URBAN AND REGIONAL DATA MANAGEMENT

PROCEEDINGS OF THE URBAN DATA MANAGEMENT SOCIETY SYMPOSIUM 2011, DELFT, THE NETHERLANDS, 28–30 SEPTEMBER 2011

Urban and Regional Data Management

UDMS Annual 2011

Editors

Sisi Zlatanova, Hugo Ledoux & Elfriede Fendel

Delft University of Technology, Delft, The Netherlands

Massimo Rumor

University of Padova, Padova and University IUAV of Venice, Venice, Italy

CRC Press
Taylor & Francis Group
Boca Raton London New York

CRC Press is an imprint of the
Taylor & Francis Group, an **informa** business
A BALKEMA BOOK

CRC Press
Taylor & Francis Group
6000 Broken Sound Parkway NW, Suite 300
Boca Raton, FL 33487-2742

First issued in hardback 2019

Typeset by Vikatan Publishing Solutions (P) Ltd., Chennai, India

ISBN-13: 978-0-415-67491-1 (hbk)

Visit the Taylor & Francis Web site at
http://www.taylorandfrancis.com

and the CRC Press Web site at
http://www.crcpress.com

Urban and Regional Data Management – Zlatanova, Ledoux, Fendel & Rumor (eds)
© 2012 Taylor & Francis Group, London, ISBN 978-0-415-67491-1

Table of contents

Part III: Environmental monitoring and assessment

Part IV: Remote sensing for urban applications

Urban and Regional Data Management – Zlatanova, Ledoux, Fendel & Rumor (eds)
© 2012 Taylor & Francis Group, London, ISBN 978-0-415-67491-1

Introduction

S. Zlatanova, H. Ledoux & E.M. Fendel
Delft University of Technology, Delft, The Netherlands

M. Rumor
University of Padova, Padova and University IUAV of Venice, Venice, Italy

Since 1971, the Urban Data Management Society (UDMS) has organised international symposia at various locations in Europe to promote the development of information systems at the local government level. At first, the focus of these symposia was mostly on urban applications, but regional and rural issues have both grown in importance over the years. Nowadays, an important aim of UDMS is to provide a forum for people to discuss new approaches, to consider new technologies, and to share practical experiences in the field of urban data management.

After 27 successful symposia, the 28th edition of UDMS was held in Delft, The Netherlands, in September 2011. This is a city and a country that many of the organizers of UDMS have called "home" for years. This year is also special for UDMS since it celebrates its 40th anniversary, which is certainly an experience not to be found elsewhere and a valuable basis on which to form planning for the future. Consequently the theme of UDMS 2011 was "Making contributions to the future".

This book contains a selection of the best papers that were submitted for UDMS 2011. Of the 42 full papers submitted, the Scientific Committee selected the 19 papers that have the highest quality and represent different aspects of the field of urban data management. It should be noticed that each paper was reviewed by at least two independent reviewers.

The topics covered in this book represent the current trends in urban and regional data management. We have organised the book into four parts: (1) 3D modelling and applications; (2) Data management for local government; (3) Environmental monitoring and assessment; (4) Remote sensing for urban applications.

3D modelling and applications. The use of 3D GIS for urban and regional planning is a growing trend that can be observed by the increasing number of tools available (for example CityGML) and by the increasing number of conferences and workshops dedicated to 3D. As in previous editions of UDMS, the third dimension continues to be one of the most important topics. This volume offers six papers related to the third dimension. The first article draws the attention to a correct and unambiguous definitions of 3D cadastral objects. Simple 2D plans to defining property rights are gradually replaced by complex 3D spatial definitions, but ambiguity in the definition of a property can lead to expensive legal disputes. The paper proposes an innovative axiomatic approach to the problem. The second paper explores the potential of Building Information Models (BIM) in urban data management. BIM models are increasingly used in construction and design and can be successfully integrated with GIS models, which offers new opportunities for urban management. The third paper presents an interesting 3D approach to represent, analyze and explore the urban rules of French local urban planning schemes. The rules (usually textual) are applied to a structured 3D geographic environment, which allows to assess their influence on the potential construction, floor area ration or to evaluate pre-process building shapes. The last three papers address different aspects of indoor navigation: extending the tagging schema of OpenStreetMap for indoor environments; a 3D data model to support individual traveller routing when using public transport, and to consider specific preferences and constraints for the pedestrian part of the journey (mostly indoor); and a framework for indoor navigation of rescuers.

Data management for local government. Data management and information sharing with citizens have been always discussed during UDMS events. Although the focus might be different, e.g. public participation systems, spatial information infrastructures, e-governments or standards for exchange of information, the purpose remains the same, i.e. supporting local governments and informing citizens with the ultimate goal to ensure sustainable development. In this volume six papers address some of these issues. The first paper presents the international project Plan4all, which aims at establishing a European network of public and private organizations involved in spatial planning and regional development. The goal of the project is the definition of a data model for the INSPIRE theme "Land Use". The second paper addresses automatic generalization. The authors argue that the requirements such as up-to-dateness may get priority over cartographic principles (but the result should remain of acceptable quality). The third paper discusses the importance of volunteers to collect geographic information and options to integrate it with spatial data infrastructures. The last three papers address issues related to improving the dialog with the citizens and engage them in the governing process by applying a systematic approach to collection, reuse and sharing of data and information. All three papers recognize the increased importance of Internet tools and community sites for exchange of information.

Environmental monitoring and assessment. This part contains four papers, which illustrate the use of urban data to monitor and analyze urban processes and phenomena. One paper focus Urban Heat Islands, which have impact on the quality of life and energy consumption in urban settlements. The paper presents a methodology for data measurement and estimation to define relationships between factors that play a role in this phenomena. Other paper presents an interesting study on the effects of population growth in Athens. The research reveals that there is lack of open spaces and green areas, although the city can be considered sustainable in terms of transportation. Two papers discuss vulnerability issues. Urban forestry fire protection model is the topic of the third paper in this part. The novelty of the model is the definition of a learning schema, which is based on different factors and aspects of fire causes, including physical, social and behavioral ones. The last paper presents a more general approach of "systemic" vulnerability evaluation, which estimates the relationship between extreme events and direct&indirect consequences of this event on the territorial system.

Remote sensing for urban applications. Although UDMS concentrates mostly on data management, at each event several high quality data collection papers are presented. The papers in this part follow again the 3D trend since these techniques are used to reconstruct and visualize 3D city models. One paper discusses a point-based rendering application, which uses high-speed spatial interpolation using a point cloud to generate VR data. This application, called LIDAR VR, is intended for real-time visualization of point clouds from urban environments. The second paper presents an innovative incremental method for 3D reconstruction from low-altitude images. The method allows first results to be obtained within a few seconds. The images are acquired by Miniature Unmanned Aerial Vehicles equipped with digital cameras. The last paper discusses a pixel-based classification approach, which extracts urban and non-urban structures from single radar imagery. In this way a global mapping of human settlements can be performed. The study is based on the German TanDEM-X coverages for two years.

This book is intended and suitable for different readers, for instance city planners, academics, students, policy makers who are involved public administration and urban planning.

We would like to thank the reviewers part of our the Scientific Committee for their time and critical remarks, which helped us select the papers with the best quality. We thank: Thomas Becker, Roland Billen, Eliseo Clementini, Volker Coors, Claire Ellul, Gilles Falquet, Sergio Freire, Ben Gorte, Gerwin de Haan, Thomas Hahmann, Andy Hamilton, Henning Hansen, Mikael Jern, Umit Isikdag, Mike Kevany, Milan Konecny, Alenka Krek, Robert Laurini, Jiyeong Lee, Bastiaan van Loenen, Martijn Meijers, Darka Mioc, Dimos Pantazis, Jacynthe Pouliot, David Prosperi, Wilko Quak, Wolgang Reinhardt, Eduard Roccatello,

Sylvie Soudarissanane, Fabio Giulio Tonolo, Edward Verbree, Marian de Vries, Monica Wachowicz and Alexander Zipf.

Last but not least, we would like to thank the authors of the submitted papers. Without their effort and contribution to the symposium it would not have been possible to achieve such a high-quality program for UDMS 2011.

Part I:
3D modelling and applications

Urban and Regional Data Management – Zlatanova, Ledoux, Fendel & Rumor (eds)
© 2012 Taylor & Francis Group, London, ISBN 978-0-415-67491-1

Modelling and validation of 3D cadastral objects

R.J. Thompson
Queensland Government, Australia
Delft University of Technology, GIS Technology Section, Delft, The Netherlands

P.J.M. van Oosterom
Delft University of Technology, GIS Technology Section, Delft, The Netherlands

ABSTRACT: As the value of land in the urban regions of the world increases, there is a trend towards the subdivision of property rights in 3D. That is to say, rights to land may be replaced by rights to the space above and below the land. This increases the complexity of the descriptions that are needed to define the regions of space to which the property rights apply. The simple plans of subdivision that are used in defining property rights on the surface of the earth are being replaced by far more complex 3D spatial definitions. An important issue in the framing of these definitions is that they must be correct and unambiguous, because an error or ambiguity in the definition of the extent of a property can lead to expensive legal disputes. This paper explores an axiomatic approach to the problem of defining what is a valid definition of a 3D Cadastral objects (parcels). This paper addresses: 1. Modelling and validation of a single cadastral parcel; and 2. Modelling and validation of a complete spatial partition.

1 INTRODUCTION

There is a strong current trend in urban areas for property rights to be defined in 3D. That is to say, rights to land may be replaced by rights to the space above and below it, increasing the complexity of the definitions of the regions of space. This paper explores an axiomatic approach to the problem of defining what is a valid definition of a 3D Cadastral objects (parcels).

The paper attempts to be neutral in terms of the storage method used, and the details of any interchange formats involved. It does, however, assume as much as possible the node, edge, face, volume paradigm (ISO-TC211 2003). It does not consider other possible representations, such as for example the Regular Polytope (Thompson 2007; Thompson and Oosterom 2011), or a set of constraints (Rigaux, Scholl, Segoufin and Grumbach 2003). It also focuses specifically on issues that arise because of the use of finite precision arithmetic such as floating point. Many of the issues addressed here do not apply if infinite precision rational numbers are used (Lema and Güting 2002), but the aim here is to achieve a similar level of rigour using the more widely and available, faster, and less storage-intensive floating point representation. Floating point arithmetic leads to accumulation of errors that must be taken into consideration if reliable and repeatable results are to be achieved.

Intuitively, the object being defined here—the parcel, is a unit of space which is the minimal unit for the definition of property rights (such as ownership). It is a generalisation of the familiar 2D land (property) parcel, or "lot". The FIG joint commission 3 and 7 working group on 3D-Cadastres (http://www.gdmc.nl/3DCadastres/) defines a 3D parcel as the spatial unit against which (one or more) unique and homogeneous rights (e.g., ownership right or land use right), responsibilities or restrictions are associated to the whole entity, as included in a Land Administration system (based on ISO 19152). Homogenous means that the same combination of rights equally apply within the whole 3D spatial unit. Unique means that this is the largest spatial unit for which this is true. Making the unit any larger would result in the combination of rights not being homogenous. Making the unit smaller would result in at least 2 neighbour 3D parcels with the same combinations of rights.

The individual land parcel should comprise a continuous unit (although some grouping of parcels into a multi-parcel will be necessary in some cases). The 3D parcel should be defined in such a way that the classical 2D parcel can be seen as a specialization (variant). In the emerging standard ISO19152, this is done by recognising that a "2D parcel" is in reality a column of space with no top or bottom defined. This reflects the legal situation in many jurisdictions—where there is no legally defined limit to the vertical extent of a 2D parcel. Individual parcels may be in contact with one another, and often share common faces, but base parcels (primary interests) are not permitted to overlap. Most jurisdictions allow "secondary interests" in land/space (e.g., a "right of way", "right of superficies") to be defined which may overlap primary interests. A further desirable attribute of a set of cadastral parcels is completeness—all the space within the area of interest must be accounted for—with no gaps of undefined space between parcels.

This paper is restricted to parcels defined by planar surfaces, but further research is planned to remove this restriction. Section 2 reviews current research in the field and positions this paper in relation to it. The rationale for validation is discussed, the precision of calculation is discussed, and cadastral issues are presented. Section 3 formulates the axioms, discussing the rationale and consequences of adopting them. Section 4 discusses practicalities and implementation issues related to validation and automatic correction of parcels. Section 5 introduces some of the issues raised by introducing partially unbounded parcels. Section 6 addresses specific algorithmic issues in more detail. An important fact when designing a system for cadastral information is that there can be significant variation in the detailed requirements between various countries/jurisdictions (van Oosterom et al., 2011), especially in the validation rules (based on laws and local practices) and the required "quality" of the representations. This is treated in Section 7. Section 8 suggests future research. The appendix gives an outline of the point in parcel test.

2 BACKGROUND

The representation of 3D objects using CAD (Computer Aided Design/Drafting) is not new, and significant work has been done on ensuring that the computer-based model is valid and is a good representation of the real-world object as it exists, or as it is to be constructed (Mortenson, 1985). The problem with the cadastral parcel is slightly different. Cadastral parcels are not "real-world" objects, although they may be associated with them. A cadastral parcel is a theoretical definition of space. One result of this is that the validation rules of cadastral parcels may have differences from the rules in CAD for physical objects; e.g., touching shell(s) in node or edge, non 2-manifold or partially bounded volume (no top). The approach taken by Oracle Spatial 11 g gets quite close to these needs (except unbounded volumes) and is described in (Kazar, Kothuri, van Oosterom and Ravada 2008). This provides a clear description of the rules for validating 3D geometries that are to be imported into the Oracle database (main rule: interior has to be strongly connected), but does not address the issue of the finite arithmetic in a computational representation. Further, for practical (implementation) reasons Oracle does not (yet) allow faces with inner rings; see Figure 1c.

Gröger and Plümer (2005) give a set of axioms that define a "2.8D" coverage—which has many of the required attributes of the Cadastral problem domain, but with restrictions. One of the restrictions—the inability to model bridges or tunnels has been removed in a later paper (Gröger and Plümmer 2011a), and extended to solid objects and a space partition in (Gröger and Plümmer 2011b). A remaining restriction is that the surfaces are required to be 2-manifold; see Figure 1a and 1b.

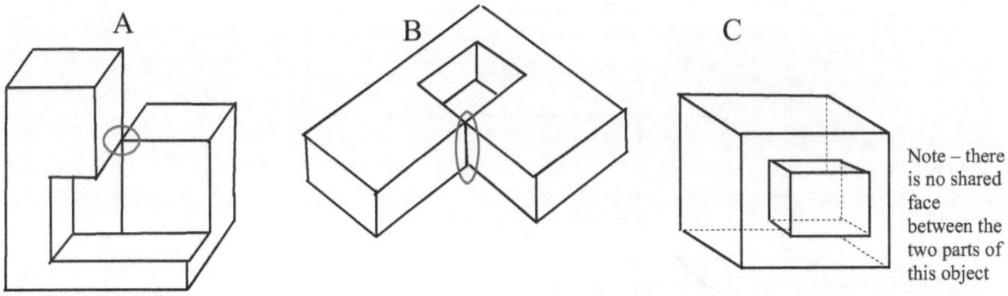

Figure 1. Cases A and B are disallowed by the axioms of Gröger and Plümer (require the surfaces to be 2-manifold). Case C is disallowed by the Oracle Spatial rules (disallow inner rings on a face).

Brugman, Tijssen and van Oosterom (2011) discuss the related problem of creating a 3D partition of space in a general context, but do not consider the issue of finite precision of the arithmetic apart from briefly discussing the difficulty of assuring planarity of faces.

2.1 *Reasons to validate spatial data*

It is important to keep in mind why validation is necessary. The action of validating data, particularly in 3D can be time-consuming and therefore costly. Unless a benefit results, it must be questioned. There are 4 aims that can be identified to justify validation rules:

Aim 1. To ensure that the definition is unambiguous. For example, in the Cadastral domain, to avoid argument in a court of law, rather the intent should be rigorously defined. As an example, if a volume is defined by surface patches which do not close, then it may be debatable whether a point in space is within or outside the "volume"; see Figure 2.

Aim 2. To allow "Programming by Contract". If data in a database is known to satisfy a set of axioms, the number of special cases that must be accommodated in an application program is dramatically reduced. E.g., given the fact that axioms "Different nodes have different coordinates" and "each edge has two distinct nodes as end points" are already satisfied, it is not necessary to test for zero length edges.

Aim 3. To ensure that data can be transferred without loss of integrity and without being rejected by the recipient. This can be difficult, as the definitions of validity checks applied by commercial software are not always clearly stated. Even worse, tight validation is seen as a selling point in commercial software, and is presented as "finding errors missed by other systems". Aim 3 can only be achieved where the validation rules are agreed, or at least well documented.

Aim 4. To prevent errors in the database. This is the least valid reason. Note that it possible to have data obtained from "good" measurements which could result in an invalid representation because it contains topological failures that are smaller than the accuracy of the data; see Figure 3.

2.2 *Finite precision*

In the 2D world of GIS, the problem of finite computation precision has been handled by the concept of normalisation (Milenkovic 1988), or some variant. This paper extends the concept into 3D. It is critical to the approach that there exists a tolerance value ε with the characteristic that arithmetic operations can be assured to give a result that is correct to an order of magnitude smaller than ε (Milenkovic uses 1/10).

This gives rise to a question of point identity ($A \equiv B$). There is a distinction to be made between points which are close together (distance $D(A, B) < d$); at zero distance apart ($D(A, B) = 0$); close but the calculated distance rounds to zero ($D(A, B)$ is calculated as zero); and points which are identified as the same point logically (e.g., the meeting of several faces). Here it is assumed that points are uniquely identified, and so the statement that points must

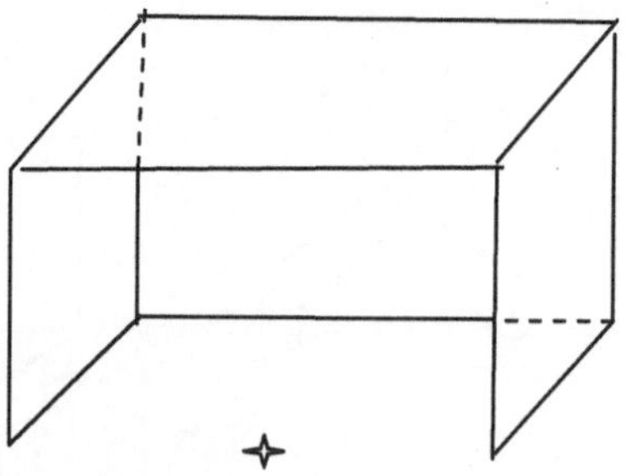

Figure 2. Is the point within the region? (Which has two faces missing).

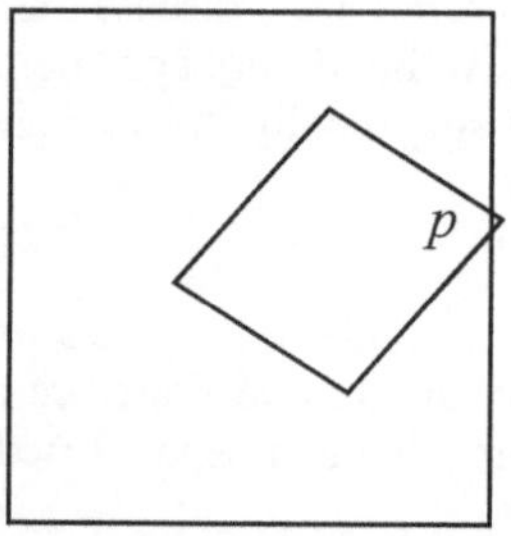

Figure 3. Invalid data (in 2D) that is "correct". The hole in the polygon crosses the outer boundary—failing validation. On the other hand, all point positions are correct within the measurement limits.

be a minimum of distance of ε apart excludes the possibility that two points are at the same 3D location.

An alternate strategy uses equality of the coordinates as a test for identity—as, for example $A = (x_a, y_a, z_a)$, $B = (x_b, y_b, z_b)$, $A \equiv B \Leftrightarrow x_a = x_b, y_a = y_b, z_a = z_b$. This may have problems if floating point coordinate values are used unless consistency of arithmetic can be guaranteed. The universal use of IEEE floating could make this acceptable (IEEE 1985; Goldberg 1991), but, for example, Java does not use IEEE floating point. A further alternative, defines a second tolerance $\iota \ll \varepsilon$. Where $D(A, B) < \iota$, they are "snapped" to one value.

2.3 *Cadastral parcels*

Although cadastral parcels in 3D have a similarity to 3D real world (physical) objects, there are significant differences.

1. A cadastral parcel is not a real world object (although it may contain or even be defined by reference to real world objects). It is a legal definition.
2. The definition of a cadastral parcel may have points, lines, or faces repeated within a single parcel, and this fact must be clear in the definition. For example, Figure 1, objects A and B could be valid definitions of cadastral parcels, and the sharing of the point/line is significant. By contrast if these objects represent part of a 3D city model for example, the parts can simply be moved (a miniscule amount) to permit a 2-manifold representation (see Section 2.4).
3. A cadastral parcel may not be fully bounded. This case arises because in many jurisdictions, the conventional 2D land parcel has no top or bottom defined. Thus, if a 3D parcel is excised from a 2D parcel, the remaining space has no upper surface or lower surface; see Figure 4.

Note that 3D parcels, when unbounded, will normally have vertical faces and missing top and/or bottom face(s). However, the unbounded character of 3D parcels might be defined in

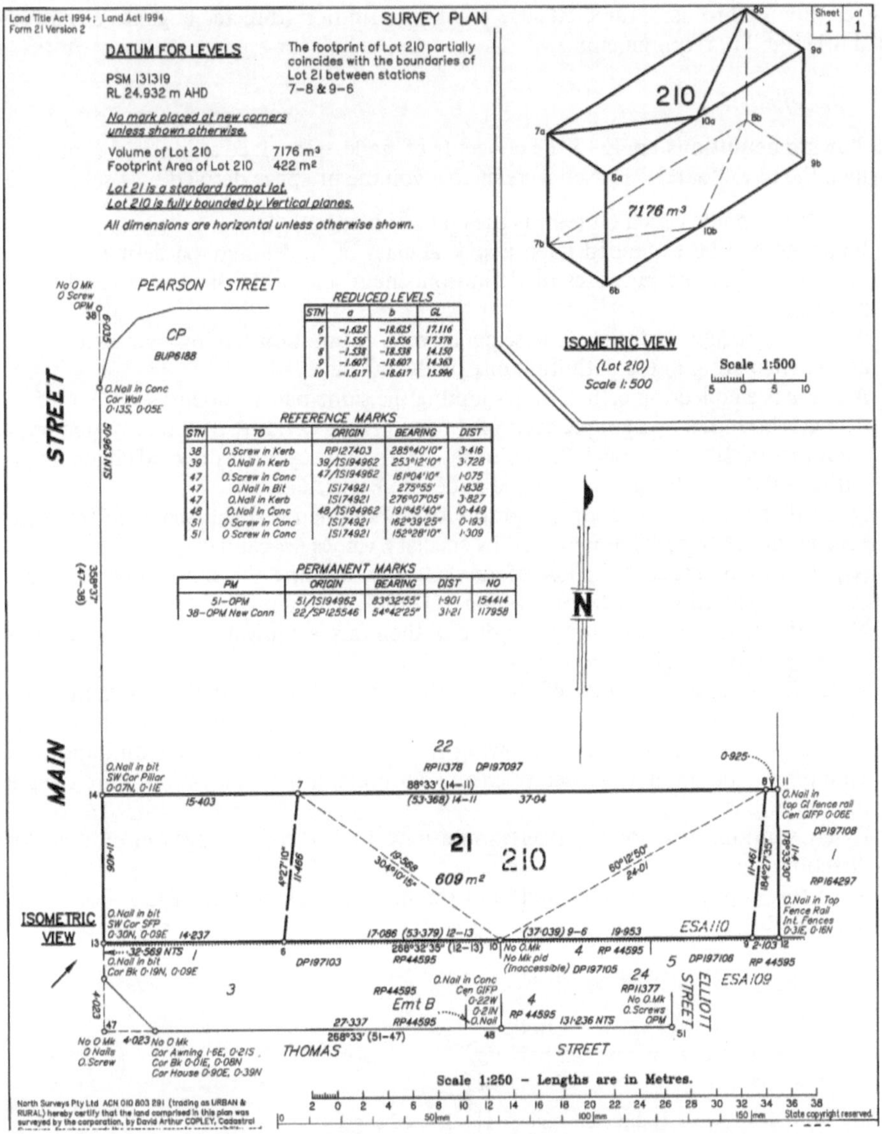

Figure 4. Bounded and partially bounded cadastral parcels. The volume of lot 210 has been excised from lot 21 (which is a conventional 2D lot). Thus lot 210 is completely bounded, but lot 21 is unbounded above and below. Note that lot 210 is part of a road tunnel and is completely below ground.

a more general manner, that is, not per se defined by the unbounded vertical faces. For example a 'viewshed ownership' in a certain direction might be defined by some of the unbounded faces being non-vertical. One could imagine this being possible as long as this unbounded volume does not overlap with any other explicitly modelled 3D Parcel. As this is quite an exotic case, for the time being, it is considered to be outside the primary requirements for valid 3D parcel representation.

3 AXIOMS

The following axioms are based on Gröger and Plümmer (2011), however they have been modified to reflect cadastral requirements, as described above. In addition, the definitions have been modified to use a directed edge form of definition (directed edge) to ensure easily verified orientability requirements.

3.1 *Basic definitions*

The following definitions apply to the concepts of node, edge, face and parcel (note that in this paper, the term "parcel" is used to refer to a volume of space defined by a set of faces):

D1. A *node* is a point which represents the end of an edge.
Note: As will be expanded later, a special class of nodes may be defined as "points at infinity" For the purposes of definition, the term "node" is taken to include such points.

D2. A *directed edge* is a straight line segment with orientation joining two distinct nodes, and participating in the definition of exactly one face.

D3. An *edge* is a collection of half-edges joining the same pair of nodes; see Figure 5.

D4. A *face* is bounded by a simple cycle of directed edges defining the outer boundary; and a number of 2D holes. Each 2D hole may be a simple cycle of directed edges, a pair of anti-equal directed edges, or a single node; see Figure 6a.
Note: If face is part of a top or bottom of unbounded vertical face, then the directed edge at infinity may be represented by special z values ($+\infty$ and $-\infty$).

D5. A *parcel* is 'bounded by' a set of faces (shell) defining the outer boundary, and a number of closed sets of faces defining 3D holes; see Figure 6b.
Note: If a parcel is open at top or bottom, then this is equivalent to an implicit face at height $= +\infty$ (or height $= -\infty$).
Note: The intention of this definition is that a parcel should define a region of space that is of use in defining interests in that region (such as ownership). The definition will be further refined by the following rules which put restrictions on its form. For example, an axiom may be used to ensure closure; another rule may require the parcel to be contiguous.

D6. By convention, faces of a parcel should have their directed edges in anti-clockwise direction, viewed from outside. (Volume calculation should produce a positive volume when this convention is followed, and the normal vector of each face points to the outside.)

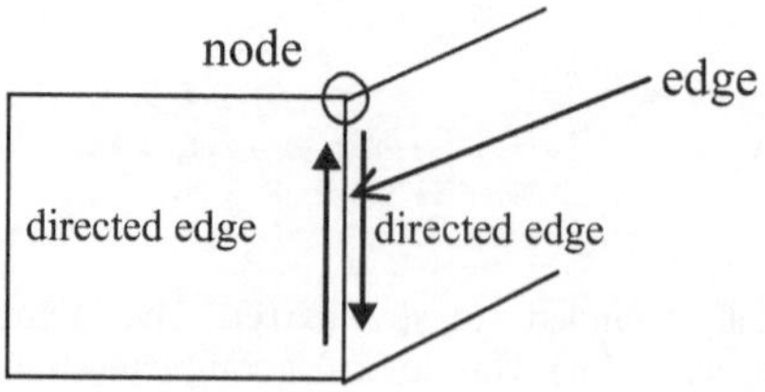

Figure 5. Node, directed edge and edge.

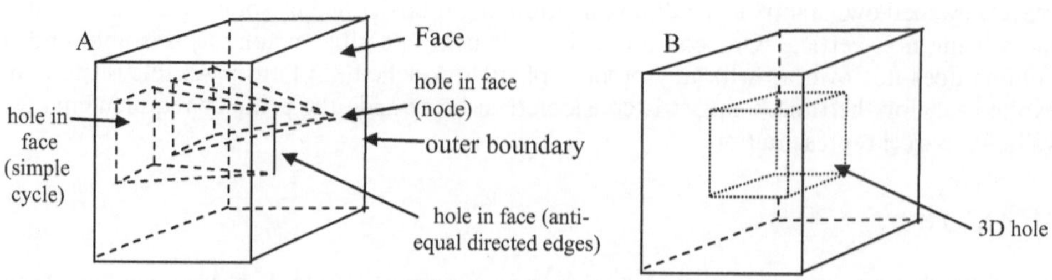

Figure 6. Holes in the faces of a parcel (A) and a 3D hole in the interior a parcel (B).

3.2 *Generic axioms*

The generic axioms for a single 3D parcel, based on polyhedral modelling are:

A1. No two nodes are closer than ε apart.
A2. Each node has at least 3 incident edges (note: this axiom is optional—see below).
A3. The faces incident at a node do not intersect one another.
A4. Edges do not intersect except at their common end point nodes.
A5. Non-intersecting edges (that is, edges not sharing a common node) must be at least a distance ε apart.
A6. Each edge is the meeting of an even number of faces such that the directed edges of those faces form an alternating sequence around the edge.
Note on A6. If an edge is viewed from either end, the faces must form a sequence radially, in pairs, with interior sides adjacent. The outside of the volume is always at the side of the normal vector (See Figure 7).
A7. The directed edges that delineate a hole in a face must have their anti-equal counterparts be part of the outer boundary of other faces.
A8. Bounded faces are planar to a tolerance of ε'.
Note the different tolerance value; see Subsection 6.1 for more details.
A9. No node is within ε of a face unless it is part of the definition of that face. (Together with A5 this implies that no edge is within ε of a face unless it is part of the definition of that face.)
A10. No edge intersects a face except at a node of that face.

3.3 *Face connectivity related definitions and theorems*

Definition: D7. Two faces F_1 and F_2 are *anti-equal* ($F_1 \rightleftharpoons F_2$) if:
$\forall e_1$: (a directed edge) $e_1 \in F_1, \exists e_2 \in F_2: e_1 \rightleftharpoons e_2$. ($e_1$ and e_2 are anti-equal).
Notes on the two degenerate cases of holes in a face (singularities): 1. Each hole of type edge in F_1 (anti-equal pair of directed edge) is defined by the same edge in F_2 and 2. Each hole of type node in F_2 is defined by the same node in F_1.
Definition: D8. Two faces *"0-meet"* if they share at least a node in common.
Definition: D9. Two faces *"1-meet"* if they share an edge in common and their directed edges at that edge are anti-equal.
Definition: D10. Two faces *"strongly-meet"* if they share an edge in common, their directed edges at that edge are anti-equal, are adjacent in the radial sequence described in axiom A6, and their interiors are adjacent—see Figure 8.
Note that 0-meet $\Leftarrow$ 1-meet $\Leftarrow$ strongly-meet (but not vice versa).

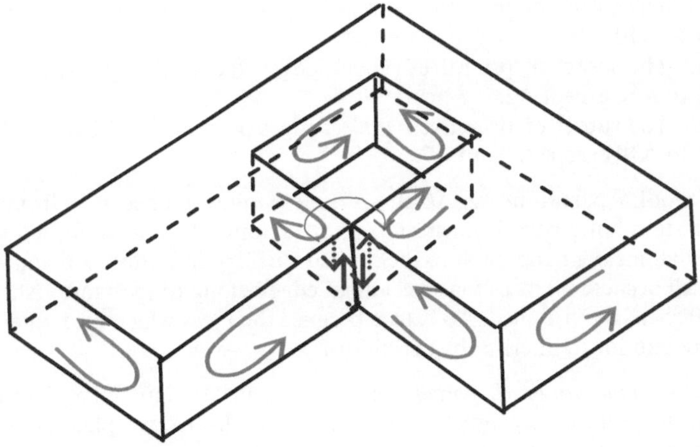

Figure 7. Alternating sequence of directed edges at an edge.

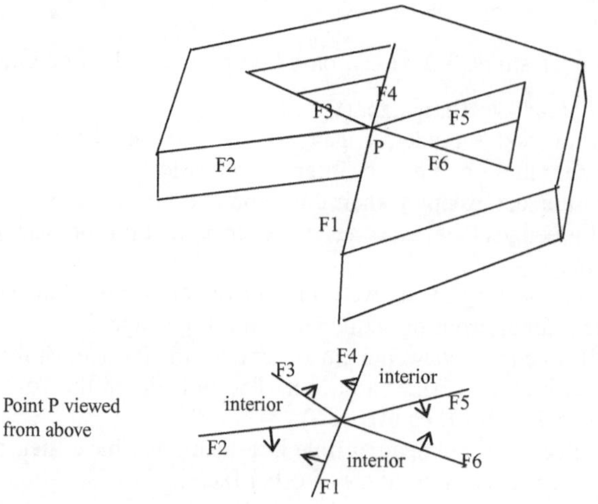

Figure 8. Meetings of faces at an edge. At point P, six faces meet. Face F1 1-meets F2, F4 and F6. On the other hand, it strongly-meets only F6, because their interiors are adjacent in the radial sequence around point P.

Definition: D11. A 0-*shell* is a subset of the faces of a 3D parcel defined by starting from a single face, and adding all faces that 0-meet that face. The procedure continues by recursively adding all faces that 0-meet any of the faces in the set, until no more can be added.

Definition: D12. A 1-*shell* is defined by a similar procedure, adding faces that 1-meet.

Definition: D13. A 2-*shell* is defined by a similar procedure, adding faces that strongly-meet.

Note that this does not imply that the 2-shell is 2-manifold.

By the definitions and axioms given so far, a parcel does not have to be a single contiguous object. So, this will be addressed in the next subsections.

3.4 *Basic parcel axioms*

The concepts of 0, 1 and 2-shells are now used to provide a definition of contiguity to be used in the definition of a valid parcel where appropriate; see Figure 9. P is a 3D object which satisfies axioms A1 to A10 (but which may be discontiguous); see Figure 10.

Theorem: Th1. The subset of primitives (nodes, edges, faces) of any 0-shell of P obeys all axioms A1 to A10.

Theorem: Th2. The subset of primitives (nodes, edges, faces) of any 1-shell of P obeys all axioms A1 to A10 except A2.

Theorem: Th3. The subset of primitives (nodes, edges, faces) of any 2-shell of P obeys all axioms A1 to A10 except A2 and A7.

Outline of proof: Apart from A2, A6 and A7, these follow immediately from the fact that the n-shell is a subset of a parcel which satisfies the axioms. A2 follows for a 0-shell because for any vertex, all incident edges are in the same 0-shell by definition. A6 and A7 are satisfied for a 1-shell because all faces that meet at an edge belong to the same 1-shell. The same applies for a 0-shell. For a 2-shell, A6 is true because the faces which meet at an edge are in pairs which are interior to interior (by definition).

Definition: D14. *Point within 3D parcel.* This is the usual Jordan Curve definition. A line is run from the point in any direction to infinity (in a horizontal plane to also treat well unbound parcels at top or bottom). If the number of times the line intersects the boundary

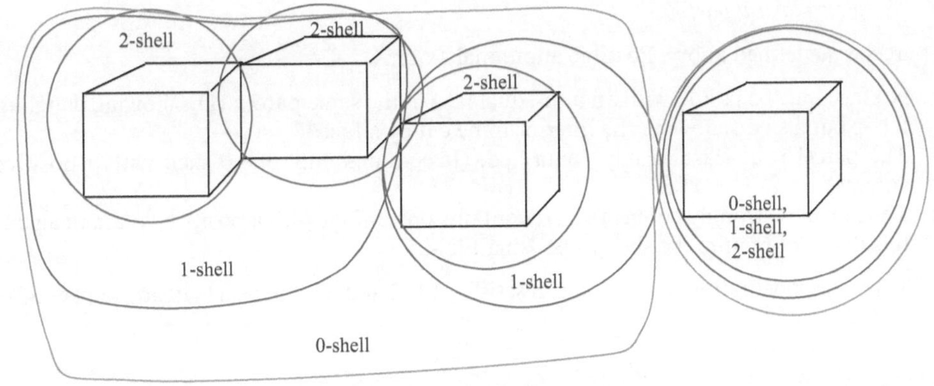

Figure 9. 0-shells (green, 2 cases), 1-shells (blue, 3 cases) and 2-shells (red, 4 cases) of a multi-part object.

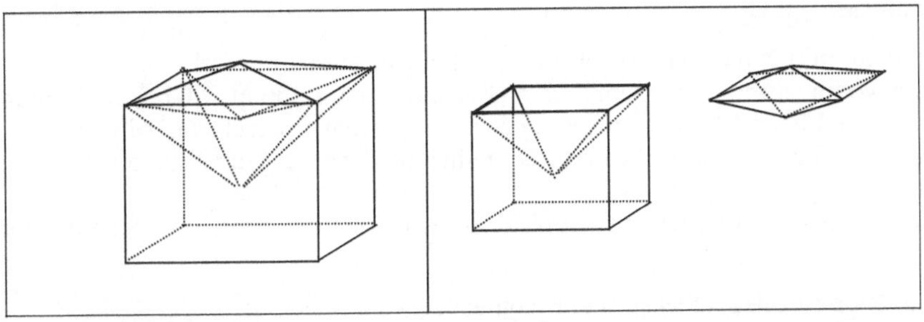

Figure 10. A parcel divided into two separate 2-shells (for clarity, the right figure shows one part displaced). The two parts are weakly connected (1-meet), but are not strongly connected. The implementation would have the option of rejecting this object as a single parcel, but accepting it as a multi-parcel.

of the parcel is odd, the point is within the parcel. See Appendix I for a safe version of this test.

Definition: D15. *N-shell within n-shell.* A n-shell S_1 is within another n-shell S_2 of the same parcel if there exists a point enclosed by S_1 that is also enclosed by S_2. Note this definition does also work for unbounded at the top (or bottom) n-shells.

Definition D16 *Positive n-shell.* Take any point enclosed by the n-shell. Apply the point in parcel test (Appendix I). If the total cutcount (of the whole parcel) returned is +1, the n-shell is positive. If the cutcount is –1, it is a negative n-shell.

Parcel axiom

A11. If a parcel is represented by 2-shells, then all negative 2-shells must fall within the single positive 2-shell. The positive 2-shell must not fall within any other 2-shell. Note: A multi parcel can also be defined, allowing one or more positive 2-shells, and any number of negative 2-shells each of which must be in a single positive 2-shell.

3.5 *Note on the axioms A2*

Axiom A2 'each node has at least 3 incident edges' prevents redundant points in edges, and redundant edges in faces, but may not always be mandated. That is, it is possible to have consistent representations (see Figure 11), which are unambiguous and programmable, but which do not respect this axiom (and can be modified to a representation that satisfies A2).

3.6 *Single cadastral parcels*

A parcel is as defined above. Possible additional axioms:

S1. No face may be paired with an anti-equal face in the same parcel. (Disallowing dangling faces, but also disallowing the interior to be cut by a face.)
S2. The parcel is at least weakly connected. (It contains only one 0-shell with a positive volume).
S3. The parcel is strongly connected. (It contains one 2-shell with a positive volume); again, this does not require the shell to be 2-manifold.

The strongest definition of a 3D Parcel satisfies S1, 2 and S3 (but as S3 already implies S2).

3.7 *Space partition*

This requires the same axioms as a single cadastral parcel. These are sufficient to prevent partial overlap of parcels, even if overlap is nicely modelled by adding intersection edges and nodes (and splitting faces). However, the existing axioms do not preclude a parcel falling completely within another (without the other having a hole). To prevent this, a further axiom may be added.

P1. No point of a parcel may be within the volume of another parcel.

This, in conjunction with axioms A1-A11 ensures that no point will belong to more than one parcel; see Figure 12. This (correctly) does not prevent a parcel from being within a void inside another, but there must be an interior shell of the outer parcel surrounding the inner parcel.

Axioms S1-S3 are optional when applied to each individual parcel, but S2 and S3 may be extended to the set of parcels:

P2. The set of parcels is at least weakly connected.
P3. The set of parcels is strongly connected.

In many cases (countries) this should not be required as the territory of the country may have exclaves (and enclaves with exclaves). Finally, a complete, non-overlapping partition of space can be mandated by adding a further axiom:

P4. Each face is paired with exactly one anti-equal face. (If the outside universal 3D parcel is not explicitly represented, this axiom does not not apply for faces at the domain boundary).

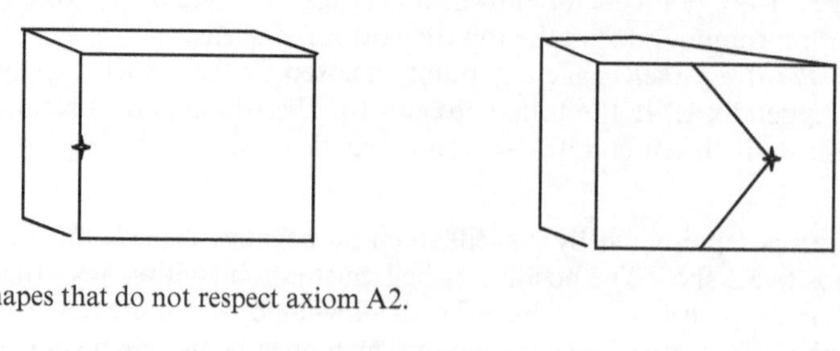

Figure 11. Shapes that do not respect axiom A2.

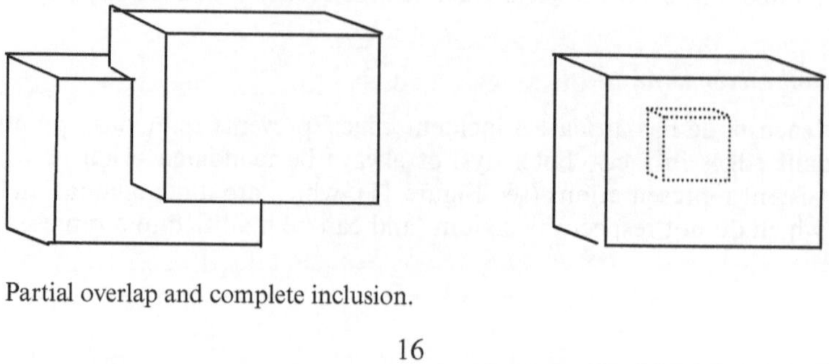

Figure 12. Partial overlap and complete inclusion.

4 VALIDATION AND CORRECTION OF 3D PARCELS

This section discusses the 'translation' from the formal axioms to a potential implementation of validation testing. It further discusses the closely related topic of (partial) automatic correction.

4.1 *Validation testing*

It is the aim of this axiomatic approach to suggest how validation tests can be applied programmatically to parcels which are of practical size and complexity (as found in actual cadastral databases), but not to insist on a particular storage structure:

- A1. Appears to be an $O(n^2)$ operation, with n being the number of points, but would reduce to 'about' $O(n \log n)$ by using methods drawn from spatial indexing (3D R-tree).
- A2. Assuming an appropriate storage structure, this would be simple to test.
- A3. Is an $O(m^2)$ operation, but the m in question is the number of faces that meet at a node. This would not usually be large in a practical implementation.
- A4 & A5. Can be tested together, requiring testing of all pairs of edges, but can be made less compute intensive using methods drawn from spatial indexing (3D R-tree). For example, if an R-Tree index is used to limit the number of edges to be tested, the complexity reduces to about $O(n \log n)$.
- A6 & A7. Assuming an appropriate storage structure, these would be simple to test.
- A8. Has some subtlety in testing (as one could argue how to exactly define a planar face and subsequently how to test this), this will be discussed in Section 6.1, but is not very compute-intensive.
- A9. Requires each node to be tested against each face. Potentially $O(n^2)$, but the complexity can be controlled by well known spatial indexing, e.g., the 3D R-tree will reduce the complexity to about $O(n \log n)$.
- A10. Requires each edge to be tested against each face. Potentially $O(n^2)$, but the same techniques that apply to A9 will work here.
- A11. Is non-trivial, but well-understood techniques such as "scan-line" algorithms extended to 3D should give acceptable response speeds.

4.2 *"Cracking" of edges and faces*

In 2D, Milenkovic (1988) discusses the "cracking" of lines to allow them to terminate at computable points. The same is not necessarily possible in 3D; see Figure 13.

For a point within ε of an edge, if that edge is cracked to meet the point, the face it belongs to may no longer be sufficiently planar (to within ε'). This may be true even if $\varepsilon = \varepsilon'$.

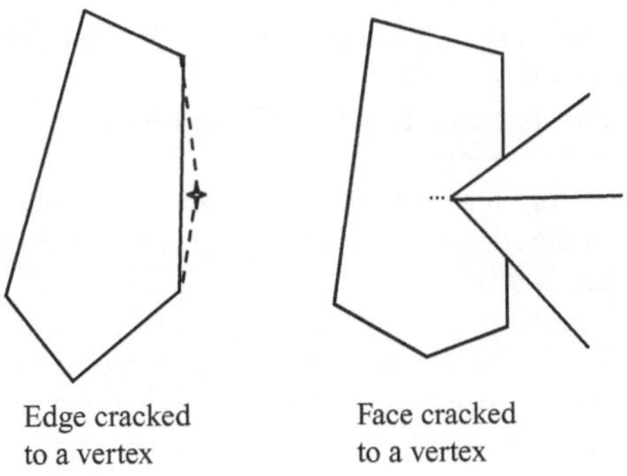

Figure 13. Cracking edges and faces.

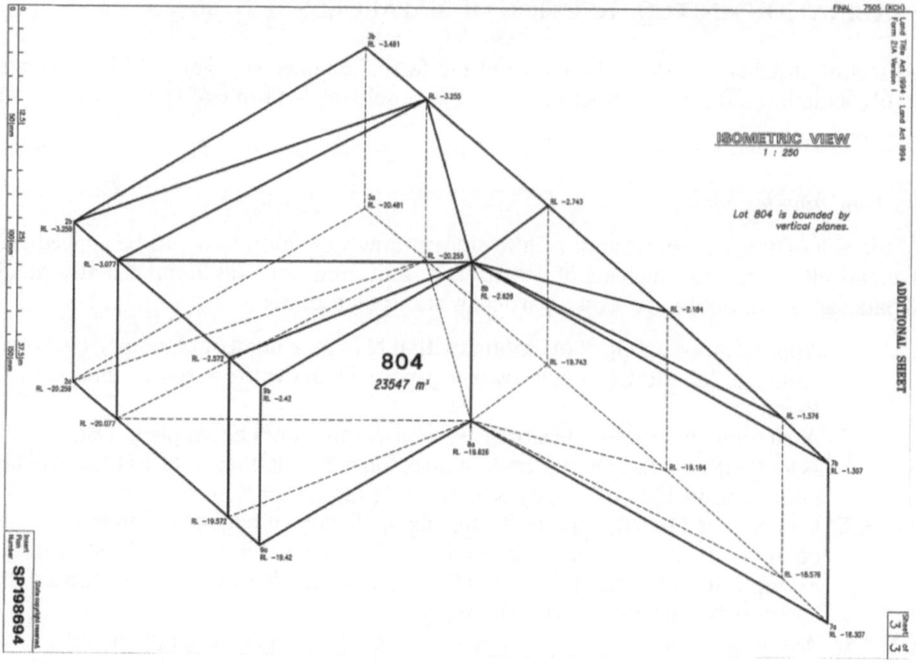

Figure 14. Cadastral parcel with vertical walls as polygons and off-horizontal faces triangulated.

Where a point is within the cracking limit of a face, the equivalent to cracking is to make it a single node "hole" in the face. Again this may not be within ε of the plane of the face.

The problem of planarity of faces is completely sidestepped if the faces are broken into triangular sub-faces. Since a triangle must be planar, a 3D parcel that is defined by triangular faces only will never fail axiom A8. On the other hand, this loses the identity of the faces, and is unnecessary for a large number of faces of cadastral parcels.

Most faces of 3D cadastral parcels (related to man-made objects) are vertical rectangular "walls" which, by definition must be planar. A further large class of cadastral faces are horizontal (constant z value). In both these cases, it is unnecessary for the face to be triangulated. There is a trend emerging in the Queensland Cadastre for parcels to be defined by polygonal faces where those faces are horizontal or vertical, but for other faces to be triangulated (for example Figure 14).

5 UNBOUNDED PARCELS

There are several ways of defining unbounded parcels. One is to use homogeneous coordinates (x, y, z, q) where the conventional point would be (x/q, y/q, z/q). In this case, letting q = 0 gives a representation of a point at "infinity"; see Figure 15.

Given any point (x, y, z, q), keeping the x, y, z constant and varying q generates a line through the origin (0, 0, 0, 1) and the point (x, y, z, q). As q → 0, the point approaches "infinity". This has application in the field of projective geometry (where the concept arises) but is not particularly applicable to cadastre.

Another approach, more appropriate to cadastral parcels, is used in the ISO 19152 concept of the "faceString" (ISO-TC211 2009; Lemmen, Van Oosterom, Thompson, Hespanha and Uitermark 2010). This defines a special direction (Z), allowing representation of edges that run to/from ±∞. Parcels which are usually considered to be conventional 2D cadastral parcels are here seen to be 3D parcels with vertical walls (known as "faceStrings"), with no top or bottom defined.

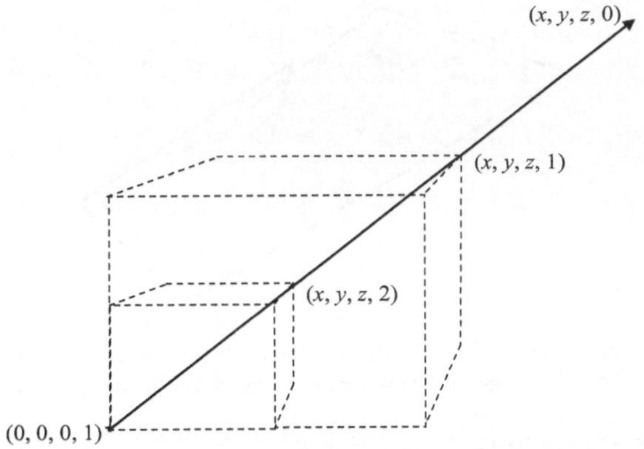

Figure 15. Points at infinity in homogeneous coordinates; Note (0, 0, 0, 1) is equal to (x, y, x, ∞).

One possible implementation of the ISO19152 approach uses a special value of the z coordinate to $+\infty$ and another for $-\infty$. (Note this will not work for tilted unbounded faces.) In order to satisfy axiom A6, there need to be faces at infinity (whether stored or inferred). This requires some care when testing "point within parcel", unless we resort to an unpleasant concept such as $(\infty + 1)$.

An alternative implementation—which is more in keeping with the spirit of ISO19152 is to make an exception to axiom A6 for directed edges at infinity so that they are not matched with any anti-equal directed edge. Note—taking this approach, the algorithm in Appendix I works even for parcels that are open top and/or bottom.

6 IMPLEMENTATION ISSUES

6.1 *Axiom 8: planar face*

In testing axiom A8 (planar face), one approach is to use a least squares or some other approximation method to determine a parametric form of the plane through the face. If this is expressed in the form $Ax + By + Cz + D = 0$, there are certain advantages in retaining these parameters (A, B, C, D). It may also be worthwhile considering these parameters to be a part of the definition of the plane—described as the "Approximated Polytope" (Thompson, van Oosterom and Pullar 2006c; Thompson 2007).

The major advantage is that once it is known that all nodes are within tolerance of the defined plane, they will remain so, and any additional nodes added will not cause difficulties. Without the parameters being stored, adding an additional node could cause new plane parameters to be calculated, invalidating previously valid points.

In addition, there is no guarantee that, even if a face has been verified to be planar within a given tolerance, an attempt to re-verify it on a different computational platform will generate the same plane parameters, leading to the possibility that the face may be rejected. Once a plane has been determined, it is necessary to ensure that the projection of the vertices onto that plane define a valid polygon; see Figure 16.

6.2 *Orientability*

Cadastral parcels, in common with real-world solid objects, must be bounded by orientable surfaces; i.e., a Moebius strip (with no thickness) cannot be part of the definition of a cadastral boundary. Note that a Moebius strip with thickness (via 'side' faces), is a valid 3D Parcel. In this set of axioms, orientability is ensured by the use of directed edges.

A Moebius band that is made up of faces defined by directed edges will always have a "seam" where the directed edges run in the same direction. This would violate axiom A6. Thus there is no need for a specific axiom to disallow un-orientable surfaces.

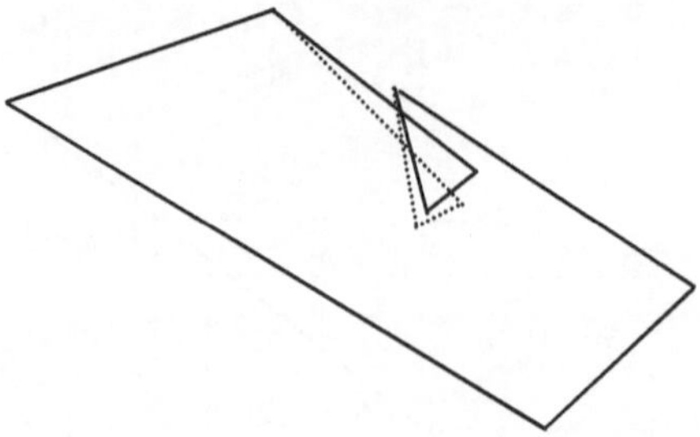

Figure 16. Although the edges of this face do not intersect or violate A5, and all vertices are within tolerance of the plane; when flattened to the plane, the face is not a valid polygon (Coors 2010).

7 VARIATIONS ON AXIOMS

This paper should be viewed as a first attempt to determine a set of axioms for the validation of 3D cadastral parcels, but there are many variants still to be discussed. For example:

- A2 could be made optional as mentioned earlier. If omitted, redundant nodes are allowed.
- S1 could be made optional. If omitted, faces and their anti-equal pair are allowed within a parcel. If this is done, there may need to be a subtle change to the definition of strong connectivity.
- In the cadastral regulations, or database rules it may be stipulated that a parcel must be at least weakly connected. These regulations/rules may be further strengthened to require strong connectivity.
- Definition D5 could be modified to disallow degenerate holes in faces such as those of parcel A of Figure 6. On the face of it, it seems unlikely that they will arise in practice, but it is a tough call to make them invalid.
- It may be mandated that all 3D parcels are within the outer shell of a 2D base parcel.
- In the paper so far, it has been mandatory that parcels do not overlap. While this is true of base cadastral parcels, it will be necessary in some jurisdictions to relax this to allow for "secondary interest" parcels (e.g., representing "right of way") to fall within or overlap parts of base parcels. These should be modelled as independent layers: primary interest layer based on space portioning parcels and secondary layer with individually valid 3D parcel (but perhaps with mutual overlap and certainly with overlap with primary 3D parcels).
- It may be mandated that all parcels that comprise a particular property (e.g., a LA_BAUnit of ISO19152) must be strongly connected or strongly connected via common property, etc.
- In the case of the cadastral parcel set, it may be desirable to relax the requirement that different parcels abut at anti-equal faces. Instead, it could be mandated that the abutting faces have anti-equal planes, but that the nodes and directed edges do not need to match.

8 CONCLUSIONS AND FUTURE WORK

8.1 *Main results*

A set of axioms have been proposed that could be used as a vendor independent and well documented basis for the validation of cadastral parcels. These axioms are presented in the context of individual parcels and space partition parcels.

A set of refinements are suggested, adding axioms making the validation stricter (depending on needs) or omitting axioms to allow more "forgiving" software. This results in a taxonomy of families of valid 3D objects (and formal definitions thereof, i.e., ontology).

A discussion of the implementation of validation and correction functionality is included. In this paper is has become clear that the needs for representation 3D parcels (non 2-manifold, partially bounded volumes, etc.) are currently not yet supported by ISO 19107 (2003). These issues have been forwarded to ISO TC211 for the revision of ISO 19107.

8.2 *Future work*

Axioms can be proved necessary by counter-examples: omitting one axiom and constructing example that, although unacceptable, passes all other axioms. The converse is not easily verified. There may, indeed be additional axioms that could be mandated, so any list must be seen as a minimum requirement only.

It should be possible to generalize our definition so it is not per se bound to the chosen type of representation (polyhedral, with topological concepts) and the same could be expressed in context of other framework; e.g., regular polytopes (Thompson & van Oosterom 2011). The approach can be extended to 3D parcels with non-flat faces/shells. Note when curved surfaces are used, then perhaps no node or edge is present in the boundary of volume; e.g., inside of a sphere.

The approach can be applied to more generic 3D objects outside the cadastral domain. The approach is not inconsistent with lower dimension primitives in same structure; for example: city model with 3D volume building, but also 2D face for a fence, and 1D edge for cable, etc.). Therefore, important aspects of this extended taxonomy of types of valid 3D objects (for different application domains) are: dangling nodes, edges, faces allowed, singularities allowed (edge or node hole in face), 2-manifold volume required, linear or curves geometries, proper orientation required, etc.

More general cases of unbounded faces, and partially bounded volumes—not limited to parcels with vertical faces (e.g., representing viewshed).

ACKNOWLEDGEMENTS

The authors of this paper want to express their gratitude to three anonymous UDMS reviewers for the constructive comments. The authors would further like to thank Elfriede Fendel for converting the text into the required format and making textual corrections. Martijn Meijers is acknowledged for proof reading of the two earlier drafts of this paper and providing very useful comments for improvements.

REFERENCES

Brugman, B., Tijssen, T. & van Oosterom, P. (2011). *Validating a 3D topological structure of a 3D space partition*. in Advancing Geoinformation Science for a Changing World (AGILE 2011): 359–378.

Coors, V. (2010). "Methoden und Metriken zum Qualitätsmanagement von 3D-Stadtmodellen" 3D-Stadtmodelle 2010 Bonn. http://www.3d-stadtmodelle.org/3d-stadtmodelle_2010/ 10_Coors_ Methoden_und_ Metriken_zum_Qualitaetsmanagement.pdf

Goldberg, D. (1991). "What Every Computer Scientist Should Know About Floating-Point Arithmetic." *ACM Computing Surveys*, March 1991.

Gröger, G. & Plümer, L. (2005). "How to Get 3D for the Price of 2-D — Topology and Consistency of 3D Urban GIS." *Geoinformatica* 9.2: 139–158.

Gröger, G. & Plümmer, L. (2011a). "Topology of Surfaces Modelling Bridges and Tunnels in 3D-GIS." *Computers, Environment and Urban Systems*.

Gröger, G. & Plümer, L. (2011b). "How to achieve consistency for 3D city models." GeoInformatica 15: 137–165.

IEEE (1985). IEEE Standard for Binary Floating-Point Arithmetic. New York, IEEE Standards Department.

ISO-TC211. (2003, 2001-11-21). "Geographic Information — Spatial Schema." *ISO 19107*, from http://www.iso.org/iso/catalogue_detail.htm?csnumber=26012

ISO-TC211 (2011). "Geographic Information—Land Administration Domain Model (LADM).". *ISO/DIS 19152*, from http://www.iso.org/iso/catalogue_detail.htm?csnumber=51206
Kazar, B.M., Kothuri, R. van Oosterom, P. & Ravada, S. (2008). On Valid and Invalid Three-Dimensional Geometries. *Advances in 3D Geoinformation Systems*. P. van Oosterom, F. Penninga, S. Zlatanova & E. Fendel. Berlin, Springer: 19–46.
Lema, J.A.C. & Güting, R.H. (2002). "Dual grid: A new Approach for Robust Spatial Algebra Implementation." *GeoInformatica* 6(1): 57–76.
Lemmen, C., Van Oosterom, P., Thompson, R.J., Hespanha, J. & Uitermark, H. (2010). *The Modelling of Spatial Units (Parcels) in the Land Administration Domain Model (LADM)*. FIG Congress 2010, Sydney, Australia.
Milenkovic, V.J. (1988). "Verifiable implementations of geometric algorithms using finite precision arithmetic." *Artificial Intelligence* 37: 377–401.
Mortenson, M.E. (1985). *Geometric modeling*. John Wiley & Sons, New York.
Rigaux, P., Scholl, M., Segoufin, L. & Grumbach, S. (2003). "Building a constraint-based spatial database system: model, languages, and implementation." *Information Systems* 28: 563–595.
Thompson, R.J. (2007). *Towards a Rigorous Logic for Spatial Data Representation*. PhD thesis, Delft University of Technology, Delft, The Netherlands, Netherlands Geodetic Commission.
Thompson, R. & Oosterom, P.v. (2011). *Integrated Representation of (Potentially Unbounded) 2D and 3D Spatial Objects for Rigorously Correct Query and Manipulation,* in Advances in 3D Geo-Information Sciences. 5th International Workshop on 3D Geo-Information, (3D GeoInfo 10) Berlin, Germany: 179–196.
Thompson, R.J., van Oosterom, P. & Pullar, D. (2006). *Robust Representation and Analysis of Geo Information*. AutoCarto, Vancouver, Washington, USA.
Van Oosterom, P., Stoter, J., Ploeger, H. Thompson, R. & Karki, S. (2011). "World-wide Inventory of the Status of 3D Cadastres in 2010 and Expectations for 2014." In proceedings FIG Working Week 2011, Marrakech, Morocco, 21 pages.

APPENDIX I

Point in parcel test

The algorithm considers a plane parallel to the x/y axes through the point (x_p, y_p, z_p)—is the plane $z = z_p$.

For each face of the 3D parcel that cuts this plane, a point is calculated where each directed edge cuts the plane. Each point is tagged +1 or −1 depending on whether the directed edge cuts in a downward or upward direction respectively.

If there is a cut point that is on or to the north of the east-west line through p (i.e., if ∃ cut point *i* (x_i, y_i, z_i) such that $y_i \geq y_p$ then the overall cutcount is incremented by the sum of the tags of the cut points with $y_j < y_p$, but only if the line from y_{i-1} to y_i passes to the west of point p.

If the total cutcount (from all faces) is +1, the point is within the 3D parcel.

It is important in this algorithm that the parametric form of the faces (A, B, C, D) is not used. If this was done, then rare special cases can occur that are very difficult to test for.

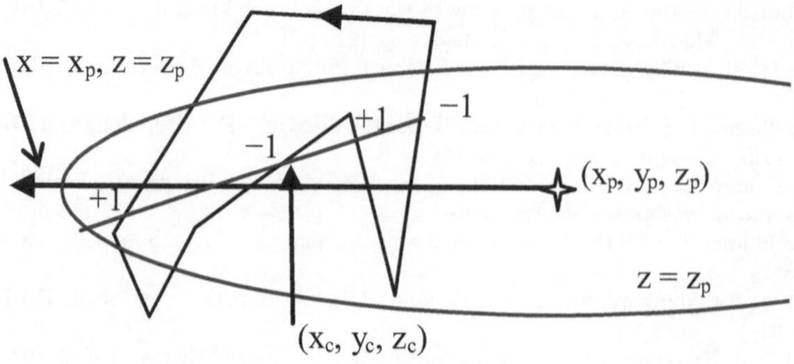

Figure 17. Point in 3D parcel test (Note—in this case the face generates a cutcount of +1).

The subtlety of this test comes about when it is realised that a face may be to the west of p, and nearly horizontal, and further that the errors in the point positions may mean that they may be scattered above and below the line $x = x_p$, $z = z_p$. Thus the face as described by the directed edges as viewed from p may resemble a birds nest, and the sum of tags to be added to the overall cut count may be ±2 or more. It is asserted here that this does not matter to the algorithm, and that the result will be correct provided that p is not within a distance of ε' of an actual face. (In the case where p is within ε' of a face, the inaccuracy of node positioning may mean that the point is determined to be on the wrong side of the face, but if this face adjoins another 3D parcel, a consistent result is still obtained with respect to that 3D parcel).

Urban and Regional Data Management – Zlatanova, Ledoux, Fendel & Rumor (eds)

An opportunity analysis on the future role of BIMs in urban data management

U. Isikdag
Beykent University, Turkey

S. Zlatanova
Delft University of Technology, GIS Technology Section, Delft, The Netherlands

J. Underwood
University of Salford, UK

ABSTRACT: In order to facilitate some urban management tasks, information regarding buildings needs to be represented in form of geospatial information. Some tasks in urban management such as indoor navigation, emergency response management, and processes in the construction life cycle e.g., site selection can be facilitated through the use of certain and sometimes high amount of, geometric and semantic information about a building. A Building Information Model (BIM) can be defined as a semantically rich shared 3D digital building model. Recent research has demonstrated that geometric and semantic information from BIMs can be transferred into the geospatial environment and CAD/GIS/BIM information can be integrated at web services level. This paper presents a comparative opportunity analysis on the future role of BIMs in Urban Data Management. The analysis presented was focused on exploring, the impact of progression in Building Information Modeling research, to the opportunities provided by the use of BIMs in Urban Data Management context.

1 INTRODUCTION

In AEC industry, tools to facilitate the design process have until the last decade been very much focused on automating 2 dimensional drawing board techniques. The emerging technologies in last 15 years focused on 3D modeling and facilitating collaborative design. In fact, most of the problems in the field have been related to insufficient interoperability and integration between different systems used, in the industry and throughout the lifecycle of an AEC project. In last 10 years Building Information Modeling have become a key research item in the global agenda to address the problems related to the inefficient sharing of information and collaboration throughout the lifecycle of an AEC project and the building (i.e., from feasibility and conceptual design stages through to demolition and re-cycling). The academic community has spent considerable time and effort in last 10 years for exploring the opportunities that these intelligent Building Information Models (BIMs) would provide to the industry. In late 1990s, the use of "Product Modeling" approach in AEC industry is defined as "Building Information Modeling" and at that time, BIMs were prescribed as a remedy for the illness of "Data Interoperability" in the industry. However, with the new interpretation of "Building Information Modeling", it is now apparent that this 'magic remedy' has evolved to cure much more 'illnesses' than it was originally prescribed for.

Today the implementation of the Building Information Modeling paradigm is achieved by using several applications supporting an intelligent digital building model defined by the agreed schema standards such as IFC and CIS/2. Well-recognized applications used in the AEC industry are today capable of importing and exporting digital building models in form of BIMs, and few of them are also capable of acquiring information from BIMs through the use of a shared resource (such as a model server database). Building Information Modeling is applied in many different areas, for example either BIMs are used as a resource to enable interoperability or Building Information Modeling has been realized as a process of managing a project through a single shared information backbone. Recent research in the area also demonstrated how 4 to N dimensional simulation applications can be facilitated using BIMs (Rebolj et al., 2010; Spearpoint, 2010). Countries such as Singapore has used BIMs to validate that building models are compliant with national code and regulations. Research has also demonstrated that Building Information Modeling can facilitate the design of energy efficient buildings towards addressing sustainability and reduction in CO2 emissions issues (Solis & Mutis, 2010; Bee Hua, 2010).

As stated in Isikdag & Zlatanova (2009) in order to effectively automate several urban management tasks, information regarding buildings needs to be represented in form of geospatial information. Several tasks of urban management such as, indoor navigation, emergency response management and some processes in the construction life cycle, like site selection (i.e., selection of the land plot for a designed project) can be facilitated through the use of certain and sometimes high amount of, geometric and semantic information about a building. Recent research efforts such as Isikdag (2006), OGC Web Services Phase 4 (OWS-4 Summary Document, 2007; Lapierre & Cote, 2008) and implementations in commercial software (i.e., IfcExplorer, 2008; Safe Software, 2009) have demonstrated that geometric and semantic information from BIMs can be transferred into the geospatial environment and CAD/GIS/BIM information can be integrated at web services level. In fact, as there are major differences in, geometric representation objects and semantics between the models in CAD/BIM and GIS domains, algorithms for seamless translation from BIMs into the geospatial environment are still in development.

In the provided context, this paper presents an opportunity analysis focusing on the future role of BIMs in Urban Data Management. The comparative analysis explained here has been completed based on two studies, first study was Isikdag and Underwood (2010) which underlined the current research directions and trends of Building Information Modeling in light of a recent edited book in the field, and the second one (Isikdag & Zlatanova, 2009) was a SWOT analysis about the implementation of BIMs within the Geospatial Environment. The comparative analysis was carried out based on a matrix defined for evaluating the impact of Building Information Modeling research trends on the opportunities provided by the use of BIMs within the Geospatial Environment. The paper starts with a background section summarizing the findings of the two studies which forms the background for the comparative analysis. The following sections provide the details on the comparative analysis performed and discuss the findings of the analysis.

2 BACKGROUND

2.1 *Research directions in Building Information Modeling*

Isikdag & Underwood (2010) defined Building Information Modeling as

> *"the information management process throughout the lifecycle of a building (from conception to demolition) which mainly focuses on enabling and facilitating the integrated way of project flow and delivery, by the collaborative use of semantically rich 3D digital building models in all stages of the project and building lifecycle"*

From this perspective a Building Information Model(s) can be defined as

> *"the (set of) semantically rich shared 3D digital building model(s) that form(s) the backbone of the Building Information Modeling process"*

In addition to its well-recognized role in facilitating the design phase of the project lifecycle, Isikdag & Underwood (2010) argued that depending on the environment in which they are used, Building Information Models can have different functions such as being a

Space Linker that links macro and micro urban spaces, an Interoperability Enabler which facilitates information sharing between various stakeholders and the software applications they use, a Data Store which stores the building information throughout the lifecycle of a building, a Procurement Facilitator that facilitates several procurement related tasks in the building lifecycle, a Collaboration Supporter through enabling the use and management of shared building information in real-time, a Process Simulator by facilitating the simulation of construction processes (i.e., nD), a System Integrator which enables the integration of several information systems across the industry, a Building Information Service which can serve real-time on-demand building information over the Internet, a Green Builder that enables advanced analysis supporting the design and construction of environment friendly/energy efficient buildings, and a Life Saver which facilitates emergency response operations.

Based on the 28 chapters of The Handbook of Research on Building Information Modeling and Construction Informatics: Concepts and Technologies (Underwood & Isikdag, 2010), Isikdag and Underwood (2010) presented a research compass for Building Information Modeling. The research compass provided the future research directions for BIM that are derived from the perspective of the 50 authors of the handbook which include academics and industry professionals (i.e., the practical users of the BIMs). The following list summarizes 7 research directions for Building Information Modeling (that are relevant for this analysis) among 12 derived from the chapters:

1. *Adoption:* The implementation of product modeling approaches and Building Information Models has been a subject of research for nearly 20 years in the AEC industry. In fact, the low interest and investment on ICT and the lack of strategic perspective on the use and implementation of ICT has prevented the successful adoption of collaborative systems and interoperable information models within the organizations. Furthermore, the move from CAD based thinking to the vision of BIM is required but difficult as it involves the shift in fundamental data management philosophy. Thus, research towards facilitating implementation and use of BIM based information management tools and methods within organizational processes, is a requirement.
2. *Standardisation:* Early building product models made use of native ISO 10303 tools and methods to represent building information. Similarly, the BIM standard of today, such as IFC, is still making use of, i. the geometric model (Part-42) of ISO 10303 in representing building elements, and ii. methods of information sharing and exchange as defined in ISO 10303. As standardisation is viewed as a key enabler and facilitator of data level interoperability in the AEC industry, this area will continue to be a focus of BIM research.
3. *BIM, Lean and Green:* In recent years the construction industry has begun to adopt lean production principles that have been applied in the manufacturing industries. This approach to construction management is known as Lean Construction. As also overviewed by Solis & Muits (2010), the aim of Lean Construction is to enable continuous improvement of all construction processes in the building lifecycle (starting from design through the demolition of the building). On the other hand, the AEC industry now is taking the initiative to build more 'environment-friendly' buildings. The use of shared building models together with collaborative environments (i.e., based on Building Information Modeling methods) during design and construction process can contribute to leaner and greener construction (and buildings) by the use of intelligent BIMs for design optimisation (i.e., in CO_2 emission analysis, etc.).
4. *Construction Process Simulation and Monitoring:* The approach to visual simulation of construction processes is known as 4D CAD. Analysis such as clash detection can now be completed using BIM software. BIMs are also used in monitoring the construction progress at the site. There is a strong research interest in developing BIMs to support and facilitate process simulation and progress monitoring.
5. *BIM based Service Oriented Architectures:* A current trend in the software industry is towards enabling interoperability over web services. The use of model servers is now increasing with open-source implementations such as BIMServer (BIMServer, 2010).

The future Building Information Modeling approaches would require the shared models in model servers to be linked with external systems in a heterogeneous environment. The BIMs will form the core data component of the future model-driven service oriented architectures of the AEC industry.

6. *Building and GeoInformation Integration:* As mentioned by Peters (2010), there is a significant value in integrating BIM and GeoInformation Systems into a single system. The differences between CAD applications and GeoInformation systems have generated barriers of integration of different representations of buildings in AEC and urban models. Van Oosterom et al. (2006), Isikdag & Zlatanova (2008) draw attention to the need of integrated geometric models and harmonised semantics between these two domains in order to tackle the interoperability problems. Song et al. (2010) explained that a range of stakeholders such as building contractors, estate agents, city management, and public sector etc. will benefit from the integration of BIM and (3D) GIS. There is a major research interest for integration of building and geoinformation, and BIMs can play a key role in these integration efforts.
7. *Supporting Emergency Response:* Emergency response operations indoors, require a high amount of geometric, semantic and state information related to the building (and building elements) in form of geoinformation. Until very recently Egress Models used in building evacuation has mainly been based on 2D floor plans. This situation is now about to change as the required level and amount of information can now be transferred from BIMs into the geospatial environment. In addition, BIMs are now used in fire simulations. Thus, the representation of semantically rich digital building models (derived from BIMs) in the geospatial environment will bring significant opportunities for supporting Emergency Response operations.

2.2 *A SWOT analysis on the use of BIMs in geospatial environment*

SWOT is an acronym for Strengths, Weaknesses, Opportunities and Threats, and is a well-known approach used prior to defining a strategy (Zach,1999). The technique is mainly used for a analyzing a company's internal capabilities (i.e., strengths and weaknesses) in relation to the competitive environment (i.e., opportunities and threats) (Avison & Fitzgerald, 2006). In the review presented in Isikdag & Zlatanova (2009), the SWOT analysis is used within a different perspective. The first stage of the analysis was focused on the technical perspective of the implementation of BIMs in the geospatial environment and evaluated the strengths and weaknesses which appear as a result of using BIMs in acquiring building information and transferring it into the geospatial environment. The second stage of the analysis, i.e., on the opportunities and threats explored the processes that can be facilitated by the implementation of BIMs within the geospatial context and the risks this implementation might bring. A list of opportunities risen by the implementation was determined as follows:

1. The implementation of BIMs in geospatial environment will facilitate the site selection process. For example, when the floor plan geometry, floor plan area (in sqm.) and number of stories is transferred into the geospatial environment, this information will help in finding out if the proposed building can be constructed over the—land plot in consideration—depending on the rules and regulations (such as, the maximum allowed value for—floor plan area/land plot area ratio-, maximum number of stories allowed for that land plot and so on.).
2. The implementation of BIMs in the geospatial environment will help in evaluation of the design proposals by answering questions such as "What windows have views of a particular spot?", "How will a design proposal affect views and shadows in an urban scene?" from multiple stakeholders' perspectives.
3. Conditions which affect both the lighting and thermal requirements of the buildings are related to the geographic location and orientation of them. Different elements of natural and built environment (i.e., hills, trees, buildings) can affect these requirements

Developing geoinformation models (enriched by detailed building information) will facilitate the analyses on energy consumption and lighting requirements.

4. Logistics operations are usually managed within a geospatial environment, and if geometric and semantic information—representing the current state of the construction-can be transformed from BIM into geoinformation models, this up-to-date information can facilitate the logistic operations in the construction sites (which occur at various stages of the construction).
5. The transfer of semantic information from BIMs into the geospatial environment will help in assessing the damage caused by a disaster (e.g., damaged electrical elements can be predicted by using the BIM in a flood situation). In addition, the assessment of the damage after a disaster will support the design of a renovation project.
6. The information transferred from BIMs can help in generation of building models with high Level of Detail (LOD) in 3D City Models.
7. The implementation of BIMs within the geospatial context will provide emergency responders with tools that will help in two aspects, i. these tools will facilitate orienting (as, the response personnel will know the geometry of the construction and possible exits in advance) and ii. these tools will also enable safer indoor navigation and evacuation as the response crew will be informed about the usage type of the different rooms (e.g.,—a room might contain flammable chemicals), and materials of the building elements (e.g., a type of flooring might get slippery when its wet).
8. Efforts towards developing 3D (indoor) geo-coding systems have started to emerge in recent years. The implementation of BIMs in geospatial context will help in developing models and algorithms for automatic 3D geo-coding.
9. The transfer of information from BIMs into the geospatial environment can aid in registration of apartment rights and rights related to different spaces (i.e., shops, shopping arcades, garages) in 3D cadastral registries.
10. The need of collaborative environments in public participation is becoming apparent. The BIMs and 3D City Models, together with integrated building and geoinformation will act as key pillars of the information infrastructure for these collaborative environments.
11. In some countries, the property tax evaluation process requires geometric and semantic information on building elements/parts and furniture, such as the precise dimensions of the rooms, and the number and type of fixtures located within a house. In addition, any structural changes in the house or property also affects the amount of the tax. The implementation of BIMs in geospatial environment will facilitate the tax evaluation process as required information can be acquired from the digital model of the building.

As mentioned in Section 2.1, the developments as a result of research related to Building Information Modeling in 7 research directions may have various impacts on, i. the use of BIMs in the context of Urban Data Management, and on ii. the opportunities provided by the use of BIMs. In this paper we focus on the latter. The following section of the paper provides an opportunity analysis, in light of an "Analysis Matrix" developed for exploring the impact of progression in each BIM research direction to the opportunities provided by the use of BIMs in Urban Data Management context. The results of the analysis are expected to provide a projection to future use and role of BIMs in the Urban Data Management scope.

3 OPPORTUNITY ANALYSIS

The Table 1 shown below provides the Analysis Matrix used for evaluating the impact of advancements in Building Information Modeling related research to various opportunities provided by the use them in Urban (Built) Environment domain. Each column in the matrix corresponds to a research direction, which means either i. the research on Building Information Modeling will support that direction/trend or ii. Building Information Models will be evolved in a way that they will support that direction or aspect of design, management, or integration. Some research directions are more fundamental (Supporting the adoption of BIM into organizational processes; Supporting the Standardization of BIMs), while others are more practice oriented (i.e., Developing BIMs that would support Lean Construction and Green Design; Developing BIMs that would enhance Process Simulation and Monitoring; Developing BIM based Service Oriented Architectures; Supporting Building and GeoInformation Integration; Developing BIMs that would facilitate Emergency Response Operations).

The rows of the matrix correspond to opportunities provided by the use of BIMs in Urban Built Environment domain. The following provides the details of the analysis.

Table 1. Analysis matrix.

		BIM Research Directions			
		Supporting the adoption of BIM into organizational processes	Supporting the Standardization of BIMs	Developing BIMs that would support Lean Construction and Green Design	Developing BIMs that would enhance Process Simulation and Monitoring
Opportunities for Urban Environment	Facilitating Site Selection	Moderate	Moderate	Low	Low
	Evaluation of Design Proposals within Stakeholders	Moderate	Moderate	High	Low
	Facilitating the analysis on energy consumption and lightning requirements	Moderate	Moderate	Major	Low
	Integration of logistics operations into large-scale 4D simulations	Moderate	Low	Major	Major
	Assessment of damage (and in support renovation projects)	High	High	Low	Moderate
	Facilitating 3D Modeling of Urban Environment	High	Major	Low	Low
	Facilitating Evacuation Activities	High	High	Low	Moderate
	3D Geo-coding	High	Moderate	Low	Low
	Registration of Ownership Rights in 3D Cadastre	High	High	Low	Low
	Public Participation	Moderate	Moderate	High	Low
	Property Tax Evaluation	Moderate	High	Low	Low

(*Continued*)

Table 1. (*Continued*).

		BIM Research Directions		
		Developing BIM based Service Oriented Architectures	Supporting Building and GeoInformation Integration	Developing BIMs that would facilitate Emergency Response Operations
Opportunities for Urban Environment	Facilitating Site Selection	Moderate	Major	Moderate
	Evaluation of Design Proposals within Stake-holders	Moderate	Major	Moderate
	Facilitating the analysis on energy consumption and lightning require-ments	Low	Major	Low
	Integration of logistics operations into large-scale 4D simulations	Low	Major	Low
	Assessment of damage (and in support renova-tion projects)	Low	Major	Low
	Facilitating 3D Modeling of Urban Environment	Moderate	Major	Moderate
	Facilitating Evacuation Activities	Moderate	Major	Major
	3D Geo-coding	Low	Major	Low
	Registration of Owner-ship Rights in 3D Cadas-tre	Low	Major	Low
	Public Participation	Moderate	Major	Moderate
	Property Tax Evaluation	Low	Major	Low

3.1 *Research direction 1 → supporting the adoption of BIM into organizational processes*

In order for BIMs to become widely recognized and used in the industry, their adoption in the organizations will be supported by the Building Information Modeling research. As organizations adopt BIMs and integrate Building Information Modeling methodologies with their organizational processes the use of, shared information models and shared databases will increase and the esteemed vision of seamless data level interoperability will become a reality for the AEC industry. As site selection, evaluation of design proposals, energy consumption analysis, logistics operations based on current status of construction, proposal evaluation by public participation, and tax evaluation processes can benefit from a wide use of—commonly recognized shared models-, high adoption rate of BIMs in organizational processes will have a moderate impact on these processes. The impact would not be high/major as these processes can also be automated when the (organizational) BIM adoption rate is not very high. In fact, other processes such as damage assessment, 3D Modeling of urban environment, evacuation activities, 3D geo-coding and registration of ownership rights in 3D cadastres, will be significantly affected by the high adoption rate of BIMs in organizations. In other words, high availability of BIMs (as a result of high adoption rate and wide-use) will meaningfully help in these activities, as it will increase the reach to geometrically and semantically rich building information that will facilitate these processes.

3.2 *Research direction 2 → supporting the standardization of BIMs*

Although BIMs are—by definition—interpreted as standard means of exchanging information, many emerging applications now term their intelligent digital building models as the BIM. So, the research is still acting as a force driving the industry towards standardization and the use of recognized standards in sharing and exchange of building information. Industry Foundation Classes (IFC) is the widely recognized standard for Building Information Modeling.

Site selection, evaluation of design proposals, energy consumption analysis and proposal evaluation by public participation can benefit from a standard BIM, but more important issue in these is the requirement for an intelligent geometrically/semantically rich building model, thus the impact of standardization of BIMs to these processes would be moderate. The requirement for 3D geo-coding is similar, so the impact of standardization would not be high for this aspect as well. Damage assessment, evacuation activities, cadastral registration, property tax evaluation require up-to-date building information to be represented accurately in applications that are used in these processes. The standardization (on the source modeling domain) ensures that the information transferred to these target domains represents—agreed semantics . In addition, as the applications on the receiving-end (target domain) would have developed tools for importing standard models, the accuracy in transferring information from a standard BIM would be higher than accuracy in transferring information from a regular digital building model. Furthermore, as the source model represents the building elements with agreed semantics, the representations on the target model can be validated by making use of the source models' schema. In summary, in the processes where information accuracy is important and model transfer is a requirement, the use of standard BIMs would be highly beneficial.

The use of standard BIMs as source models would have a major impact on 3D Modeling of urban environment, as standardization of the source models (i.e., BIMs) would serve to quick and batch transfer of information from multiple BIMs to geospatial environment, as it would be easy to develop interfaces (APIs) or Web Services for transforming information from a standard BIM (i.e., IFC) to a standard 3D City Model (i.e., CityGML). Otherwise, when the source models are regular digital building models (i.e., with different object schemas), a need for developing adapters and translators from each of them to the target Modeling domain would become inevitable, and this would negatively affect the performance/accuracy of information transfer from BIMs into the geospatial environment. Finally, the process simulation activities can also be performed by using non-standard BIMs. Thus, BIM standardization efforts will have a low impact on logistics operations (which are performed based on current status of construction), as building information in these operations will be obtained from process simulation applications but not from BIMs directly.

3.3 *Research direction 3 → developing BIMs that would support lean construction and green design*

Currently AEC industry is focused on enabling and facilitating, lean construction (i.e., by applying the lean production principles to construction) and design/construction of environmentally friendly and sustainable buildings. BIMs—as being facilitators of all information management processes in the lifecycle of building—should also support lean construction processes and construction of environmentally friendly buildings. The geographic location and positioning of the building will have an impact on the energy consumption and this impact is seriously considered in evaluation of design proposals. Thus, BIMs providing more information for energy consumption calculations/analysis would be highly beneficial in evaluation of design proposals (with some stakeholders or with public participation). As aimed in this research direction, when BIMs contain more information regarding energy requirements and consumption, this would have a major impact for facilitating energy consumption analysis. Lean construction depends on lean supply and Just-in-Time (JIT) delivery of materials to the construction site. When BIMs are developed for providing support to enabling leaner processes (i.e., by use of process simulation tools) and (JIT) logistics operations, this would have a major impact on facilitating the logistic operations performed based on 4D simulations. Other opportunities in the matrix (not mentioned here) will not be much affected by the developments in this research direction (i.e., the impact level is considered as low for all other opportunities).

3.4 *Research direction 4 → developing BIMs that would enhance process simulation and monitoring*

Process simulation, visualization of construction based on process simulation, model based monitoring of construction activities are currently being accomplished by the use of semantically rich building models, and the use of BIMs within these simulation and monitoring applications is highly encouraged. This type of simulations is generally termed as 4D/nD CAD. If an emergency situation occurs during the construction of a building, information acquired from the 4D/nD Model (which is in-sync with the actual construction) can serve for assessing the damage in the building, and can assist to navigation and evacuation activities within the building, by providing information based on the current state of the construction. Information acquired from the 4D/nD model would enable routing and wayfinding within the building (based on the current status of the construction). For example, the navigation path from room A in first floor to room B in second can naturally be different depending on the current state of the construction (i.e., a staircase connecting floors might not have built yet at the time the emergency situation occurred, although it is represented in the architectural model of the building). Thus, information regarding current state of the building is of a key importance for evacuation processes and for accurately assessing the damage when a disaster occurs during the construction, and when required this information can be acquired from applications supporting process simulation and monitoring. For this reason, developments in Building Information Modeling towards enhancing process simulation and progress monitoring will have a high impact on, facilitating evacuation processes and damage assessment exercises. The major impact of developing BIMs that would enhance Process Simulation and Monitoring would be facilitating, the 4D simulations, and in consequence the logistics operations that acquire information from these simulations. Other opportunities in the matrix (not mentioned here) will not be much affected by the developments in this research direction (i.e., the impact level can be regarded as low for all other opportunities).

3.5 *Research direction 5 → developing BIM based service oriented architectures*

Although Building Information Modeling is interpreted as a model-based interoperability and integration paradigm, the use of model-driven service oriented architectures is highly encouraged by researchers and technology providers in the field. In the proposed model-driven web service architectures, a shared BIM forms the core of a web service, where multiple clients (or systems) can interact with—to acquire information from the shared model—. The use of BIM based web services would facilitate interactions on the 3D visual representation of the BIM in web-based collaborative environments. The use of BIMs in such environments would have a considerable impact on facilitation of, i. the processes in site selection, evaluation of design proposals and proposal evaluations by public participation and ii. the communication and coordination in emergency evacuation processes. Furthermore, as shown in the OGC Web Services Phase 4 (OWS-4 Summary Document, 2007) service oriented architectures can have an impact on information integration towards representation of buildings together with (or within) a 3D City Model. The impact of this research direction to other opportunities mentioned in the matrix can be considered as relatively low when compared with the areas mentioned previously.

3.6 *Research direction 6 → supporting Building and geoInformation integration*

BIMs are generated by using CAD systems. The approach to Modeling in CAD Systems and GIS are significantly different, and in consequence, the resulting information models have differences in representation of the buildings (Isikdag & Zlatanova, 2009). CAD systems are developed to model objects that do not exist prior to the generation of the model, and are designed for representing the maximum level of detail in terms of, geometry and attributes of the model. On the other hand, GIS are developed to represent objects that already exist around us, and geospatial data models are defined for representing the objects in a way that they can be recognized most easily. In addition, GIS highly require and make use of the geo-referenced geometries in representing the real-world objects. The differences in the representations of buildings in both systems (CAD/GIS) and in both information domains (AEC/Geo), together with the need for representing highly detailed building information in geospatial environment, have ignited the efforts towards enhancing BIMs for supporting building and geoinformation integration. This integration would have a major impact for facilitating all the processes (opportunities) mentioned in the impact analysis matrix (as all of them would be highly facilitated as a result of a building and geoinformation integration).

3.7 *Research direction 7 → developing BIMs that would facilitate emergency response operations*

A digital building model designed to support emergency response operations should provide, i. detailed 3D geometric representation of building elements (which should be geo-referenced and compatible with GIS representations), ii. semantic information regarding the building elements (i.e., including a clear definition and naming of objects and their relations), iii. properties of each building element (such as material of the walls, opening directions of the doors, doors used as exits) and iv. states of them (e.g., door can be 'open' or 'closed' at a certain point in time). In the first stage BIM based information management methods can be enhanced to provide geo-referenced and GIS compliant geometric representations of building elements when required. Once this achieved, the transfer of geometric information from BIMs into the geospatial environment will be a straightforward task. Furthermore, information regarding the object states needs to be stored in the BIMs. In this case, the (geometric/semantic) information transferred from BIMs (which also provide object's states) into the geospatial environment will positively affect, the site selection activities, evaluation of design proposals (by stakeholders and public participation), and representation of 3D building information in city models. In fact, the major impact of this research direction would be on facilitating the evacuation activities—especially when the object states are stored in BIMs along with geometric/semantic information and transferred into the geospatial environment—. In fact, we view the impact of this research direction to other opportunities mentioned in the matrix as relatively low.

4 CONCLUSIONS

In recent years the vision of model-driven management of AEC processes is becoming a reality. Shared use of semantically rich building models (BIMs) which are defined by standardized schemas currently referred as Building Information Modeling. In fact, recent research and industry visions also regard Building Information Modeling as an information management process and strategy that spans along the lifecycle of a project and facility. In this respect, the shared information models (i.e., BIMs) forms the core of this new emerging information management strategy. This new strategy focuses on enabling seamless integration and interoperability between different applications used in automating the tasks in the lifecycle of a facility/building by the use of shared digital models.

Although this new strategy and shared digital models has been developed for facilitating the various tasks in the AEC industry, the use of them is now positively influencing several processes related.to Urban Data Management. A SWOT analysis recently carried out by the authors has outlined the tasks that would be facilitated by this positive impact of implementation and use of the BIMs in Urban Data Management Systems. The results of the SWOT analysis demonstrated that activities including site selection, evaluation of design proposals, emergency response and post-disaster damage assessment, registration of ownerships in 3D cadastre and even tax evaluation will benefit from this implementation.

The analysis explained in this paper was focused on exploring, the impact of progression in Building Information Modeling research, to the opportunities provided by the use of BIMs in Urban Data Management context.

The analysis results indicate that BIM research towards supporting Building and GeoInformation Integration would have a major impact on facilitating all processes/ opportunities in focus. Other research directions that would have a considerable impact on all opportunities in focus were found as, i. research towards facilitating organizational adoption of Building Information Modeling and ii. research towards facilitating standardization of BIMs. Developing model-driven service oriented architectures and developing BIMs that would facilitate emergency response operations will have an impact on half of the BIM related opportunities in focus. On the other hand, the impact of i. research towards, developing BIMs for supporting lean construction and green design, and ii. developing BIMs that would enhance process simulation and progress monitoring, found as relatively low. Further research will focus on exploring how this 7 research directions related to BIM research will have an impact on the opportunities related to UDM by via a series of surveys and one-to-one interviews with participants from the AEC industry.

REFERENCES

Avison, D. & Fitzgerald, G. 1996. Information Systems Development, McGraw-Hill, Berkshire, UK.

BIM Server, 2010. *Open Source BIM Server Available* online at: http://www.bimserver.org Last Accessed: 05 Jan 2010.

Hua, G.B. 2010. "A BIM based application to support Cost Feasible 'Green Building' concept decisions" in J. Underwood and U. Isikdag (eds.) *Handbook of Research on Building Information Modeling and Construction Informatics: Concepts and Technologies,* IGI Global.

IFC Explorer, 2008. *Tool for viewing and conversion of IFC models* http://www.iai.fzk.de/www-extern/index.php?id=1040&L=1 [last accessed 11-2010].

Isikdag U. & Zlatanova, S. 2008. Towards defining a framework for automatic generation of buildings in CityGML using Building Information Models, *Submitted to 3D Geoinfo 08.*

Isikdag, U & Underwood, J. 2010. A Synopsis of the Handbook of Research on Building Information Modeling, In *Proceedings of CIB 2010 World Building Congress* http://www.cib2010.com/ [last accessed 01-2011].

Isikdag, U. & Zlatanova, S. 2009. A SWOT analysis on the implementation of Building Information Models within the Geospatial Environment., In A. Krek, M. Rumor, S. Zlatanova & E.M. Fendel (eds.) *Urban and Regional Data Management. UDMS Annual 2009*, CRC Press.

Isikdag, U. 2006. Towards the Implementation of Building Information Models in Geospatial Context, *PhD Thesis*, University of Salford, UK.

Lapierre, A. & P. Cote, 2008. Using Open Web Services for urban data management: a testbed resulting from an OGC initiative offering standard CAD/GIS/BIM services, In Coors, M. Rumor, E. Fendel & S. Zlatanova (eds.): *Urban and Regional Data Management; UDMS Annual 2007*, Taylor and Francis, London, pp. 381–393.

OWS-4 Summary Document, 2007. *OGC Document 07–037r4:Summary of the OGC Web Services, Phase 4 (OWS-4)* Available online at: http://www.opengeospatial.org/ projects/initiatives/ows-4 [last accessed 11-2007].

Peters, E. 2010. BIM and Geospatial Information Systems. In J. Underwood and U. Isikdag (eds.) *Handbook of Research on Building Information Modeling and Construction Informatics: Concepts and Technologies*, IGI Global, 2010.

Rebolj, D., Babic, N.C. & PodBreznik, P. 2010. Automated Building Process Monitoring, in J. Underwood & U. Isikdag (eds.) *Handbook of Research on Building Information Modeling and Construction Informatics: Concepts and Technologies*, IGI Global.

Safe Software, 2008. *FME Desktop Translator/Converter Software* http://www.safe.com/products/desktop/formats.php [last accessed 11-2010].

Solis, J.L.F. & Mutis, I. 2010. The Idealization of an Integrated BIM, Lean, and "Green model (BLG)" in J. Underwood & U. Isikdag (eds.) *Handbook of Research on Building Information Modeling and Construction Informatics: Concepts and Technologies*, IGI Global.

Song,Y., Boghdan, J., Hamilton, A. & Wang, H. 2010. Integrating BIM with Urban Spatial Applications: A VEPS perspective, in J. Underwood & U. Isikdag (eds.) *Handbook of Research on Building Information Modeling and Construction Informatics: Concepts and Technologies*, IGI Global, 2010.

Spearpoint, M. 2010. Extracting Fire Engineering Simulation Data from the IFC, in J. Underwood & U. Isikdag (eds.) *Handbook of Research on Building Information Modeling and Construction Informatics: Concepts and Technologies*, IGI Global.

Underwood, J. & Isikdag, U. 2010. *Handbook of Research on Building Information Modeling and Construction Informatics: Concepts and Technologies*, IGI Global, 2010.

Van Oosterom, P., Stoter, J. & Janssen, E. 2006. Bridging the worlds of CAD and GIS, In: Zlatanova & Prosperi (eds.): *Large-scale 3D data integration—Challenges and Opportunities*, Taylor & Francis, Boca Raton, pp. 9–36.

Zach, M.H. 1999. Developing a knowledge strategy, *California Management Review* 41(3): 1.

Urban and Regional Data Management – Zlatanova, Ledoux, Fendel & Rumor (eds)
© 2012 Taylor & Francis Group, London, ISBN 978-0-415-67491-1

Towards a 3D geographic information system for the exploration of urban rules: Application to the French local urban planning schemes

M. Brasebin & J. Perret
COGIT laboratory, IGN, Saint-Mandé, France

C. Haëck
IGN Conseil, Saint-Mandé, France

ABSTRACT: This paper presents the use of 3D GIS to represent, analyse and explore the most common urban rules described in French local urban planning schemes. As these rules are only expressed in free formed texts, it is difficult to assess their impact on the evolution of the urban environment. A schema is proposed in order to allow the capture and management of urban rules. These rules are applied to a structured geographic environment in order to automatically determine and represent their influence by 3D volumes. Three different applications of the proposed model are explored in this article: (1) to asses or emulate the influence of urban rules in terms of potential constructions (by visual analysis), (2) to determine their influence on the Floor Area Ratio (by spatial analysis), (3) to evaluate pre-project building shapes.

1 INTRODUCTION

The development of the French territory is worked out through large and complex sets of documents, plans or schemes (Territorial Coherence Schemes[1], local urban planning schemes[2], Urban mobility plans[3], etc.). These documents illustrate strategies of urban and territorial development but they are also the expression of these strategies throughseveral constraints that projects are subjected to. The underlying logic refers to several domains (construction, housing, transport, urban planning, etc.) for which the relevant level of detail is different. These difficulties and complexities highlight several issues and for planners' needs such as new techniques to better understand, evaluate, manage and design the city (Bain, Maujean & Theys 2008).

In the past, several methods were proposed in order to model the urban environment at different scales through different properties during the design process. Such declarative modelling provides tools for architectural or urban design (Donikian 1992; Desmontils 1995), to assess building volumes according to urban restrictions (Faucher 2001; Viet-Tung 2003; Shen & Kawakami 2004) or to take into account visual accessibility (Nivet 1999). (Murata 2004) proposed the use of a 3D GIS to visualise the state of a city, the capacity of urban blocks and to assess a construction project. (Gilles & Metral 2005) proposed a 3D urban knowledge visualisation tool and the prototyping of projects to improve communication and to involve non-experts. Finally, (Coors, Hunlich & on 2009) propose the constraint-based generation of building volumes where the constraints are adapted to the german urban planning regulations.

[1] Schéma de Cohérence Territoriale (SCoT) in French.
[2] Plan Local d'Urbanisme (PLU) in French.
[3] Plan de Déplacement Urbain (PDU) in French.

The goal of our work is to propose tools that integrate the specificities of urban knowledge linked to the urban rules of French urban planning schemes. Graphic representation and use of relevant symbols to represent urban regulation rules is another aspect of this work. Similar efforts are being made (Caillaud, Denegre, Barbier & Denis 2006) for the representation of such schemes on 2D maps.

2 A TOOL TO DISPLAY AND TO EXPLORE URBAN RULES

In this article, we propose three possible applications of our 3D GIS for the exploration of urban rules (*cf.* Figure 1):

1. checking whether a set of urban rules for existing or future buildings is respected on a parcel,
2. visualizing constraints inferred from a set of rules by means of automatically generated volumes,
3. computing volumes respecting a set of rules and a 2D footprint or a proposed building.

In Figure 1, the different necessary inputs are presented. Indeed, in order to study urban rules on a specific area, two pieces of information have to be modelled and integrated: the rules (*cf.* section 2.1) and the urban environment (*cf.* section 2.2).

2.1 *Urban rules*

The Local Urban Planning Scheme (LUPS) is a document laying down regulations based on a general approach to develop an area. This approach is the expression of urban project, including restructuring and development plans for the area,with constant collaboration with its citizens. Developed at the level of municipalities or inter-municipalities, LUPSs might eventually cover the entire French territory. Each LUPS specifies a zoning plan where each zone is assigned to a type. The typology of zones includes urban zones (type *U*), zones to be built up (type *AU*), agricultural zones (type *A*) and natural or forested zones (type *N*), which can be subdivided into subtypes (type *UA* for residential urban zones for instance). For each type (or subtype) of zone, 14 articles have to be defined. These articles have the same titles for all LUPSs (for example, article number 10 always concerns the height of buildings) but their contents are free so that each municipality can express its own set of customised regulations. Therefore, there is a great diversity in the expression of the rules of the same article in different LUPSs and one cannot pretend to formalise all possible LUPS rules.

As a matter of fact, only a selection of articles has been considered in this paper. This selection is twofold: it concerns the nature as well as the granularity of the objects the rules relate to. The first selection is linked to the focus of this study: buildings' location. Therefore, articles concerning car parks or green spaces were filtered out. The second selection comes

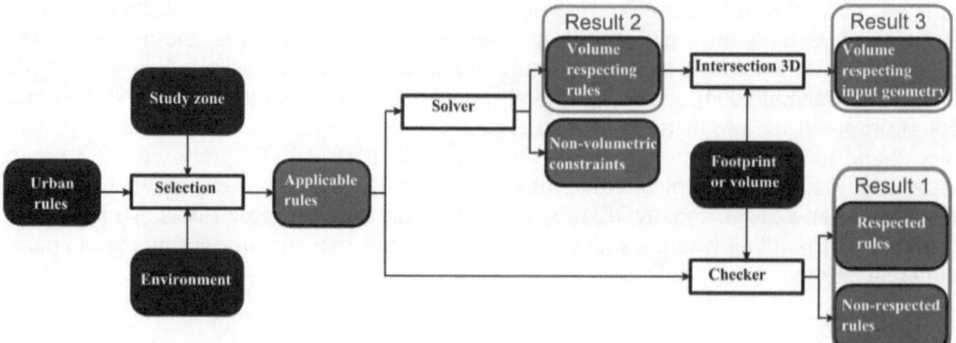

Figure 1. Global architecture of the application.

from the limitations of the available databases. Indeed, rules relating to elements of buildings such as balconies were kept out of the study because of the difficulty of acquiring and maintaining reliable data at such granularity on large areas.

Finally, we propose a schema for expressing the rules. This formalisation limits the range of the rules that can be represented. Indeed, only the most common expressions of these rules were kept. Further work is required to explore the impact of this limitation and the extension of the formalisation to include a wider range of rules. Table 1 presents the main articles and rules that were selected. LUPS refer to geographic information and relationships which have to be implemented in the system in order to utilize the different rules. For instance, in some rules, the distance of a building to a road border may depend on the width of this road: the knowledge of the width of roads is necessary to use this rule.

Selected rules can be formalised from the different elements identified in the study. This formalisation transcribes the logic of the administrative partition: a set of rules is defined for each type (or subtype) of zone. Once defined, a rule can be applied to a parcel associated to the type (or subtype) of zone.

A rule is composed of a condition and a set of constraints (as a consequence of the condition). Thus, if no condition is defined for a rule, the constraints always apply (i.e., for all parcels associated with the considered type of zone). Otherwise, the condition must be met for the constraints to apply. For instance, let us consider the following rule (the condition is in bold and the consequence is in italics):

If **the width of a road is smaller than 7 m** *then distance between the borders of this road and adjacent buildings must be higher than 3 m* will only apply to parcels with an adjoining road wider than 7 meters.

We classify the different elements of rules as follows:

Conditions

Conditions are divided into two categories:

- conditions concerning the properties of the road adjoining the parcel,
- conditions concerning the properties of the constructions on the parcel.

Constraints

Constraints are divided into four categories:

- *volumetric constraints* concerning limitations on volumes in which constructions are allowed on the parcel,
- *surface constraints* concerning the surface of the constructions in the parcel (either of the footprint or the total area of the buildings floors),
- *shape constraints* concerning specific shapes of certain elements of the constructions on the parcel, especially roof slopes,
- *appearance constraints* on the appearance of the constructions in the parcel (mostly concerning materials).

Table 1. Rules implemented in our prototype.

Title of the article	Constraints considered
Forbidden or allowed land uses	Appearance of the building's facades
Land uses subjected to particular restrictions	Building functional use
Building location in relation to roads	Distance between a building and a road
Building location in relation to parcel borders	Distance between a building and a parcel Buildable 2D surface
Building location in relation to other buildings	Minimal distance between 2 buildings
Building footprint	Footprint max area
Building maximal height	Maximal height of a building Difference of height between 2 buildings
Exterior appearance	Material choice of roofs and walls Roof slopes
Floor Area Ratio (FAR)	Limit values of the FAR

Volumetric constraints allow the creation of the maximal buildable volume. In the rest of this article, we focus on estimated constraints particularly relevant to 3D analysis:

Maximal height of a building: The aim of this constraint is to limit the maximal height H of buildings on a parcel (*cf.* Figure 2(a)). This means that the building geometry should be included in the solid processed from the extrusion of its parcel with a value H.

Floor Area Ratio (FAR): The FAR is calculated from the total floor area of buildings (surface of all the floors deducted by the surface of technical units, insulation, etc.) on a parcel divided by the parcel surface.

Distance between features (adjoining road, neighbouring parcel or building): These constraints limit the distance between a construction and another feature. Two kinds of distance are included in our model (*cf.* Figures 2(b) et 2(c)): the *Euclidean distance* and the *prospect distance*. The Euclidean distance indicates the distance at which a construction can be implanted from an other feature. The prospect distance is expressed through an affine equation whose maximal height limit depends on the distance from a feature: $D = s \times H + H_{ini}$, where s is a positive coefficient (the slope) and H_{ini} the maximal height to respect at the border of the feature.

2.2 *Geographic schema*

In order to draw conclusions from on a set of urban rules, geographic features, mentioned in LUPS, are organised in our approach according to a conceptual schema (*cf.* Figure 3). This schema only contains the required information for the proposed application to building location (objects such as natural areas or electrical wires are not included). For instance, evaluating the FAR requires information about buildings a parcel contains. As this schema is quite generic, several datasets can be used to describe the urban environment.

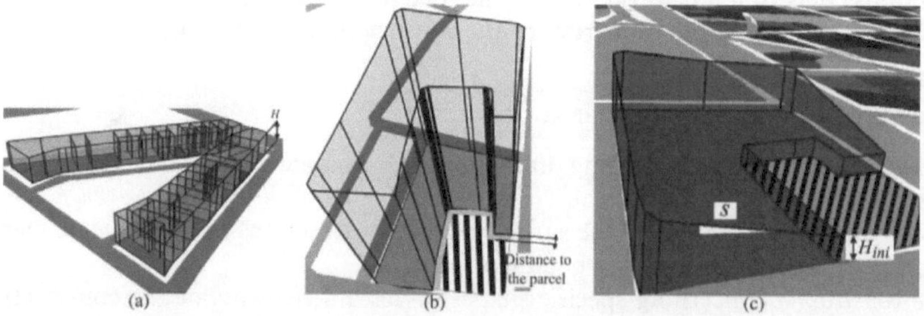

Figure 2. (a) Limitation of a maximal height H for a set of parcels. (b) Minimum distance to the hatched parcel. (c) Prospect distance from the hatched parcel with a slope s and height against it $Hini$.

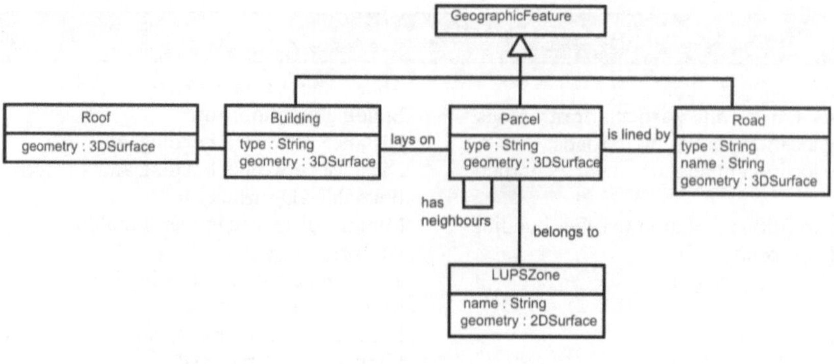

Figure 3. Schema of the studied objects of the urban environment.

3 IMPLEMENTATION

In order to implement such a system, it is necessary to create tools to format the inputs as described in our internal schemas. That refers to rules (*cf.* section 3.1) and the geographic environment (*cf.* section 3.2). The technical choices of the system are described in this part (*cf.* section 3.3).

3.1 *Graphic user interface dedicated to rule creation*

A graphic user interface is implemented in Java in order to capture and to export rules according to our schema. The JAXB (Java Architecture for XML Binding) library is used to implement our conceptual schema from a schema file (.xsd) and to save the corresponding data in XML format. The interface is the main window for filling in information about the LUPS zoning plan. It is possible to add or delete rules. Another window enables rule editing (to add conditions and consequences and to set their parameters).

Furthermore, in order make it easy to use, a text description of each rule element is generated. The exported XML file contains this description for a better understanding of its content. Any other XML file compliant with our conceptual schema can be imported into our system and uitilized. Since our rule model is based on a conceptual schema, it is possible to extend it in order to include new functions for the definition of new rules (adapted to other local urban planning schemes).

3.2 *Integration and data loading*

A pre-process (carried out during data loading) initializes the relationships between a parcel and the buildings it contains (a building belongs to only one parcel), between neighbour parcels and between a parcel and its surrounding roads.

In order to determine the relationships between roads and parcels, it is necessary to know the geometry of the borders of the roads. Nevertheless, in some databases, roads are modelled with a linear geometry (the road axis). In this case, the geometry of the borders is approximated by applying a buffer whose size corresponds to the attribute *width* of the roads. The link between a building and its roof is directly extracted from imported data. The application can directly represent and query relationships in the 3D navigator in order to check if the pre-process is successful. In this article, we propose a generic schema to represent relevant features and some of their attributes and relationships. This schema is extendable in order to include new geographic features, attributes and their relationships.

3.3 *System*

An urban environment and a set of rules (as described in the previous section) are used as input in the system. The process is then applied on a set of parcels selected by the user. Functions presented in this article are developed on the open-source GIS platform *GeOxygene* (http://oxygene-project.sourceforge.net/http://oxygene-project.sourceforge.net/) (Brasebin 2009). This platform implements the ISO 19107 specifications that describes a standard geometric schema (ISO 2003). It provides tools to load data (Shapefile, CityGML, etc.), 3D geometric functions (boolean operators, extrusions, etc.) and a viewer (which was used to produce most of the figures of this article).

4 RESULTS

Different results of the use cases of our application are presented in Figure 1.

4.1 *Presentation of the studied area*

In order to illustrate our results more practically, the different processes presented in the next section are applied on a specific area. The studied area is located in the centre of Paris with a high-density of buildings (Figure 4.2a). The relevence of this area is to determine its potential in terms of buildable area to assess the possibilities of densification in accordance with the LUPS regulation. This LUPS is expressed through several atlases: a zoning atlas, a general height atlas, a risk atlas and a synthesis atlas, that contain specific planning processes. The zoning plan of Paris is composed of a general urban zone (UG), an urban zone of services and natural zones. Our study area is located in the UG zone. The maximum building height is expressed by two atlases: the general height atlas, that defines the maximal height of a district, and a 1/2000 map that conveys information about the prospect distance on parcels according to adjacent roads (Figure 4.2b).

For most of the results presented in this article, the following rules are used:

- Minimal distance between 2 buildings on a single parcel is 6 m
- Minimal distance to the parcel border is 0 m
- Maximal FAR value is 3
- Prospect distance is applied according to the 1:2000 map
- Maximal height is 25 m

The study area was described by integrating 3 different datasets from IGN (*cf.* Figure 5):

- **buildings**, are from BATI3-D® dataset. It contains 3D buildings (with roofs) in the CitYGML format,
- **parcels**, are loaded from the BD PARCELLAIRE® (RGE) dataset. This dataset represents French cadastral parcels,
- **roads**, are imported from the BD TOPO® (RGE) dataset. It covers the French territory at the 1/25,000 scale. In this dataset, road geometries are described as 3D lines.

Since, the FAR of a parcel is not described in our database, it needs to be evaluated. Empirically, to calculate a parcel's FAR, the total floor area of its buildings is approximated by using 80 percent of the surface of all the floors of the buildings. Furthermore, the surface of floors of all the buildings is evaluated using the surface of the buildings' footprints and their number of floors (approximated by dividing the height of the buildings by the empirical height of a level: 3 m).

4.2 *Verification of the rules of a LUPS*

The first result presented (*cf.* Figure 6) is a check-up of the rules from a LUPS on a set of parcels. The system tests for a collection of parcels if the rules concerning the buildings (real or emulated) of each parcel are met or not. On the scale of a parcel, this enables to verify if a new project respects the different constraints imposed by the LUPS. On the scale of a district or a city, this feature is useful to visualize the less respected rules while anticipating a modification of the LUPS.

Results are two-fold. Each parcel is automatically associated with a list of rules that are respected and another list of rules that are not respected. Moreover, a layer containing

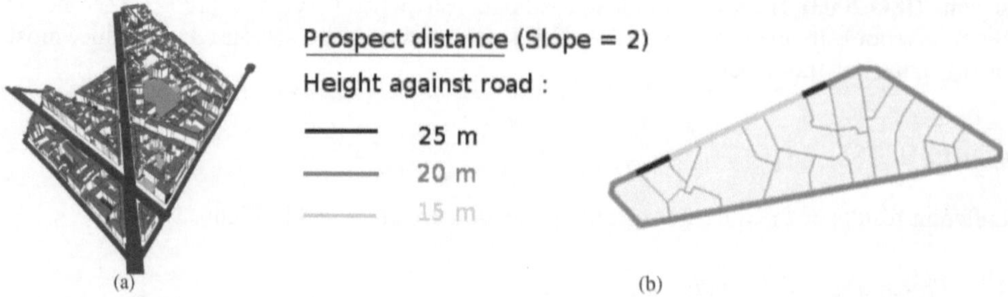

Figure 4. (a) 3D representation of the studied area. (b) Height plan of the studied area and its legend.

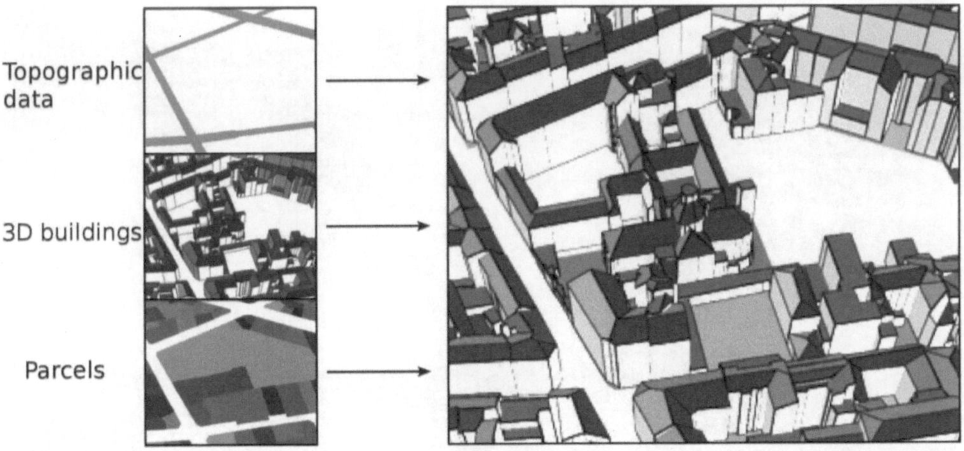

Figure 5. Layers composing the urban environment.

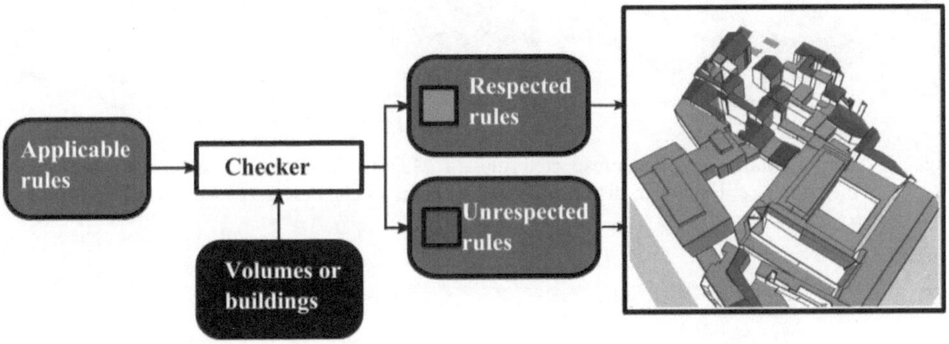

Figure 6. Verifying a rule that concerns the roof angle.

localized rule violations is created. This layer enables the user to zoom on a particular violation and to visualize it thanks to an adapted representation. For each type of rule, a specific representation is proposed.

Different tests have been carried out to determine the representation of unrespected rules. Violations of rules linked to the prospect distance or to the maximal height are represented by a plane whose altitude is set according to the maximal height allowed (*cf.* Figures 7(a) et (b)). This representation enables to easily distinguish the parts of the volume that respect the rule and the parts that do not. Violations of rules that concern the Euclidean distance between features are represented by a volume with a certain coefficient of transparency corresponding to a non-buildable zone (*cf.* Figure 7(c)). It is thus possible to visualise the parts of a building that do not respect the rule by observing the color of the building sides. The textual display presented in Figure 7 (d) represents the respect of the FAR (a dark text with a white halo if the FAR is not respected or a white text with a black halo if it is).

Some results of rule verification on the studied zone are presented in Figure 8. A 3D extract of our scene after verification processing is shown in Figure 8(a). In this figure, three different types of unrespected rules are represented:

- maximal height,
- prospect distance,
- maximal FAR: this rule is generally unrespected (*cf.* Figure 8(b)). This is explained by the fact that, recently, the maximal FAR decreased from 3.5 to 3.0, long after most of the buildings of this area were built.

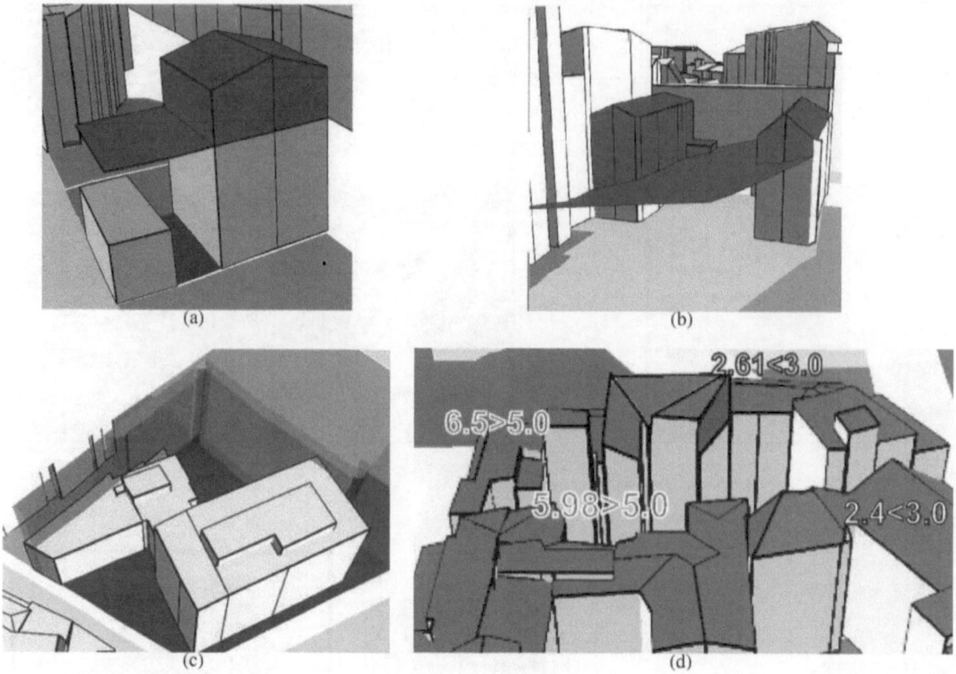

Figue 7. Representation of some unrespected rules. (a) Maximal height. (b) Prospect distance from the border of the parcel. (c) Euclidean distance from the border of the parcel. (d) Respect of the maximal FAR value.

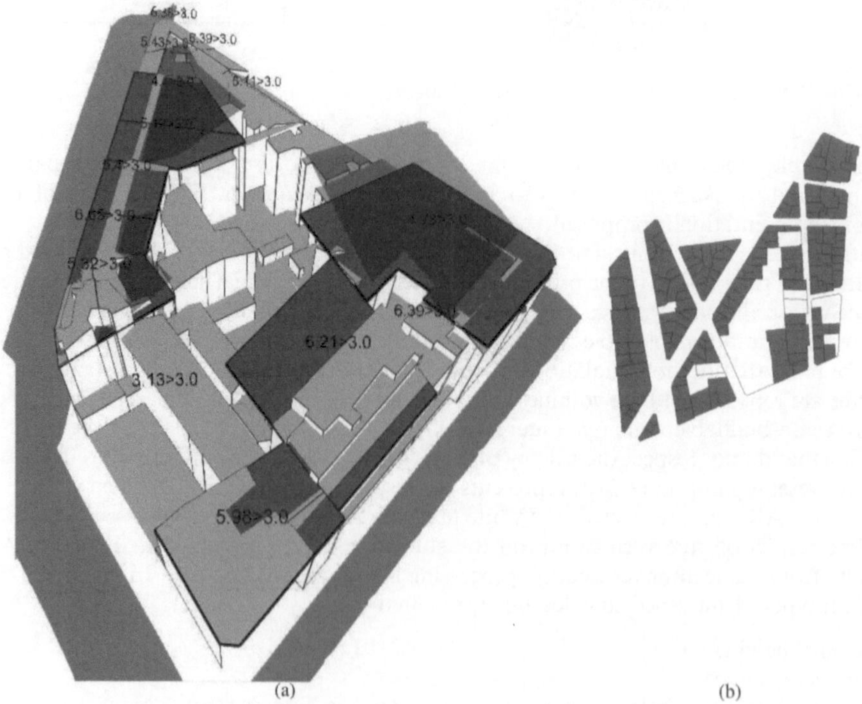

Figure 8. (a) Representation of unrespected rules on an extract of the studied zone. (b) Dark parcels have a higher than 3 FAR.

4.3 *Computation of a volume with respect to urban rules*

The second function presented is automatic evaluation, from a set of applicable urban rules, of volumes that respect inferred geometric constraints. Here, we only consider the volumetric constraints. aint that immediately modify the buildable maximal volume of a parcel. The other constraints do not permit us to directly deduce their influence on the maximal buildable volume (such as, for example, the roof angle). The application can handle four types of non volumetric constraints: roof angle, material used, FAR and footprint area limitation.

This function described in Figure 9 enables us to visually take into account the impact of a LUPS or of its variation on unbuilt parcels.

Computation is processed through a solver (*cf.* Figure 10) that uses geometric operators. Some necessary 3D functions, such as 3D intersection, are time consuming. This is why a strategy was developed in order to limit the repetition of such operations.

The first step consists in sorting the different constraints that can be applied on a parcel. Constraints are sorted in to four categories and are processed in this order: 2D constraints, height rules, rules about prospect distances and non geometric rules. *2D constraints* limit the buildable surface of a parcel in 2D (Euclidean distance between features and imported buildable areas). They are processed together to define the buildable part of a parcel (which is similar to the Baufenster found on German zoning maps (Coors, Hunlich & On 2009). Let us specify that the surface of existing buildings on the parcel is considered as non-buildable. If the result is a collection of simple polygons and not a single polygon, each polygon (each buildable part of a parcel) is treated separately. *Height constraints* limit the height of buildings and the difference of height between buildings on the same parcel. dable volume generated. *Prospect constraints* generate a volume that corresponds to the mathematical definition of these constraints (*cf.* section 2.). Several volumes can be generated on the same parcel if this kind of rules is defined for the different features from the urban environment. Currently, *Non volumetric constraints* do not directly influence the final result of this work as they do not enable one to produce only one solution.pects these rules (Notably, footprint area and FAR limitations).

In order to generate a final maximum buildable volume, first volumes are obtained by the extrusion of the 2D buildable areas of the parcel according to height rules. Theses volumes are then intersected with a set of volumes obtained by prospect constraints to process the final volumes. Figure 11 presents the different steps of the generation of a final volume

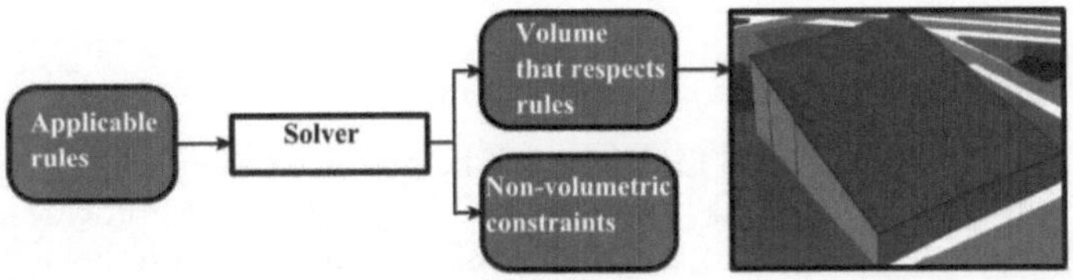

Figure 9. Creation of a volume that respects a set of rules.

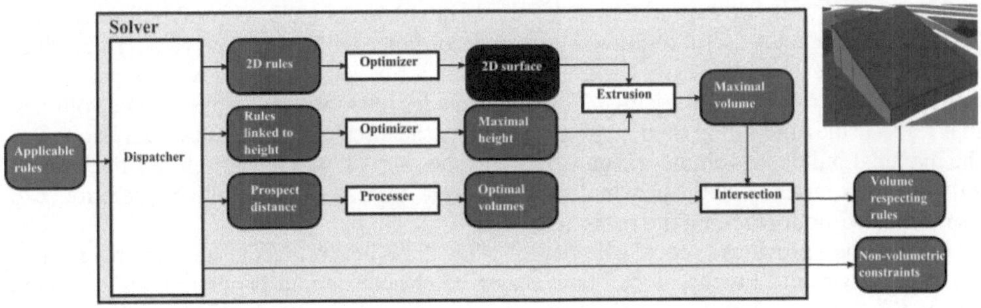

Figue 10. Solver organisation.

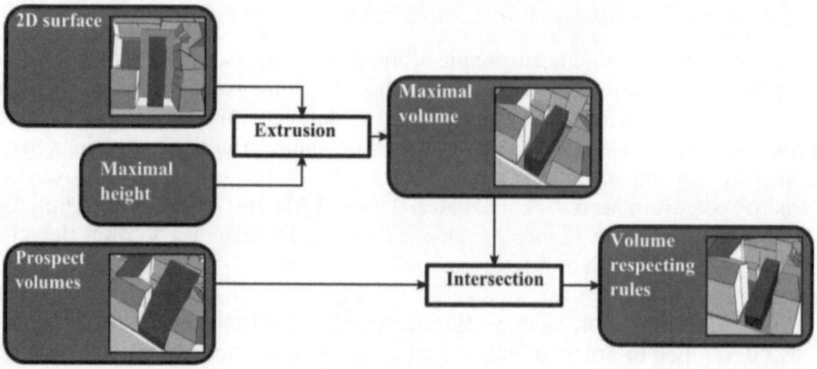

Figure 11. Steps to generate the maximum buildable volume of a parcel.

with the following set of constraints: 15 m as maximal height, prospect distance with 5 m as maximal height againstthe road and a coefficient (s) of value 2, a minimum distance to the parcel borders of 4 m and a minimum distance to the road of 2 m.

This process was applied on the studied area the selected rules presented in 4.1. Results are presented in Figure 12, they only integrate volumetric constraints. They provide a way to visualise the potential impact of adding new buildings in this area.

These results can be used better together with a map of buildable areas (Figure 13). 3D processes enable one to produce this map thanks to the FAR assessment and enable one to determine which areas have sufficient potential to receive new buildings. Thus, with these two results, it is possible to know automatically which parcels have a building potential, an assessment of the cumulated floor area that can be added on a parcel and the volume that must contain new construction. In our area of study, some of the buildable volumes spread on parcels with a sufficient buildable potential, this result will be used in the next section to include propositions of new constructions. We can notice the presence of a parcel (at the bottom right of the area) with a sufficient buildable potential whose buildable volume is void: its FAR is lower than the maximum limit but no construction can be added due to urban planning rules (notably distance between buildings on the same parcel). Nevertheless, it seems possible to raise the buildings of this parcel to densify it.

4.4 *Use of a maximum volume in a pre-project phase*

The results of the fore-mentionned functions can be perceived as rough, they process some results whose shapes do not look like plausible buildings. In order to provide some help for the pre-project phase, there is a possibility to cross the information on the maximum buildable volume with propositions from users. Two options are available: the use of building footprints or the import of a 3D building. Thus, our application proposes a final solid that respects the input geometry and a set of applicable rules (*cf.* Figure 14).

In our case study, we captured a set of footprints from our 2D map of buildable areas (Figure 15 (a)), that is useful to draw shapes in accordance with our urban rules. To produce 3D volumes, the system assesses the maximum number of levels allowed with respect to the FAR limitation (considering 1 level = 3 m) and extrudes the footprints. To end, final volumes, that respect rules and initial shapes, are produced by the intersections of these extrusions and the maximal buildable volumes (Figure 15 (b)). The map of buildable area can be updated with our area enriched by the new buildings. An large decrease in the buildable potential can be notified without reaching the maximum FAR.

This function provides a global overview of what could be a project in such an area regarding these constraints. Furthermore, theses shapes can be considered as possibilities to explore during the creative conception work of the city planner as mentioned in (Edmonds & Candy 2002).

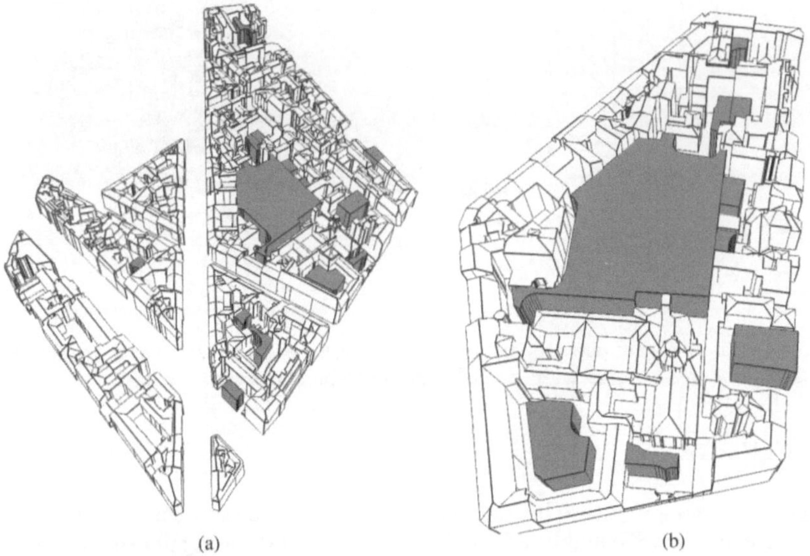

Figue 12. Views of buildable volumes (in dark gray) according to the study case. (a) Global view of the zone. (b) Zoom on an urban block.

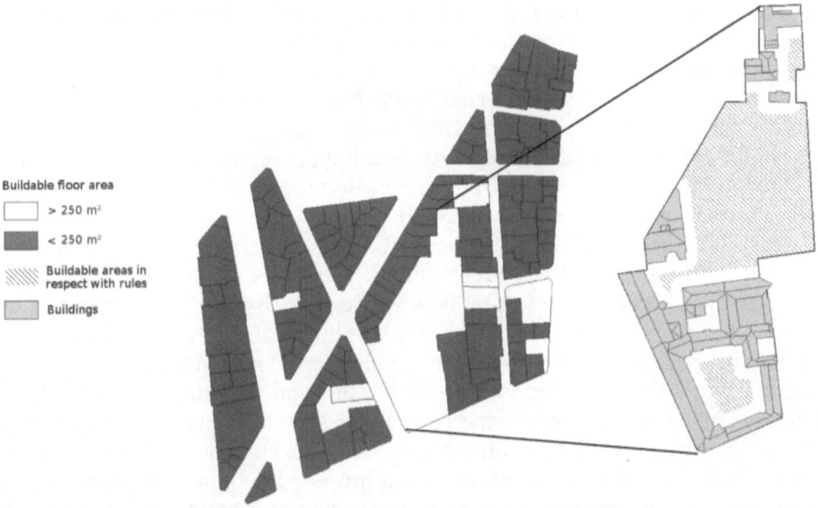

Figure 13. Buildable areas with respect to FAR limitation: map of parcels and extract of a parcel with its buildings and its buildable volume in 2D.

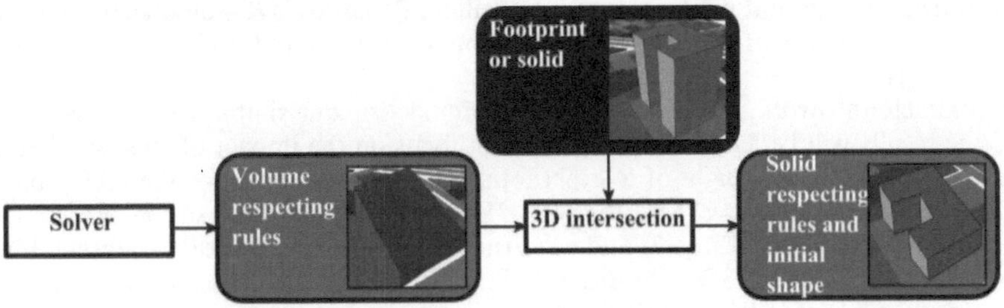

Figure 14. Optimal volume calculation from a building footprint.

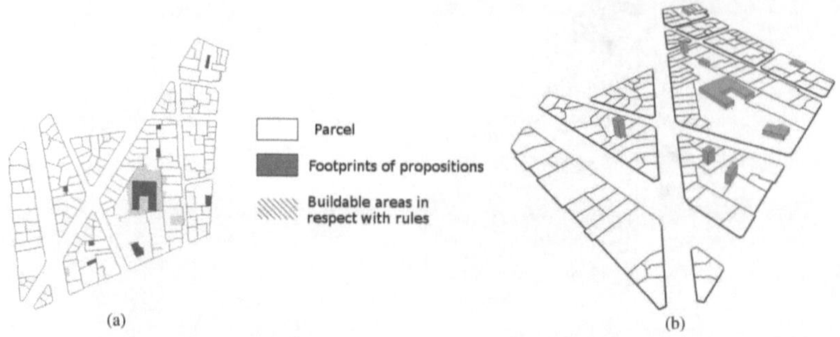

Figure 15. (a) Footprints of new proposition of building shapes on the 2D map of buildable volumes. (b) Buildable floor area enriched with new propositions of buildings with 3 m levels.

5 CONCLUSION

This paper proposes a model to include the most common rules of the French Local Urban Planning Schemes (LUPS) and to utilize them within 3D GIS using 3D spatial analysis. The proposed system allows 3D visualisation of the impact of urban rules, checks if they are respected and provides the integration and evaluation of new projects with regard to urban regulations. The 3D GIS is used as a relevant tool to better comprehend the urban environment (especially regarding height rules) and regulation constraints.

Our aim is to provide a tool that can help during the design process, discussion and decision phases in the context of urban projects. The LUPS is a reference document in France and such a tool cannot be substituted by current administrative procedures. Nevertheless, our process requires the LUPS rule formulations and offers interesting experiments about the definition of the rules to detect possible ambiguities. Eventually, the development of these rules in time should enable one to study their impact on the evolution of a territory on a platform for urban dynamics simulation (Perret, Boffet Mas & Ruas 2009; Curie, Perret & Ruas 2010; Perret, Curie, Gaffuri & Ruas 2010).

This work is currently being implemented as a prototype. Information about non-geometric constraints have not been utilized. Nevertheless, this information could be integrated with the checker. Thus, many perspectives are possible, for example in order to obtain more detailed buildings. For instance, non-photorealistic rendering techniques and textures could be used in order to have more visually-appealing results.

The approach and the tool proposed are tested through the integration of some LUPS, and validations of the domain with experts (planners, municipalities, geographers etc.) are forthcoming. Tests have shown that some rules such as the alignment of buildings on a zone are missing. These rules should be added to the proposed system. Furthermore, the addition of new rules can become very relevant if more accurate semantic or geometric data are available (for example, the presence of balconies, chimneys, functions of buildings etc.). Terrestrial imagery acquired from vehicles (Devaux, Paparoditis, Precioso & Cannelle 2009) and information extracted from them (such as wall openings (Hammoudi, Dornaika, Soheilian & Paparoditis 2010), ground marks or signs (Soheilian, Paparoditis & Boldo 2010) opens new perspectives for the use of 3D GIS for conception, discussion and decision making around urban projects.

To complete this work, a reflexion about the impact of our simplifications on the quality of the results will be led. This study will first focus on the impact of the use of the $3m$ value to approximate the height of a level (notably for the assessment of the FAR), but also on the choice of our building (does the IGN 3D dataset for buildings supply more relevant information than the 2D dataset?) and road (notably for the road width) datasets. Finally, a more global work about 3D semiotics in GIS will improve the visual effectiveness of our system for the understanding of urban regulations.

ACKNOWLEDGMENTS

This work is partly funded by clusters *Advancity* and *Cap Digital* through the *TerraMagna project*. It was realised in collaboration with *Bionatics* and the *IGN*. The authors would like to thank Sébastien Mustière and Christiane Weber for their guidance and support as well as the reviewers for their helpful comments. Finally, the authors would like to thank Iwona Mc Kenzie for her corrections.

REFERENCES

Bain, P., Maujean, S. & Theys, J. (2008). Agora 2020, vivre, habiter, se déplacer en 2020: quelles priorités de recherche ? Technical report, MEEDDAT/DRAST.

Brasebin, M. (2009, June). Geoxygene: An open 3d framework for the development of geographic applications. In 12th *International Conference on Geographic Information Science (AGILE'09)*, Hanovre, Germany.

Caillaud, D., Denegre, J., Barbier, M. & Denis, P. (2006, October). Proposition de sémiologie pour l' édition des plu à partir d'un sig. Technical report, Groupe de travail "informatisation des PLU", CNIG. pdf.

Coors, V., Hünlich, K. & On, G. (2009). Constraint-based generation and visualization of 3D city models. In J. Lee & S. Zlatanova (Eds.), *3D Geo-Information Sciences*, Lecture Notes in Geoinformation and Cartography, Chapter 23, pp. 365–378. Berlin, Heidelberg: Springer Berlin Heidelberg.

Curie, F., Perret, J. & Ruas, A. (2010, May). Simulation of urban blocks densification. In 13th *AGILE* International Conference on Geographic Information Science.

Desmontils, E. (1995, September). Les modeleurs déclaratifs. Research report RR-IRIN-95, Institut de Recherche en Informatique de Nantes, Nantes.

Devaux, A., Paparoditis, N., Precioso, F. & Cannelle, B. (2009, May). Face blurring for privacy in street-level geoviewers combining face, body and skin detectors. In *IAPR Machine Vision Applications*.

Donikian, S. (1992). *Une approche déclarative pour la création de scènes tridimensionnelles: application à la conception architecturale*. Phd thesis, Université de Rennes 1.

Edmonds, E. & Candy, L. (2002, October). Creativity, art practice, and knowledge. *Commun. ACM 45*, 91–95.

Faucher, D. (2001). *UrbanLab modélisation déclarative des enveloppes urbaines réglementaires*. Phd thesis, Université de Nantes.

Gilles, F. & Metral, C. (2005, June). Integrating urban knowledge into 3d city models. In 1*st International Workshop on Next Generation 3D City Models*, Bonn. ISPRS WG III/4, EuroSDR, DGPF, and University of Bonn. pdf.

Hammoudi, K., Dornaika, F., Soheilian, B. & Paparoditis, N. (2010). Extracting outlined planar clusters of street facades from 3d point clouds. In *IEEE/CRV—Seventh Canadian Conference on Computer and Robot Vision*.

ISO (2003). *Geographic information—Spatial Schema, ISO Draft International Standard 19107:2003*. International Organization for Standardization.

Murata, M. (2004, August). 3d-gis application for urban planning based on 3d city model. In 24th *Annual ESRI International User Conference*, pp. 9–13. pdf.

Nivet, M.-L. (1999). *De Visu: un logiciel pour la prise en compte de l'accessibilité visuelle dans le projet architectural, urbain et paysager*. Phd thesis, Université de Nantes. École d'architecture de Nantes.

Perret, J., Boffet Mas, A. & Ruas, A. (2009, November). Understanding urban dynamics: the use of vector topographic databases and the creation of spatio-temporal databases. In 24th *International Cartography Conference (ICC'09)*.

Perret, J., Curie, F., Gaffuri, J. & Ruas, A. (2010, September). A multi-agent system for the simulation of urban dynamics. In 10th *European Conference on Complex Systems (ECCS'2010)*. to appear.

Shen, Z. & Kawakami, M. (2004). Visualisation of usable building space according to planning permission ordinances for public participation in district plan in japan. In J.P. van Leeuwen & H.J. Timmermans (Eds.), *Recent advances in design and decision support systems*, pp. 85–98. Springer.

Soheilian, B., Paparoditis, N. & Boldo, D. (2010, July). 3d road marking reconstruction from street-level calibrated stereo pairs. *ISPRS Journal of Photogrammetry and Remote Sensing 65*(4), 347–359.

Viet-Tung, L. (2003). Modélisation des volumétries autorisées de construction du bâtiment basées sur les réglements d'urbanisme. Masters thesis, École d'architecture de Nancy.

Urban and Regional Data Management – Zlatanova, Ledoux, Fendel & Rumor (eds)
 ISBN 978-0-415-67491-1

Extending OpenStreetMap to indoor environments: Bringing volunteered geographic information to the next level

M. Goetz & A. Zipf
Chair of GIScience, Department of Geography, University of Heidelberg, Heidelberg, Germany

ABSTRACT: Extensive and high-quality geographic data sources are important for any kind of spatial analysis or application, especially in the field of urban data management. In the last couple of years, Volunteered Geographic Information (VGI) has increasingly gained attractiveness not only to amateur users, but also to professionals in the geoinformation industry. Different VGI communities evolved and especially the OpenStreetMap (OSM) community became very strong. OSM provides very detailed information about the landscape, the street network and also more and more buildings are mapped. However, until now, this building mapping is mainly related to the outer shape of the ground space of the building and there is hardly any information about the inner structure available. This paper presents an approach of extending the OSM tagging schema to indoor environments. A 3D Building Ontology targeted to VGI communities is presented for describing different information aspects about buildings and their inner structure. Based on this ontology, the OSM extension is developed and explained. A proof of concept is given by applying the developed extension in mapping a use case building.

1 INTRODUCTION

When dealing with spatial and geographic phenomena in urban regions and performing analysis and computations about them, it is always important to have enough high-quality data. In the past, urban data management used to be performed by professional cartographers, public authorities or commercial data providers. Nevertheless, in the last couple of years a new and different trend for data collection has evolved, describing the collaborative and volunteered collection of geographic data. According to Elwood (2009) there are many different terms for this trend, but one of the most popular terms is Volunteered Geographic Information (VGI). VGI comprises the effect that an ever-expanding range of users creates, assembles and disseminates geographic and spatial data in a collaborative and volunteered manner (Goodchild 2007a). That is, individual persons or groups collect and create geographic data based on their personal measurements (via GPS etc.) and their knowledge about their surroundings and furthermore share that information with others through open web platforms. VGI is a combination of elements of the Web 2.0 and a collective intelligence, and there is an enormous potential arising from six billion humans acting as remote sensors (Goodchild 2007b). That is, VGI can also be considered as a new opportunity of systems and sensors for monitoring urban and regional environments. Particularly in urban environments the coverage is very good, because many humans results in many potential sensors and therefore the usage of VGI in urban management increases (Song & Sun 2010). Furthermore, public participation in urban planning and data management gets increasingly important (Cf. (Hansen 2004), (Bugs et al., 2010)).

Diverse online mapping platforms such as OpenStreetMap (OSM) or Wikimapia have been initiated, allowing users to contribute and collaboratively edit spatial data. The increasing number of participants leads to a variety of different spatial data and information about geographic phenomena such as the street network, cities, POIs, buildings, landuse etc. At UDMS 2009, VGI has also been shown to be useful for urban data management by providing new sources of information that even can be integrated into a 3D platform (Schilling et al., 2009).

However, most of current mapping activities are related to the outdoor environments (e.g., rough building structures, information about type of businesses, playgrounds, footpaths etc.) and there is hardly any information about indoor environments available. Since there is an increasing need for mature indoor navigation solutions and other indoor location based services (Goetz & Zipf 2010) and data providers are hardly able to commercially capture indoor data for large areas, there is an enormous potential within VGI communities for capturing and providing information about indoor environments which are open to the public (e.g., airports or shopping malls). For that purpose, it is essential to provide clear and understandable methodologies for mapping data about indoor environments. Therefore the main contribution of this paper is an extension to indoor environments for bringing VGI to the next level.

The remainder of this paper is organized as follows: First, there is an introduction to OpenStreetMap, especially focusing on its data model and data acquisition methods. Afterwards, there is a brief overview about related research. Thereafter, an extensive ontology for 3D building models with detailed information about indoor environments is presented. Subsequently, there is a description of how to extend the existing OSM tagging schema to indoor environments according to this ontology. The presented methodology is demonstrated on a use-case building and the last chapter summarizes the presented research and discusses future work.

2 THE OPENSTREETMAP COMMUNITY

One very popular (or even the most popular) VGI community is the so called OpenStreetMap project. OSM follows the peer production model (Haklay & Weber 2008) that created Wikipedia and aims for the provision of free to use and editable map data. Since 2004, the project grew rapidly and by November 2010 there were more than 320.000 registered users and more than 2.000 millions tracked points in the database (OSM 2010a). The data in OSM is created in different ways. The most important way is the acquisition of original data, manually captured by users via GPS devices. However, people can also contribute data based on aerial images (e.g., by Bing or Yahoo) or by contributing their local knowledge about the region they live in.

Basically, OSM consist of differently tagged nodes, i.e., a point with distinct coordinates. For defining lines or simple polygons (i.e., a polygon without holes), users can create so called ways which consist of several nodes. For defining a polygon, this way needs to be closed, i.e., the start point equals the end point. For mapping complex polygons or describing existing relationships between different elements, there are furthermore so called relations. These contain ways, nodes and also other relations.

In conjunction with user-generated content and collaboratively collected data, there is always a question about the accuracy and quality of the provided data. By comparing OSM data with data provided by commercial vendors like Teleatlas, it became evident that VGI is able to compete against commercial providers (Cf. (Zielstra & Zipf 2010), (Haklay 2010), (Ludwig et al., 2010)). That is, data from OSM can be considered as a real alternative data source for spatial and geographic data in urban environments.

The data of OSM mainly focuses on outdoor environments and objects (e.g., streets, landuse etc.). When considering buildings, some information can be available within OSM, but these do only refer to very basic things like the location, outer shape or the height of the building. The latter mentioned information can be applied in the 3D visualization of city models as within the OSM-3D project (Over et al., 2010). There are also some discussions about indoor mapping (Cf. next chapter), but until now they cannot be regarded as mature.

3 RELATED WORK

When investigating new models for the built environment provided by volunteers in a crowdsourcing approach, it first needs to be assessed, what the state of art of existing and professional building models looks like. There are some basic topological data structures and models such as Molenaar (1990), Zlatanova & Tempfli (1998), Coors (2003), Billen & Zlatanova (2003), Holweg et al. (2004) or Zlatanova et al. (2004), but they do not focus on semantic aspects, thus they are not applicable for the intended purpose.

The term Building Information Modeling (BIM) describes the process of generating and managing building data (Ashcraft 2007), whereby typically 3D modeling approaches are utilized. A commonly used format for BIM is the Industry Foundation Classes (IFC), describing a neutral and open specification which is also registered as ISO 16739 (IFC 2010). IFC is an object-oriented model which consists of hierarchically organized classes (often hundreds or thousands). Examples for such classes are *ifcWall* or *ifcWindowType*, describing on the one hand a particular part of the building and on the other hand adding additional information about the type of the window. The classes in IFC can be divided into many different groups or concepts (for a detailed description refer to IFC (2010)), which on the one hand makes the whole framework very extensive and powerful, but on the other hand very complicated and confusing. The latter mentioned fact is also the reason why IFC is good for professional purposes, however for involving the broad public and average hobbyists, IFC is way too detailed and complex.

A closely related initiative to IFC is the City Geographic Markup Language (CityGML) which is an Open Geospatial Consortium (OGC) standard based on GML3. It is usable for describing geometric, topologic and semantic aspects of urban spaces and city models in a three-dimensional way (Kolbe et al., 2005). Thereby it does not only focus on buildings, but also on other objects (e.g., vegetation). Furthermore, semantic and spatial aspects are structured into five different Levels of Details (LoD), where LoD0 describes a coarse model of the city and LoD4 (i.e., the most detailed one) even contains information about indoor environments. That is, CityGML can be utilized for a detailed and fine-grained description of buildings and their interiors. Nevertheless, CityGML is rather considered as a storage and exchange format for 3D city models and is mainly used by professionals. It is likely that CityGML is too complex for an application within VGI. But when new models for community based urban modeling tasks are proposed, these existing standards will provide some useful hints that shall ease the future combination of models from both worlds.

When trying to model 3D buildings, it is essential to share a common understanding of what kind of information is relevant and important. Therefore, different ontologies for buildings and indoor environments have been developed in past research efforts. Anagnostopoulos et al. (2005) developed an extensive indoor navigation ontology, which is usable for path searching as well as for the presentation tasks of the navigation system. However, despite the many details about the indoor space, the presented ontology does not provide any information about the building itself. For modeling, visualization and other purposes, such information is nevertheless very important and therefore should be considered in a 3D building ontology.

A different approach is presented by Yuan & Zizhang (2008). Based on BIM, the authors propose an indoor navigation ontology which captures different parts inside the building such as corridors, elevators, stairways, doors etc. However, this ontology does also not provide information about the overall building (e.g., façade color, roof type etc.) and furthermore details about concrete measures and positions of the rooms etc. are also not covered by this ontology.

Dudas et al. (2009) present ONALIN, an ontology which considers the special needs and requirements of different groups and individuals, whereby a strong focus lies on the limitations of different users (e.g., visually impaired, wheelchair driver etc.). ONALIN is a "*comprehensive ontology that considers the needs of all potential people traversing in buildings based on the ADA (Americans with Disabilities Act) standards*" (Dudas et al., 2009).

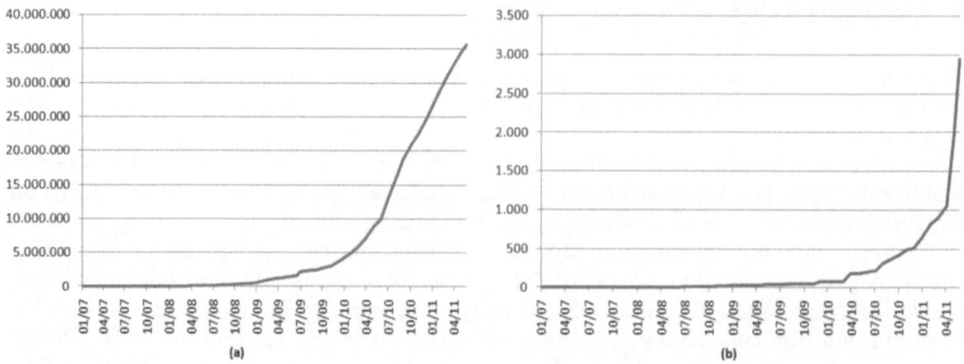

Figure 1. Usage of the *building* key (a) and the *indoor* key (b) between 2007 and 2010.

Other ontologies are furthermore presented by Andersen & Vasilakis (2007) and Lee et al. (2008). However, these are either not detailed enough for both outer appearance and inner structure, or on the other hand they are too detailed, so it is likely that the average OSM user won't be able to capture all required data.

Besides scientific work on 3D building modeling frameworks and ontologies, there is also some related work within the OSM community. In (OSM 2010b) it is described how to map buildings. By utilizing the key-value *building* = *yes*, users are able to describe the ground shape of a building with a set of individual nodes. Furthermore the keys *height* and *building:height* can be utilized for adding information about the (estimated) height of the building. Currently there are almost 36 million buildings available and about 636.000 of them contain height information. Furthermore there are some discussions about how to map indoor environments (OSM 2010c). Thereby the key-value *indoor* = *yes* has been proposed, indicating that the corresponding OSM object is inside a room or building. Furthermore the key *level* shall be used for describing in which building level the corresponding OSM object is located. Currently there are about 3.000 indoor objects available, so compared to the whole dataset, the coverage of the *indoor* key is negligible. Nevertheless, there is a strong trend towards building and indoor mapping activities. In Germany currently over 100.000 new building outlines are added to OSM every week. The development of the usage of the keys *building* and *indoor* is depicted in Figure 1 (a) and Figure 1 (b). There is an accelerating increase of the usage and it is likely that this development will increase even further in the near future as it has been the case with other object types in OSM before. Especially when more and more outdoor spaces are completely captured, it is likely that the community will commence with buildings and indoor spaces.

4 3D BUILDING ONTOLOGY

Indoor spatial information and data allows transferring and extending popular existing outdoor applications and services (e.g., routing, navigation or mapping) to indoor environments. In doing so, diverse indoor services (e.g., navigation in a shopping mall or train station) can be provided to the public, thus assist the users in difficult situations and therefore increase their quality of life. As mentioned above, a commercial data collection for broad indoor areas is not likely and therefore it is a good opportunity to include and involve voluntarily acting hobbyists in the data acquisition process. This however leads to kind of a cleavage, because on the one hand a scientific correctness and completeness is required for proper applications, but on the other hand the OSM community is likely to not accept and utilize too complex methodologies. That is, there is a trade-off between the comprehensiveness and the usability of the invented methodology.

Trying to be able to share a common understanding of building information and fulfilling a scientific correctness and completeness, a 3D Building Ontology (3DBO) has been developed. Thereby, the ontology is designed as simple as possible, having in mind that not only professional users will use the extension but mainly normal users without professional skills. The ontology is suitable for describing the outer appearance and characteristics of the building as well as the inner building structure and it is likely that all relevant data can be captured by the community. So on the one hand, the ontology is kept as simple as possible, but on the other hand it contains enough information for different purposes such as a quite realistic visualization of the building or the provision of different indoor applications and services.

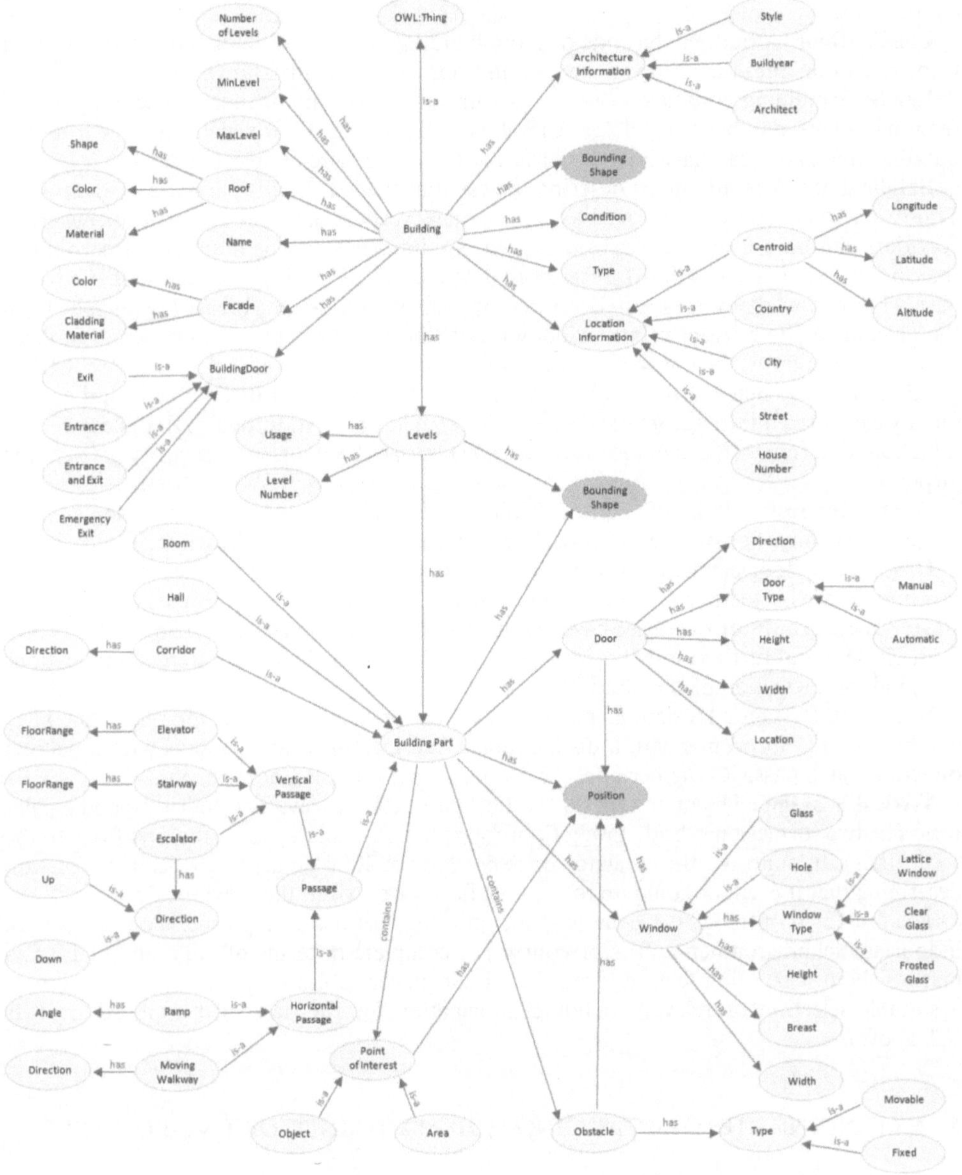

Figure 2. 3D Building Ontology for describing the inside and outside of a building.

The ontology has been developed in OWL/RDF using Protégé. It is visualized in Figure 2 and a human-readable documentation is given below:

Building: The concept of a building is the main part of the developed ontology. A building has diverse characteristics which describe the outer appearance of the building. One example is the roof of the building which can be described according to its shape (e.g., flat, gabled etc.), color (e.g., black, red etc.) or material (e.g., aluminum, wood etc.). A building also has distinct shape, considering the ground area. The façade of the building can be described by a color and the cladding material (e.g., glass, concrete etc.). There are also additional semantic characteristics describing the building such as its condition (e.g., new, renovated etc.), its type (e.g., public, academic etc.), its name (e.g., "BST 48") or its total number of levels. Architectural information such as style or buildyear can also serve as additional information. Furthermore the location of the building can be described by an address and/or the exact coordinates.

BuildingDoor: A building has one or more building doors, describing a direct connection through a door from the inside of the ground floor to the outside and vice versa.

Levels: A building contains a distinct amount of levels (one or more), whereby each of them has a level number (e.g., -1, 2). A level has a bounding shape which can vary from the building area and it can have a particular usage (e.g., commercial, residential, etc.).

BuildingPart: A buildingpart describes any spatial element that is part of a distinct level. Those can be generally categorized as rooms, halls, vertical passages or corridors, whereby corridors can have a distinct direction. Such buildingparts can be described according to their bounding shape and they furthermore have a concrete position on the floor. They can have windows, whereby these are described by their width, height, breast (i.e., height from the ground to the bottom of the window), position and type (e.g., lattice window, clear glass etc.). Similar to the concept of windows, buildingparts also have doors, whereby these are described by their width, height and position. Doors have a distinct type and can be mainly categorized into automatic doors and manual doors. Additionally, doors can have a direction, describing that the door is only usable in one direction (e.g., security gates at the airport).

Point of Interest: A Point of Interest (POI) describes a distinct area (e.g., smoking area) or object, which might be of interest or usefulness for a user (e.g., a landmark for navigation).

Obstacle: A floor element can contain an obstacle, whereby the obstacle has a distinct position. These obstacles can either be a movable (e.g., an office desk) or fixed (e.g., a raised stage). These groups of obstacles have in common that they are characterizing features of a floor element and furthermore it is important to be aware of them for different applications (e.g., indoor navigation of the blind).

Horizontal Passages: Horizontal passages are areas or elements inside floor elements. They can be divided into ramps with a distinct angle and into moving walkways with a distinct direction. Such passages furthermore have a position inside the floor element.

Vertical Passages: The main types of vertical passages are elevators, stairways and escalators. The two former ones both have a floor range (e.g., a stairway from the first floor to the sixth floor). In contrast, the escalator does not have a floor range, but a distinct direction describing that it is either going up to the next floor or down to the previous floor.

The presented ontology is a generic spatial ontology for describing generic buildings with indoor spatial environments. It is of course not complete in terms of containing all information about a building (e.g., information about power cables etc). However it represents a suitable selection of relevant building information for the described application areas (Cf. above).

5 EXTENDING THE OSM TAGGING SCHEMA TO INDOOR ENVIRONMENTS

The fundamental basis of the OSM schema extension is the above presented 3DBO. In most cases, buildings in OSM are mapped as a closed way or, in the case of a complex building shape, as a relation. The principle of the developed extension is to map the building as a relation. For every level inside the building there is one relation member (which also will be a relation) and one additional relation-member per building door. These doors are mapped as single nodes on the bounding shape with key *building:entrance* and values *yes, entrance, exit, emergency, both*. The relation itself has diverse key-value pairs for adding (semantic) information about the building (Cf. 3DBO). The proposed keys with exemplary values are described in Table 1. The blue colored ovals with dashed contour in the 3DBO (Cf. Fig. 2) illustrate all building parts which actually need to be mapped in JOSM—all other information is added as key-value pairs.

As mentioned before, each level is also defined as a relation respectively. The members of these relations are other relations, ways or single nodes, representing the different elements (rooms, halls, corridors, passages etc.) in this level. An additional relation-member (way or relation) is introduced for mapping the bounding shape of this distinct level. The relation itself again has diverse key-value pairs for adding additional information. The proposed keys for the OSM extension are depicted in Table 2.

The different elements of a distinct level are modeled as relations, ways or single nodes (depending on their requirements and complexity). For example, windows or doors are mapped as single nodes which are part of the corresponding element shape. That is, the bounding shape is mapped with a closed way and at every position where a window is located, there is a node with the tag *buildingpart:window*. Furthermore, there is additional information attached to them (e.g., height or width). In a similar way, rooms, halls or corridors are mapped as (closed) ways and tagged with *indoor* = *yes*. The before described procedure is valid for any kind of level element (Cf. 3DBO). The possible and reasonable keys for additional information are listed and described in Table 3. The position of the different elements is implicitly provided by their nodes.

For mapping obstacles inside buildingparts (e.g., a strut in a hall) there are two possibilities. If the obstacle belongs to the building and is not movable, it is the best to map

Table 1. Proposed relation tags for general building information.

Key	Description	Exemplary value(s)	Tag count*
building	it's a building	yes	35.670.028
building: levels	number of levels	4	305.603
building: min_level	minimum level	−1	1504
building: max_level	maximum level	6	
building: roof	roof shape	flat, pitch, hip	21.445
building: roof: color	roof color	black	781
building: roof: material	roof material	cardboard	28
name	building name	Federal hospital	475.935
building: cladding	façade material	glass	6.876
building: facade: color	façade color	yellow	60
building: facade: image	URL to façade image	http://url.de/image.gif	
building: architecture	architecture style	modern	270
building: buildyear	buildyear	1987	14
building: architect	architect	Neumann	2
building: height	building height	25 m	27.804
height	building height	25 m	608.877
building: condition	building condition	renovated	46
addr: country	country	Germany	1.075.942
addr: city	city	Munich	1.396.895
addr: street	street	Luisenstraße	2.158.250
addr: housenumber	house number	5	2.298.275

*if this key is already available in OSM, the number in the column indicates the total usage amount (based on personal analysis).

Table 2. Proposed relation tags for general level information.

Key	Description	Exemplary value(s)	Tag count
indoor	it's inside the building	yes	2.950
name	name of the level	ground floor	
level: usage	usage of the level	academic*	
level: height	height of the level	4 m	
building: level	actual level number	1	2.884

* some values of the OSM key *amenity* are also reasonable for *level:usage.*

Table 3. Proposed tags for buildingparts (nodes, ways or relations).

Key	Description	Exemplary value(s)
buildingpart	what type of buildingpart	room, hall, corridor
name	name of the buildingpart	Room 101
buildingpart: corridor: direction	corridor direction	one-way
buildingpart: height	height	3 m
buildingpart: window	it's a window	yes, glass, hole
buildingpart: window: type	window type	lattice window, frosted glass
buildingpart: window: height	window height	1.2 m
buildingpart: window: width	window width	1 m
buildingpart: window: breast	breast of the window	1.3 m
buildingpart: poi	point of interest	object, area
buildingpart: door	it's a door	yes, manual, automatic
buildingpart: door: height	door height	2 m
buildingpart: door: width	door width	1 m
buildingpart: door: direction	door direction	one-way
buildingpart: horizontalpassage	it's a horizontal passage	ramp, moving walkway
buildingpart: horizontalpassage: angle	horizontal passage angle	5 degree
buildingpart: horizontalpassage: direction	horizontal passage direction	one-way
buildingpart: verticalpassage	it's a vertical passage	elevator, stairway, escalator
buildingpart: verticalpassage: floorrange	vertical passage floor range	1–5
buildingpart: verticalpassage: direction	escalator direction	up, down

it as a hole in a polygon (considering the hall shape as a polygon and the strut as a hole). In contrast, if the obstacle is not part of a building (e.g., an office desk), it is possible to map it as a closed way or single node with the keys *obstacle = yes* and *obstacle:type* with value *movable* or *fixed.*

6 EXAMPLARY USAGE OF THE PROPOSED SCHEMA

For demonstrating the OSM indoor extension, the building of the Chair of GIScience of the University of Heidelberg (Figure 3) has been mapped. It has four different levels (one basement and three above the ground) and one main entrance. It mainly contains offices and three bigger rooms for lectures. The levels are connected with two different staircases. The location of the building was gathered via GPS. Information about the specific dimensions

Figure 3. Use case building.

and the inner structure of the building were gathered by CAD floor plans, kindly provided by building authorities.

The whole building was mapped as a relation and several keys (e.g., *building* = *yes*, *building:cladding* = *concrete*, *height* = 14.5 *m*, *building:roof* = *flat* etc.) were attached to it. There are a total of five relation-members, whereby one is a single node (indicating the building entrance) and the other four members are again relations for describing the different levels of the building. For every single level there were different keys (e.g., *level* = *yes*, *building:level* = 1 etc.) attached to the corresponding relation. Several closed ways (as relation-members) were utilized for mapping the bounding shape and the buildingparts of the respective level. Every buildingpart was tagged with the key *indoor* = *yes* and additional semantic information was added by several key-value-pairs (as described in chapter 5). Windows were mapped by adding additional nodes to the bounding-shape ways and adding different keys e.g., *buildingpart:window* = *glass* or *buildingpart:window:breast* = 1.2 *m* to them. Doors were mapped in a similar way by adding additional nodes to the corresponding buildingpart ways. The two stairways inside the building were mapped as a closed ways, whereby it has been decided to model the stairs on every level (it is also possible to simply map the stairway on one level and to leave it blank for the other levels). The key *buildingpart:verticalpassage:floor-range* = −1 *to* 2 was utilized for describing the floor-range of the respective stairways.

The size and shape of the different buildingparts were gained from the CAD files and manually mapped in the OSM editor JOSM. Thereby, all levels have been mapped individually and afterwards manually combined by editing the OSM-XML file. This manual code-editing is required, because JOSM (and other OSM editors) lack visualization possibilities for multi-storage buildings with indoor environments and it is hard to distinguish between several overlapping ways. Another problem has been encountered when uploading the mapped building into the OSM database. For some reasons, the OSM renderer does not recognize buildings mapped as relations, whereby the main-relation is tagged with *building* = *yes*. Instead, it is required that a member of the relation (a closed way) is tagged with *building* = *yes* key. So at first, the use-case building was not displayed in the OSM map. For overcoming this problem, it has been decided to additionally tag the bounding-shape of the ground floor with the key *building* = *yes*. Utilizing the bound-shape of the ground floor was obvious, because it describes the building shape on the ground but in general also the other bounding-shapes could have been utilized. Nevertheless it must be said that these problems occur in current editors and renderers because they are not intended for this purpose, but the presented methodology shall be considered as an impulse for improvements.

7 CONCLUSIONS AND FUTURE WORK

In this paper, an extension of the existing OSM tagging schema for mapping indoor environments has been proposed. An extensive 3D building ontology has been presented, whereby the ontology suits both the description of the outer appearance and characteristics of the building as well as the detailed description of the inner building structure. Based on this ontology, the existing OSM tagging schema has been extended and methodologies for mapping indoor spaces have been developed. By applying these extensions, members of the OSM community are now able to map indoor spaces, thus provide more detailed information about buildings and their inner structures. Both the ontology and the tagging schema extension have been kept as simple as possible, so it is realistic that the OSM community will understand and accept them.

As described above, existing OSM editors are currently not able to provide an easy-to-use functionality for the presented extensions. Therefore it is important to develop an OSM editor which provides such functionality, so it is possible to map inner building structures in an easy-to-use manner. The use-case demonstrated that existing OSM renderers need refinements, so that relations tagged with *building = yes* are also recognized in the map. Furthermore it will be interesting to see, how relevant data can be captured. Different information such as façade color or roof type can be extracted from images or visual experience, but for measuring distances or positions inside buildings, different methodologies (such as Lidar or pedometers) must be applied. Also future cell phones are likely to contain more sensors for different measurements. Moreover, importing existing standards like CAD, CityGML or IFC into OSM is wishful but a challenging idea for future research, because this would allow an integration of existing floor plans into OSM. For achieving this goal, the presented 3DBO needs to be compared with IFC and CityGML, so possibilities for transformations can be discovered.

From a different point of view, privacy aspects for indoor spaces are also an important factor. Not every kind of information is appointed to be of interest for the public and providing those bears some kind of security risk, so this field also needs to be further investigated and explored.

ACKNOWLEDGEMENTS

The authors would like to thank the anonymous reviewers for providing valuable comments towards the improvement of this paper. We would also like to thank the building authorities of the University of Heidelberg for providing the floor plans of our use case building. Furthermore we would like to thank all members of the Chair of GIScience for their proofreading and helpful hints.

REFERENCES

Anagnostopoulos, C., Tsetsos, V., Kikiras, P. & Hadjiefthymiades, S.P. (2005). OntoNav: A Semantic Indoor Navigation System, in: W. Mansoor, et al. (Eds.), *First International Workshop on Managing Context Information in Mobile and Pervasive Environments (MCMP '05)*, Ayia Napa, Cyprus. pp. 123–131.

Andersen, O. & Vasilakis, G. (2007). Building an Ontology of CAD Model Information in: G. Hasle, et al. (Eds.), *Geometric Modelling, Numerical Simulation, and Optimization Applied Mathematics at SINTEF*, Springer, Heidelberg. pp. 11–40.

Ashcraft, H. (2007). Building information modeling—A Framework for Collaboration, *American Bar Association—Forum on the Construction Industry*, Newport, Rhode Island, United States.

Billen, R. & Zlatanova, S. (2003). 3D spatial relationships model: a useful concept for 3D cadastre? *Computers, Environment and Urban Systems* 27(4): 411–425.

Bugs, G., Granell, C., Fonts, O., Huerta, J. & Painho, M. (2010). An assessment of Public Participation GIS and Web 2.0 technologies in urban planning practice in Canela, Brazil. *Cities: The International Journal of Urban Policy and Planning* 27(3): 172–181.

Coors, V. (2003). 3D-GIS in networking environments. *Computers, Environment and Urban Systems* 27(4): 345–357.

Dudas, P.M., Ghafourian, M. & Karimi, H.A. (2009). ONALIN: Ontology and Algorithm for Indoor Routing, *First International Workshop on Indoor Spatial Awareness (ISA 2009)*, Taipei, Taiwan. pp. 720–725.

Elwood, S. (2009). Geographic Information Science: new geovisualization technologies—emerging questions and linkages with GIScience research. *Progress in Human Geography* 33(2): 256–263.

Goetz, M. & Zipf, A. (2010). Open issues in bringing 3D to Location Based Services (LBS)—a review focusing on 3D data streaming and 3D indoor navigation *Proc. of 5th 3D GeoInfo Conference*, Berlin, Germany. pp. 121–124.

Goodchild, M.F. (2007a). Citizens as sensors: the world of volunteered geography. *GeoJournal* 69(4): 211–221.

Goodchild, M.F. (2007b). Citizens as Voluntary Sensors: Spatial Data Infrastructure in the World of Web 2.0. *International Journal of Spatial Data Infrastructures Research* 2: 24–32.

Haklay, M. (2010). How good is volunteered geographical information? A comparative study of OpenStreetMap and Ordnance Survey datasets. *Environment and Planning B-Planning & Design* 37(4): 682–703.

Haklay, M.M. & Weber, P. (2008). OpenStreetMap: User-Generated Street Maps. *IEEE Pervasive Computing* 7(4):12–18.

Hansen, H.S. (2004). Public Participation GIS: An Important Tool in Environmental Planning, *24th Urban Data Management Symposium (UDMS 2004)*, Chioggia, Italy.

Holweg, D., Zlatanova, S. & Coors, V. (2004). Geometrical and Topological Models for Real-time GIS, *24th Urban Data Management Symposium (UDMS 2004)*, Chioggia, Italy.

IFC, (2010). building SMART technical resources. http://buildingsmart-tech.org/. (Accessed 11/30/2010).

Kolbe, T.H., Gröger, G. & Plümer, L. (2005). CityGML—Interoperable Access to 3D City Models, in: Oosterom, et al. (Eds.), *International Symposium on Geo-information for Disaster Management*, Delft, Netherlands.

Lee, J.-S., Min, K.-M., Lee, Y.-S., Kim, J.-H. & Kim, J.-J. (2008). Building Ontology to Implement the BIM (Building Information Modeling) Focused on Pre-Design Stage, *25th International Symposium on Automation and Robotics in Construction—ISARC-2008*, Vilnius, Lithuania. pp. 350–354.

Ludwig, I., Voss, A. & Krause-Traudes, M. (2010). Wie gut ist OpenStreeetMap? Zur Methodik eines automatisierten objektbasierten Vergleichs der Straßennetze von OSM und Navteq in Deutschland. *GIS.Science* 4: 148–158.

Molenaar, M. (1990). A formal data structure for 3D vector maps, *EGIS'90*, Amsterdam, The Netherlands. pp. 770–781.

OSM, (2010a). OSM Stats. *http://wiki.openstreetmap.org/wiki/Statistics*. Accessed (11/23/2010).

OSM, (2010b). Key:building. *http://wiki.openstreetmap.org/wiki/Key:building*. (Accessed 11/22/2010).

OSM, (2010c). Indoor Mapping. *http://wiki.openstreetmap.org/wiki/Indoor_Mapping*. (Accessed 11/22/2010).

Over, M., Schilling, A., Neubauer, S. & Zipf, A. (2010). Generating web-based 3D City Models from OpenStreetMap: The current situation in Germany. *Computers, Environment and Urban Systems* 34(6): 496–507.

Schilling, A., Over, M., Neubauer, S., Neis, P., Walenciak, G. & Zipf, A. (2009). Interoperable Location Based Services for 3D cities on the Web using user generated content from OpenStreetMap, *27th Urban Data Management Symposium (UDMS 2009)*, Ljubljana, Slovenia.

Song, W. & Sun, G. (2010). The role of mobile volunteered geographic information in urban management, *18th International Conference on Geoinformatics Beijing*, China. pp. 1–5.

Yuan, L. & Zizhang, H. (2008). 3D Indoor Navigation: a Framework of Combining BIM with 3D GIS, *44th ISOCARP Congress 2008*, Dalian, China.

Zielstra, D. & Zipf, A. (2010). A Comparative Study of Proprietary Geodata and Volunteered Geographic Information for Germany. *The 13th AGILE International Conference on Geographic Information Science (AGILE 2010)*, Guimarães, Portugal.

Zlatanova, S., Rahman, A.A. & Shi, W. (2004). *Topological models and frameworks for 3D spatial objects Computers & Geosciences* 30(4):419–428.

Zlatanova, S. & Tempfli, K. (1998). Data Structuring and Visualization of 3D Urban Data, *AGILE 1998*, Enschede, The Netherlands.

Urban and Regional Data Management – Zlatanova, Ledoux, Fendel & Rumor (eds)
© 2012 Taylor & Francis Group, London, ISBN 978-0-415-67491-1

Towards a 3D geo-data model to support pedestrian routing in multimodal public transport travel advices

J. Schaap
9292 REISinformatiegroep bv, Utrecht, The Netherlands

S. Zlatanova & P.J.M. van Oosterom
Delft University of Technology, GIS Technology Section, Delft, The Netherlands

ABSTRACT: Current web-based multimodal travel planners are able to generate adequate travel advises for public transport journeys using timetable information, but they lack detailed pedestrian routing for the walking parts of a journey. This paper describes an approach to support pedestrian routing in multimodal travel advises. It focuses on a 3D data model to support finding of an optimal route for the individual public transport traveller taking into account his specific preferences and constrains (e.g., wheelchair) for the pedestrian part of the journey. Existing standards and models are explored to investigate approaches to develop such a 3D geo data model. An important characteristic of this 3D model is that no difference is made between indoor and outdoor spaces. The model is a combination of networks and surfaces with corresponding attributes and behaviour. The possible pedestrian connections are modelled by combining so called *NodeSurfaces* and *LinkSurfaces* to compute the possible routes. Attributes of *LinkSurface* such as the direction (including up/down in 3D), the accessibility and pedestrian duration for a specific traveller, support the computation of the optimal routes. A prototype using this new conceptual geo data model was implemented in Oracle 11 g and results were visualized in Google Earth. Oracle Spatial 11 g network shortest route capabilities for logical network are used for finding the optimal route for individual travellers with specific speed and accessibility properties. The developed prototype and performed tests have proven that this approach is very beneficial for enhancing the traditional journey timetables with routing for the walking part of the journey.

1 INTRODUCTION

Many countries offer travel planning websites for the public transport, e.g., the Dutch Door-to-Door Journey Planner (www.9292ov.nl), the German DELFI (www.delfi.de) the international EU-Spirit (www.eu-spirit.com). Such travel systems provide a travel advice with the most appropriate connections, given a departure/arrival time (Figure 1a). Some planners provide also a map for the walking part to a station or between stations (Figure 1b). Two deficiencies can be observed in such planners: 1) the journey planners use standard times for the walking part and 2) the possible routes inside stations are not indicated. The time is usually computed on the basis of an average walking speed. Many factors may influence the speed of walking and thus the time to move between public transports vehicles. The condition of the traveller (age, disability, etc.) and the geometry of the connection places are some of the most critical factors. During transfers, travellers have to find their way from the place of arrival to the place of departure of other means of transport. If the traveller is not familiar with the complexity of the transfer place, he/she might have difficulties with finding the correct exit (or correct platform), might get lost and finally miss the connection. Most commonly the traveller needs to claim stairs, wait for elevators, etc. when walking inside building, which also influences the moving time. The need for adequate information at the planning stage of a trip

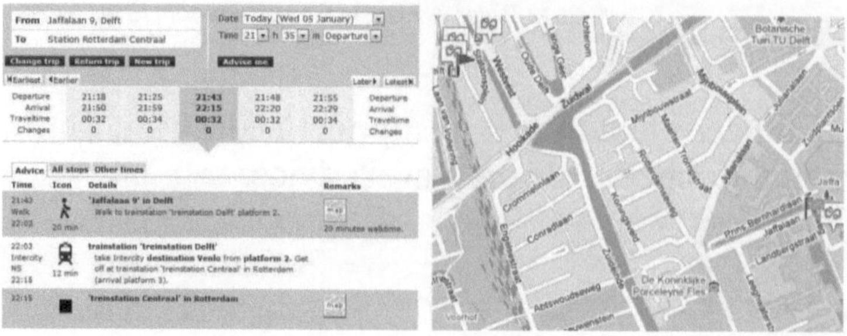

Figure 1. Result of travel advice of the Dutch Door-to-door journey planner: a) text, b) map for the walking part.

is of greater importance for a traveller using public transport than for a car. The planning stage of a public transport trip is an important stage to get familiar with transfer places especially when different kinds of transport modalities are used (e.g., bus and train) (Grotenhuis, et al., 2007).

To overcome the above deficiencies various studies have been completed and models and standards have been developed. A very good example of these developments is the Identification of Fixed Objects in Public Transport model (IFOPT) accepted as European standard in 2007. As discussed in IFOPT, 2007, a good information system for Public Transport makes use of at least three groups of information about: fixed objects (bus stops, roads, stations), mobile objects (vehicles) and events (accidents, construction works). All this information is usually available by different providers and its up-to-date status, accuracy and completeness creates problems. IFOPT provides a model for the fixed objects. A lot of research has been conducted on modelling and wayfinding for pedestrians in transportation networks. Raubal and Worboys, 1999 developed a model that reflects the goal-driven reasoning chain that leads to taking a decision and performing an action. Hochmair, 2004 investigated 15 criteria that influence multi-modal travel, but the author concludes that a smaller set would be also sufficient. Rüetschi, 2007 and Rüetschi and Timpf, 2005 proposed a model for wayfinding in railway stations.

This work builds on the concept of (public transport) *Network Space* and (pedestrian) *Scene Space*. Important in this model is that the two spaces are defined and distinguished from each other by a set of specific properties, including differences in the level of scale, the wayfinding processes, the possibility to plan ahead, the role of time, and most important the intrinsic architectural building structure. It claims that any model of Scene Space must depend on Network Space and vice versa, because all pedestrian movement along routes in Scene Space is governed by spatial and temporal constraints established by Network Space and its public transport timetable schedule. This paper adopts these concepts but develops a different model to represent Scene Space for planning ahead pedestrian routing.

Clearly, the knowledge about the individual travellers and the specific geometry of transfer areas has to be taken into consideration in the individual travel advice. In this paper we argue that 3D information about the transfer place will be of a great benefit for the communicating explicit directions to the travellers. Realistic 3D visualisation showing walking directions added to travel advices gives the traveller the opportunity to prepare better himself for a walk in an unknown situation.

Three-dimensional (3D) GIS data models are studied for various applications. In research for emergency response in urban areas, evacuation and routing 3D data models are studied (Lee, 2004, Meijers et al., 2005, Miller 2006, Stoffel et al., 2007, Hagendoorn et al., 2009, Yuan and Schneider, 2010). Navigation inside and outside buildings is of vital importance in emergency situations and therefore various 3D data models are studied to support 3D routing. In this 'emergency response' research it is found that the 3D topological and graph models are best suitable for 3D routing. Construction of these topological and graph models is described in literature (Lee 2004, Stevens and Choi 2006).

However, research on 3D GIS data models in relation to public transport network models is still new. Visualisation of pedestrian routes in 3D GIS data models can improve travel advice from public transport information websites. Applications using 3D data models can help to prepare the traveller to find faster the way to the next public transport vehicle. A traveller could train himself for a transfer in an unknown situation with a simulated walk in a 3D visualisation. Adequate information in this preparation stage of a public transport vehicle exchange can substantially contribute to the overall satisfaction of multimodal public transport quality.

This paper is organised as follows: the next section defines the requirements on a 3D geo-data model to provide the information needed for pedestrian routing in multi-model transport context. In section 3 existing models are explored on the suitability to accomplish these requirements. No model was found applicable for our research, therefore an improved 3D geo-data model is proposed in section 4. Section 5 describes how this new 3D geo data model is implemented in a first prototype for pedestrian routing in our research. Finally conclusions are drawn and an outlook is given in Section 6.

2 REQUIREMENTS TO THE 3D MODELLING OF THE SCENE SPACE

The traveller has different mental images when he is preparing a travel with a public transport compared to when he is planning a walk in a building or on the street (Rüetschi and Timpf 2005, Rüetschi 2007). When thinking of public transport, the traveller imagines a (public transport) network. This environment is dominated by connections made by transport lines (links) and the locations (nodes) where you can enter and leave or change the vehicles. It's the world of timetables, line maps, fare structures, alternative journey routes and sometimes delays. The public transport environment is called *Network Space* because it is based on the public transport node/link structure (Rüetschi 2007). This environment is created by strategic planners of public transport. The result of network design, planning and timetabling is designed to optimize travel times for the public. Very often it is influenced by historical, social, and economical processes.

The environment a traveller knows when he walks has a different origin. It is created by infrastructure and urban design. This walking environment is called a *Scene Space* (Rüetschi 2007). It is an environment for way finding. It has another type of structure, which is dominated by nested open spaces. Environments like stations (with its halls, squares, platform areas, etc.), airports, harbours, shopping malls, or public parks are samples of these walk environments. These environments are generally available in 2D but they have to be captured in a 3D geo-data model.

An effective and well-structured model of the 3D Scene Space is needed to support preparation of walking routing as part of a travel advice. The functional requirements for such a model can be specified as follows:

- Provide a sufficient number of objects to be included in the Scene Spaces
- Store spatial characteristics of objects relevant for pedestrian use of Scene Spaces
- Define accessibility constrains of Scene Spaces
- Combine indoor and outdoor Scene Spaces in one geo data model
- Connect Scene Space with the Network Space
- 3D visualisation of Scene Spaces.

In such a 3D model two kinds of structures can be created: *network* and *hierarchical*. 3D spaces when defined can be connected to each other creating a network topology of possible walking connections in Scene Space. For example an entrance is connected to a hall. A hierarchical topology of Scene Spaces exists because of the nesting of open spaces, e.g., a platform is part of a station. The network topology of Scene Spaces is important for accessibility and routing for pedestrians. The hierarchical topology is important for identification and visualisation of Scene Spaces to represent this nesting of open spaces.

The most import function of a journey planner is to help the traveller in preparation of his/her journey. Thus the traveller must be able to prepare himself for the pedestrian parts of a public transport trip. In the planning phase of a public transport trip the 3D geo-data model should be able to support this preparation. Consequently, several 'traveller's' requirements can be identified as well. An important element in his preparation is to identify the origin, the destination and the accessibility of the pedestrian parts. The pedestrian parts are needed to recognize the locations where egress of public transport vehicles in a journey takes place. Entrances and exits of public transport locations as part of a pedestrian route are important landmarks. Therefore a 3D geo-data model should have a structure to support the following functionality:

- Locate the entrance(s) and exit(s) of a public transport place in relation to the surroundings and the entry and exit points of public transport.
- Get detailed routing for access, transfer and egress of public transport.
- Get detailed routing inside and outside a public transport place.
- Get detailed walk time and accessibility information for routing.
- Get proper 3D visualisation of routing.

The representation of 3D Scene Space must be also able to accommodate large amounts of data (assuming it will be built for cities, countries or groups of countries) and to be practical to support the preparation of walking. These are implementation recommendations to arrive at a usable, well-structured geo-data model. In our research we have proposed the following guidelines:

- The geo-data model should be implemented in a spatial database.
- The 3D model should follow the semantic developments worldwide.
- Apply when possible (Open or industry) standards in design and implementation.

3 ANALYSIS OF EXISTING 3D MODELS

The above specified requirements were used to evaluate the existing 3D models and approaches in their suitability for modelling Scene Spaces. Some models are focusing on indoor navigation (Choi and Lee 2010) or interior construction details (buildingSMART), while other are having more focus on outdoor like CityGML (Gröger et al., 2008).

3.1 *The CityGML and IFC models*

CityGML LOD4 has information for indoor but is lacking detailed semantic objects relevant for pedestrian navigation for public transport. CityGML is extensible and adding these elements could form a good basis for pedestrian navigation related to public transport. Building Information Models (BIM) are used in construction work, these models are focussing on the indoor environment and on the construction details of facilities. BIM is a digital representation of physical and functional characteristics of a facility. Industry Foundation Classes (IFC, ISO PAS 16739) are a standard in this field. IFC represents a data schema for sharing construction and facility management data across various applications used in the building construction industry. It is a common exchange format in the field of architecture, engineering, and facilities management. It is an object-oriented data schema based on class definitions representing the objects (such as building elements, spaces, properties, shapes, etc.) that are used by different software applications used in construction or facility management project. There is also research on mapping elements from IFC to CityGML (Benner, et al., 2005, Isikdag and Zlatanova, 2009).

3.2 *The IFOPT model*

For Scene Space the IFOPT standard is very appropriate. The Fixed Object Model has a relationship to other standards describing the geographical features of a country, but it is not a GIS standard. The IFOPT Stop Place Model is strong in defining the semantic aspect, organizing, identifying and naming, of spaces relevant to support pedestrian navigation in relation to public transport areas. In the Stop Place model the entities from the passenger, vehicle and equipment domains are considered relevant to travellers:

- Vehicle area's locations like tracks and stopping positions
- Equipment e.g., ticket machine, lockers
- Spaces which are publicly accessible (e.g., platforms, boarding positions, entrances)
- Spaces (stores, kiosks, public toilets)
- Buildings/ part of buildings and indoor spaces
- Connecting elements between levels (steps, escalators, elevators, stairways ramps)

A weak point of the Stop Place Model is that it is a complex standard with a lot of classes, which however are lacking completeness. A "light" version of the standard which could be extended with local relevant elements would be easier to implement. However even in the current complex standard not all relevant items for travellers are included. For example signs are not included in the model. The signs are actively used for orientation. The location of these signs in relation to the footpath of the traveller must be in line with the descriptions of directions used in travel advises. Capturing all information from signs and their locations and direction in relation to the footpath will improve the information in travel advises. Furthermore 3D geographical aspects of objects are ignored in this standard.

3.3 *The Slingsby model*

In the model of Slingsby (Slingsby and Raper 2008) the pedestrian space combines geometric, semantic and navigation aspects. In this model 3D spaces are represented only by their lower surfaces. The objects described by the lower surfaces can have existence, accessibility, pedestrian direction and construction material attached as attributes. Some of these attributes are time-depended, while others are pedestrian-dependent. This model consist of 'barriers' (walls), 'portals' (doors and windows), 'teleports' (lifts) and 'spaces' (specific delineations of space). Persistent information is provided through lists of unique or recurring time periods (e.g., some barriers only exist at certain times of day). Using this information, the model attempts to incorporate some of the micro scale details of pedestrian access. A pedestrian has a step-height he or she is able to negotiate, which would be zero or very small for a wheelchair user. Although this model gives a good basis for a pedestrian model its semantics containing 'barriers' (walls), 'portals' (doors and windows), 'teleports' (lifts) and 'spaces' (specific delineations of space) is not in line with semantics used in the field of public transport. However, the concept of lower, walking surfaces is adopted in our approach.

4 A 3D MODEL FOR SCENE SPACE

As illustrated above existing 3D GIS models are not directly suitable to support pedestrian routing to improve travel advice on public transport information websites. Some models focus on storing geometrical data while others on visualisation of this data, some models are a strong on the semantic aspects, others on supporting navigation. Therefore a new 3D data model to represent Network Space and Scene Space is proposed.

4.1 *The main concepts in the model*

The following concepts are taken into consideration (see Figure 2):

- *PublicTransportJourney* is the spatial and temporal constraints established in the Network space and in the Scene Space. It defines the entrance and access time and location to the public transport vehicles in a planned public transport trip.
- *Single space* is the smallest representation of a 3D indoor or outdoor public accessible area. A Single space can represent a real world space limited by physical borders or a logical part of a space limited for a specific purpose by unseen imaginary borders. Scale, time, pedestrian purpose and architectural structure of a Single space are influencing the borders. Single space can be limited by temporal or moving objects, e.g., doors, elevators or lifts. A *NodeSurface* models the lower surface of a Single space.

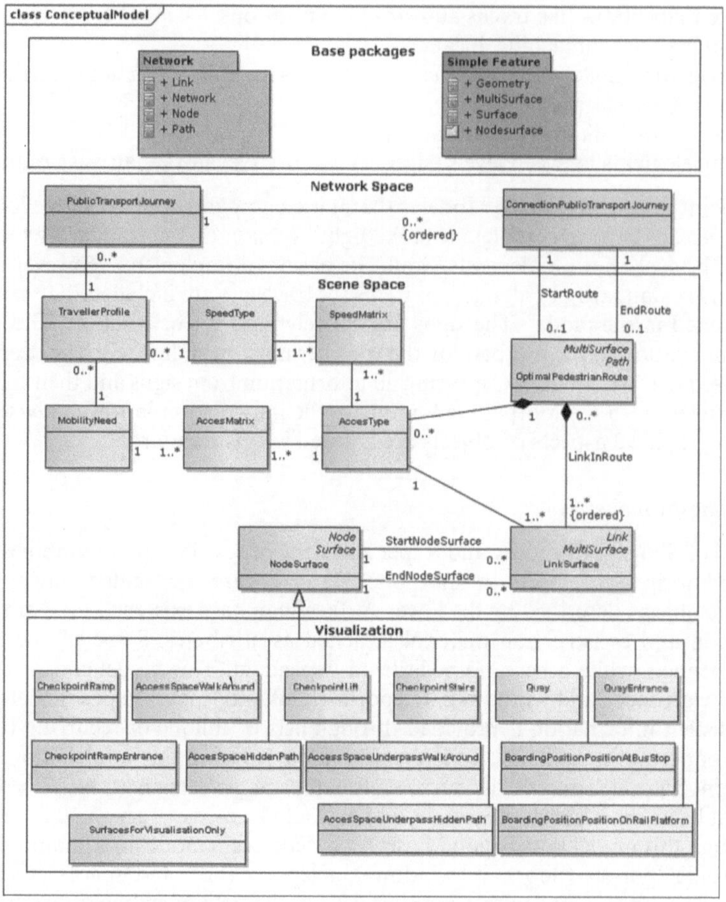

Figure 2. UML Overview class diagram of the conceptual model for pedestrian routing.

- The pedestrian connection of single spaces named *LinkSurface* defines how pedestrian entrance or exit of single space takes place. These connections define possible pedestrian movement (border crossings) in 3D between single spaces. They are describing the possible directions and accessibility of a pedestrian area and are modelling the accessible Scene Space.
- Pedestrian routes are the ways a traveller walks through a collection of connected Single spaces. The order of Single spaces determines the direction of pedestrian movement. The term pedestrian route is used to describe a path for all pedestrian movement followed by all kinds of individual or assisted travellers' locomotion by e.g., foot, wheelchair, dragging luggage, sleeping in a baby buggy. The representation of pedestrian routes is a *MultiSurfacePath* in the model.
- *TravellerProfile* is the selection of preferences on accessibility restrictions and speed of pedestrian movement chosen by traveller for a specific situation. These preferences are time and situation, public transport journey depending, various aspects are influencing these selection e.g., luggage, time pressures, physical efforts, safety, comfort, personal characteristics, habits, health concerns. This traveller profile defines the possibility to use a connection of single spaces and duration of pedestrian movement passing single spaces.
- *OptimalPedestrianRoute* is an accessible pedestrian route optimal in time (fastest) using the time and accessibility preferences from the travellers profile.

The conceptual model is organized around the separation of Network Space and Scene Space and at the same time much attention is given to the visualisation. In Network Space only the spatial and scheduled temporal constraints created by planned public transport journeys are considered in this model, other important aspects in Network Space e.g., public transport line names, delays, disturbances or vehicle properties are covered in existing journey planners and not considered in this model. Scene Space is limited to 3D representation of accessible pedestrian areas relevant in the planning process of a walk as part of a public transport journey. To model this idea we separate core functionalities for Public transport, Pedestrian and Visualization. Other functionalities, Network and Simple Feature are considered base functionalities used as foundation in the conceptual model.

Figure 2 shows the UML class diagram of the conceptual model. In Base package the base classes of Network and Simple Feature are captured. The Pedestrian functionality is related to storing the geometry and finding optimal pedestrian routes for pedestrians. The pedestrian functionality uses the Simple Feature Implementation Specification for SQL (OGC 1999) as standard to store the spatial data. The other base functionality is Network. Network is used to model the pedestrian connections of Single spaces. The classes *Link*, *Network*, *Node* and *Path* in the Network functionality are needed to model a pedestrian route.

4.2 *The three packages in the model*

The Public transport package represents the public transport journey. Important are the locations were boarding and egress of the public transport vehicles takes place. Class *ConnectionPublicTransportJourney* models the locations where connecting vehicles entrances and exits locations to Single spaces take place.

The Pedestrian package captures the information related to the travellers. The traveller's preferences are given in the in the class *TravellerProfile*. The class *MobilityNeed* represents accessibility preferences to select in a profile. From the class *SpeedType* the preferences for modification of pedestrian movements speed are selected (e.g., Fast, Normal, or Slow). The class *NodeSurface* models the lower surface of a Single space. The *NodeSurface* is the generalization for all kind of Single spaces. The representation of pedestrian a connection is given with the class *LinkSurface*. *LinkSurface* represents the concept of a border between two Single spaces. It also specifies the direction of pedestrian movement over this border. The pedestrian connection of Single spaces is important for the accessibility of the next Single space in pedestrian movement. A classification of accessibility can be added to the combination of representation of Single spaces in a *LinkSurface*. This accessibility classification is captured in the class *AccesType*. To link this accessibility to the accessibility preferences in the travellers profile, the class *AccesMatrix* is defined. It holds the knowledge which preferred *MobilityNeed* are influencing the border that can be crossed by a pedestrian. The pedestrian connection of Single spaces is also important in the pedestrian speed moving over the next Single space. The class *SpeedMatrix* is linked to the *SpeedType*, which gives information which speed must be used to calculate the walk duration to the next a Single Space.

The Visualization package encompasses the objects relevant for the visualisation of the pedestrian navigation, combining visual and textual information and using the semantics derived from the IFOPT standard for the travel advices. The visualisation is used to distinguish between the different Single spaces and their purposes. Single spaces with different purpose are named differently. When applicable exiting names are derived from the IFOPT standard to further define a specific single space:

- *BoardingPosition* is a Single space where a traveller waits for boarding to a public transport vehicle.
- *AccessibleSpace* is a passenger area such as a concourse or booking hall, immigration hall or security area that is accessible by passengers, but without a direct access to vehicles Single space, may be a room, hall, concourse, corridor, or bounded open space.
- *Quay* is a Single space, such as platform or quayside where passengers have access to public transport vehicles, taxi cars or other means of transportation.
- *Checkpoint* is a Single space, which may potentially create an extra time penalty on top of the walk duration that should be allowed for when journey planning. Checkpoint is used

IFOPT for processes like ticket control or immigrations. In this model it is also used for stairs, ramps, lifts and escalators.

4.3 *Refinement of a Single space*

To express the differences in Single Space between a real world space limited by physical borders and a virtual logical part of a space, a new terminology is created. The term *Walk around* is used for the Single space were a space is limited by real word visible borders and no specific pedestrian area is defined. With *AccesSpaceWalkAround* it is possible to explore the whole space, e.g., in a room of a museum or in square in a shopping area. Figure 4 illustrates the lower surface of one Single space modelled this way.

The term *Hidden* is used for a Single space with invisible borders in the physical world. If the purpose of a logical part of space is to define a logical walking area the single space is called *AccesSpaceHiddenPath*. The term '*hidden path*' is used because a logical walk area ('path') is supposed to exist in an open space, but this walkway is imaginary (or hidden). Figure 5 illustrates the lower surfaces of two Single spaces modelled this way. Please note that modelling Single spaces is an abstraction process and borders of this kind of Single spaces are invisible in the real world. In this process scale is important and cartographic generalisation in 3D influences this process.

A similar approach to Slingsby and Raper 2008 is used to model the lower surfaces of accessible 3D spaces. However, the representation of logical links is different. In our model the logical links between a surface and its entrances and exits are modelled as the combined surface of the origin (*StartNodeSurface*) surface and the destination (*EndNodeSurface*) (see Figure 3).

4.4 *Linking network space and scene space*

A link, called *LinkSurface* is a (derived) *multiSurface* geometry. The order of the surfaces is important in this *multiSurface* because it defines the direction of the link. The model restricts

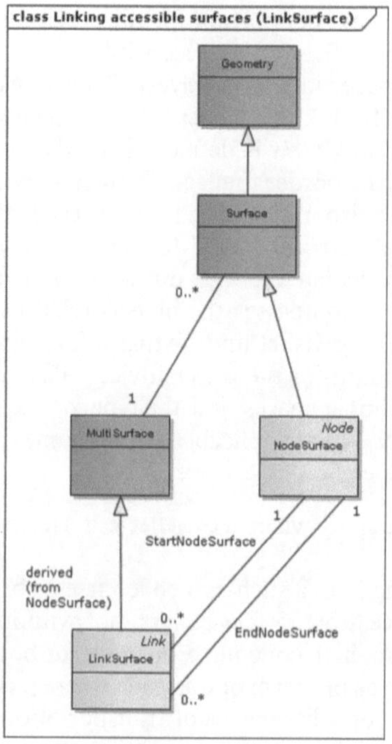

Figure 3. Class diagram showing LinkSurface as multiSurface and the start and end node roles.

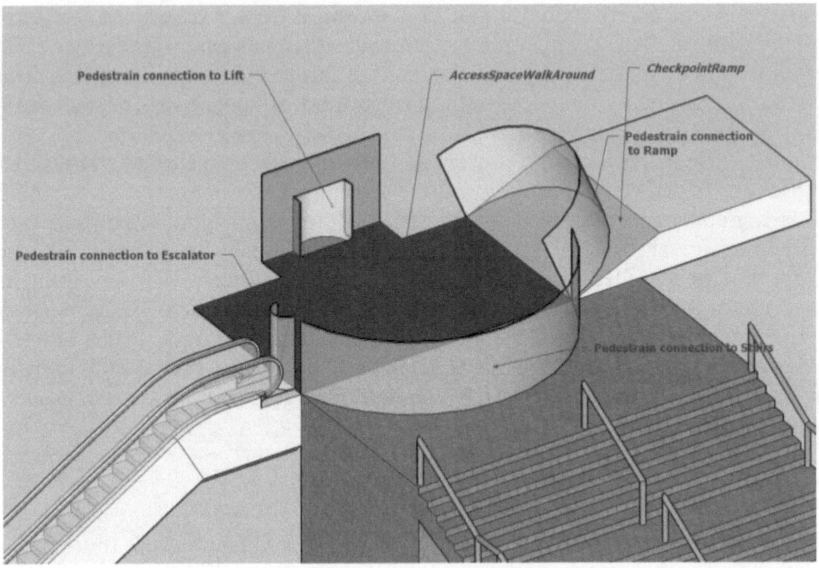

Figure 4. *AccessSpaceWalkAround* (the area at front of the lift), *CheckpointRamp* (the area towards the Ramp) and possible pedestrian connections to other *NodeSurfaces*.

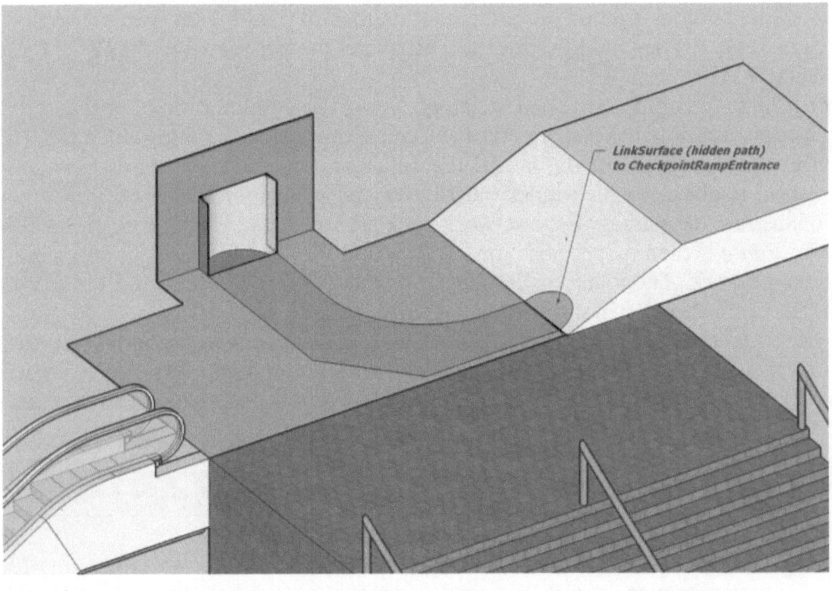

Figure 5. *Hidden path* and *hidden ramp entrance*.

this *multiSurface* to a surface consisting of only two surfaces with only one connecting border. No disconnected surfaces are allowed in this *multiSurface*. This is done because route choice in this model is based on selecting relevant directions to unique entrances and exits of *NodeSurfaces*. Figure 4 is an example of a route choice after leaving the lift. After leaving the lift a central floor is reached (a *NodeSurface* indicated as *AccesSpaceWalkAround*). The possible directions are shown when leaving the *AccesSpaceWalkAround* to Ramp, to Escalator and to Stairs. These directions are shown as 'curtains' to represent better the possible pedestrian connectionss in Scene Space after leaving the *AccesSpaceWalkAroun*. To model the direction, a new surface is created by spatial union of the central floor (*AccesSpaceWalkAround* as *StartNodeSurface)* and the ramp surface (*CheckpointRamp* as *EndNodeSurface)*. This new surface (*LinkSurface)* models the pedestrian connection to the ramp from the central floor. This *LinkSurface* explicitly connects both surfaces and as such gives the walking (lower) surface of this pedestrian connection.

The model is able to supports the concept of walking freely on a surface, *Walk around* and a more strict way of walking on *Hidden paths*. This manner of representation combines both ways of cartographic generalization of pedestrian routing. In Figure 4 the route a traveller follows when exiting the lift and walking to the ramp is not defined. The model supports the concept that the traveller walks freely to the ramp. This concept of free walking makes it possible to use only visible borders in the real world. For applications where more detail is needed, the model also supports the concept of logical walking routes in the Single spaces, i.e., *Hidden Paths*. When modelling *Hidden paths*, the *NodeSurfaces* are split in more detailed *NodeSurfaces*. The resulting *LinkSurface* is showed in Figure 5. Both ways of walks can be represented in the model.

4.5 *Accessibility restrictions and the individual traveller*

The *LinkSurfaces* are categorized in different access types. The visualisation class of the *EndNodeSurface* in combination with the relative height of two *NodeSurfaces* determine the acces type of the *LinkSurface*. In Figure 5 the ramp is located relative higher than the *NodeSurface* and is connected to the lift therefore the type of the *LinkSurface* is 'Ramp-Up'. These access types are used as accessibility restrictions for the *LinkSurface*. Furthermore they are used to add pedestrian duration as attribute to the *LinkSurfaces*.

These accessibility restrictions in are direct mapped to possible individual traveller needs as shown in Table 1. The semantics for these attributes are derived from the IFOPT standard (IFOPT 2007).

All *LinkSurfaces* compose the pedestrian network. The combination of access restrictions and pedestrian duration makes it possible to calculate optimal walk routes with a given transfer point in public transport. In the model the pedestrian accessible surfaces (*NodeSurfaces*) are modelled as objects, which inherit attributes and behaviour from nodes in a network. Figure 6 illustrate the classes of the network package, i.e., *Path*, *Link* and *Node*. The *Optimal Pedestrian Route* is then a specialisation of *Path* and the *LinkSurface* is a specialisation of *Link* in the network. The relations between *NodeSurface*, *LinkSurface* and the navigational path for the optimal pedestrian route are shown as well.

The necessity to connect public transport with pedestrian routing is illustrated by the possibility to connect *ConnectionPublicTransportJourney* and *OptimalPedestrianRoute* classes. Connecting boarding and egress locations of public transport vehicles, in a public

Table 1. Individual needs (*MobilityNeed*) related to accessibility (*AccesType*) of a *LinkSurface*.

	No restriction	Single step (up or down)	Ramp up	Ramp down	Stair up	Stairs down	Lift	Escalator up	Escalator down
Wheelchair	Y	N	Y	Y	N	N	Y	N	N
Step free	Y	N	Y	Y	N	N	Y	Y	Y
Averse to lifts	Y	Y	Y	Y	Y	Y	N	Y	Y
Escalator free	Y	Y	Y	Y	Y	Y	Y	N	N
Luggage heavy	Y	N	N	N	N	N	Y	Y	Y
Luggage medium (no stairs ramps up)	Y	Y	N	Y	N	Y	Y	Y	Y
No down stairs/ ramps/escalators	Y	N	Y	N	Y	N	Y	Y	N

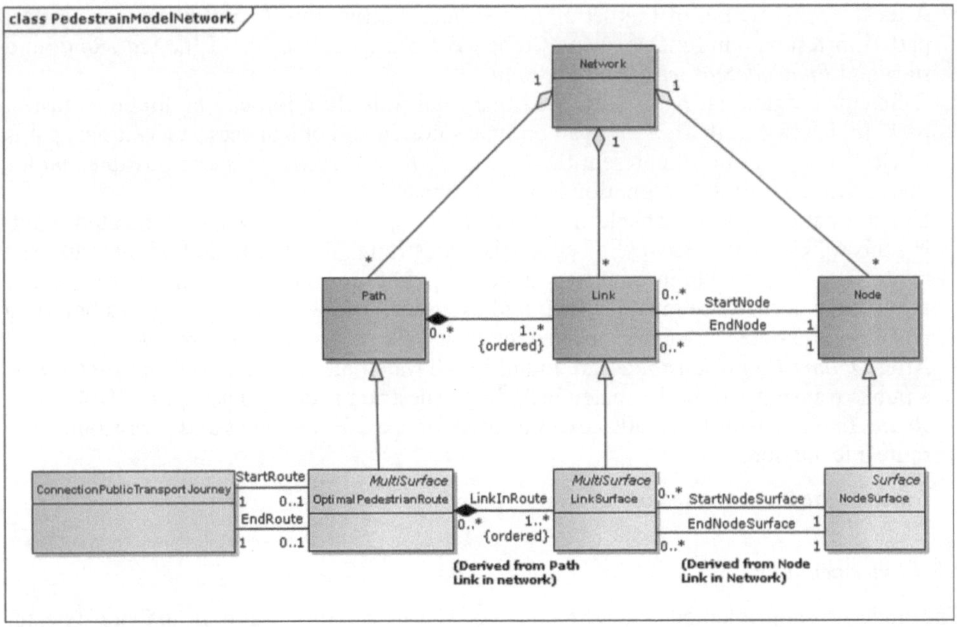

Figure 6. Class diagram surfaces are modelled as objects inherit attributes and behaviour from nodes in a network.

transport advice, to an optimal pedestrian route is done though the start- en end- *NodeSurfaces*. An *OptimalPedestrianRoute* is an ordered collection of *NodeSurfaces*. The start *NodeSurface* is defined as the first *NodeSurface* in the sequence of an optimal route and the end *NodeSurfaces* as the last *NodeSurface*. More details on the model and the implementation can be found in (Schaap 2010).

5 PEDESTRIAN ROUTING WITH THE MODEL

Using this model and the classes related to the *TravellersProfile* (see Figure 2), an optimal pedestrian route can be calculated.

5.1 *Computing the optimal path*

To enhance a standard public transport travel advice with walk routing for a traveller having a specific travel profile the following steps must be taken:

1. Optimal navigational path calculation starts with deriving the start en origin from the walk parts from a public transport travel advice. These start and origin points must be matched on *NodeSurfaces* describing these points. (E.g., IFOPT Boarding positions, entrance or public places). Connecting boarding and egress locations of public transport vehicles is done though the start- en end- *NodeSurfaces* in an optimal pedestrian route.
2. The *TravellerProfile* is composed. The walk *SpeedType* in combination with the individual *MobilityNeed* from the traveller is defined for this traveller.
3. With the individual *MobilityNeeds* from the *TravellerProfile*, a sub-selection of all *LinkSurfaces* is made containing all *LinkSurfaces* this traveller can possibly use in an optimal path.
4. For the set of *LinkSurfaces* selected for the *TravellerProfile*, the walking durations are assigned to each *LinkSurface* according to the profile. This duration is calculated for each *LinkSurface*.

5. After the construction of this set of *LinkSurfaces*, fitting this *TravellerProfile* the fastest path (shortest path in time) is calculated between *StartNodeSurface* of the *StartRoute* and start and *EndNodeSurface* of the *EndRoute*.
6. Different navigational paths with the same total walk duration can be found as fastest path. In this case both are seen as an optimal solution and one of these navigational paths is selected at random to represent the *OptimalPedestrianRoute*. It is also possible that for this individual user the destination is not reachable.
7. If a destination is not reachable, a new *PublicTransportJourney* must be generated avoiding this walking part. Because of pedestrian constraints the destination of the walk part might be not reachable in time to connect to public transport. Also in this case a new public transport advice must be generated using the new departure time at this boarding position. The steps 1 till 6 are repeated for each walk part of a *PublicTransportJourney*.
8. After *OptimalPedestrianRoutes* are found for all walk parts of a *PublicTransportJourney*, a public transport advice is created including pedestrian routing. The *OptimalPedestrianRoutes* found for this travel advice can be used for textual walk times and accessibility and route information.

The steps are implemented in our prototype.

5.2 *Implementation*

The model was implemented Oracle Spatial 11 g. The visualisation was done in Google Earth. Relevant data for the model were manually created in Google SketchUp and exported to KML files. Figure 7 shows the Google SketchUp model for the test location (i.e., the railway station of Wageningen, the Netherlands).

This dataset was constructed to illustrate and validate the proposed geo-data model with Scene Space from a transfer location with indoor and outdoor Single Spaces, Multi model complexity (bus and train stops) and 3D Scene complexities with specific access constrains (Underpass, Stairs, Ramp and a Lift). The tested transfer location is not very complex. The evaluated public transport journey assumed transfer from bus to train. For a traveller with local knowledge and no restrictions, an optimal route can be straightforward found.

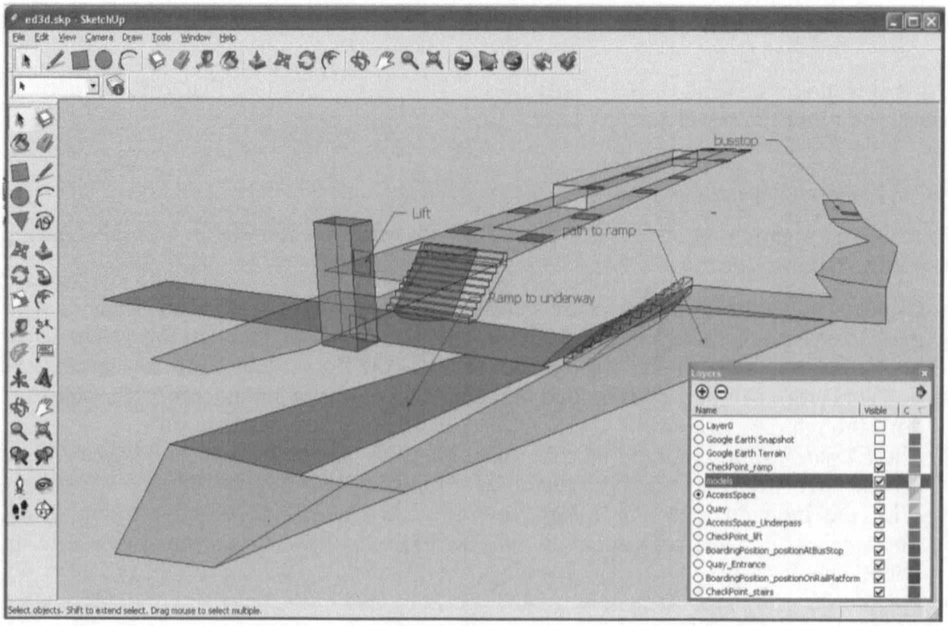

Figure 7. *NodeSurfaces* for studied location Ede-Wageningen in Google SketchUp.

The restricted traveller needs to use a ramp and/or a lift (based on his traveller profile). In the test model, people not able to use stairs have to consider the ramp down to the under passage at the left hand when entering the south entrance. Furthermore there is a lift in the underpass to the platform. This lift is located opposite of the stairs to the platform in the underpass (Figure 7).

Our first test focussed on traveller's profiles with different pedestrian movement speed (Fast, Normal and Slow). The second test included accessibility needs (No specific needs, Wheelchair, Step Free, Averse to lifts, Escalator free, Luggage heavy, Luggage medium and No down stairs/ramps).

5.3 *Testing*

The first pedestrian routing without accessibility restrictions revealed some limitations in the prototype when the Scene Space was modelled with only *WalkAround* Single spaces. The used distance calculation algorithm in the prototype, based to half the perimeter of the *NodeSurface*, was not precise enough to be useful for optimal pedestrian routing (but quite crucial in optimal route computation). Furthermore wrong pedestrian connections were created at certain locations along pedestrian routes due to the limited 3D intersection capabilities of Oracle Spatial 11 g. The Oracle Spatial 11 g offers only ANYINTERACT between two surfaces in 3D. The operator SDO_RELATE with various masks to test for the different kind of spatial relations is available only for 2D geometries (Murray 2010). In further evaluation of the prototype the problematic Single spaces were redrawn using *Hidden* path and *NodeSurfaces* to overcome these limitations.

The overall results of tests were promising: it was possible to find optimal pedestrian routes with respect to the tested traveller's profiles. Figure 8 illustrates the optimal route between the bus station and the needed train platform. Restricted pedestrians use the ramp as indicated in Figure 9. They have to use ramp down to underpass and the lift up to Quay.

Figure 8. Walking route between bus stop and entrance of a train.

Figure 9. The optimal route between the bus stop and platform (and entrance) of a train for travellers who cannot use stairs, heavy luggage or wheelchair.

6 CONCLUSION AND OUTLOOK

This paper describes a novel 3D geo-data model of accessible pedestrian spaces in 3D to support pedestrian routing in multimodal public transport travel advices.

6.1 *Main results*

Innovative aspects in this geo data model are:

- The model manages information about the walk duration preferences for specific travellers by the traveller's profiles. This information addresses the first disadvantage of journey planners, i.e., the standard time used for walking.
- The model to represent the Scene Space incorporates accessibility and 3D spatial characteristics. This model allows a connection between the Scene Space and the Network Space. Thus this approach provides a solution to the lack of information about transfer places in current journey planning systems.
- The concept of *NodeSurface* and *LinkSurface* to represent Scene Spaces is a simple, effective and yet well-structured way to represent complex spaces used by pedestrian routes in a geo data model.

Furthermore the model supports the concept of walking freely on a surface, i.e., *Walk around* and the more strict way of walking on *Hidden paths* over surfaces. This can be used to have differences in conceptual generalization when modelling pedestrian routing in visualisations at different scale.

The overall results of the prototype were promising especially when modelling *Hidden paths*. It was possible to implement the geo data model in Oracle Spatial 11 g. Oracle Spatial 11 g network shortest route capabilities for logical network turned out to be suitable for finding the optimal route for individual travellers with specific speed and accessibility properties.

6.2 *Future work*

The tests are still of limited complexity and the model needs further research towards the feasibility and suitability of this new model for different transfer points. Currently no tests with escalators and complex lift situations are completed. The algorithms should be further elaborated to define more efficiently hidden path areas is useful to limit labour when modelling complex public transport places.

Visualisation of the results could also be improved. Better visualisation of pedestrian routing is needed including underground objects in a clear way. As known, Google Earth does not provide underground visualisation yet. Integrating sign information in the pedestrian model and improved visualisations are needed to be included on travel information websites.

Websites with travel information for public transportation must be able to generate travel advises fast and in large amounts. The prototype did not focus on performance and scalability aspects. The separation of logical steps (i.e., link assigning according to traveller needs, network loading and route calculation, storing and visualisation) need to be reconsidered when developing a scalable and fast website applications. Performance could be improved by keeping the pedestrian network in memory for all optimal route requests. A dedicated software program specific written for this functionality could optimize this process and outperform the Oracle network search capabilities used in the described prototype.

Applications using the proposed geo data model are not limited to applications for trip preparation. Future applications on mobile devices with adequate indoor and outdoor positioning could use the proposed geo data model to facilitate travellers during a transfer. Showing directions in real-time on mobile devices could guide travellers to the next public transport vehicle. These applications could be very sophisticated. An application might show a virtual guide, e.g., a guide dog, projected over an image of the actual environment: augmented reality based (similar to Layar for outdoor applications). The symbolisation of speediness of the virtual guide in such application could inform the traveller to hurry to catch the next vehicle. The symbolisation of speediness could be based on the real-time departure times of vehicles and the expected walk distance in the geo data model. Lack of adequate indoor positioning in mobile devices and detailed 3D datasets of pedestrian locations is currently limiting these developments, but it is most likely that these (indoor positioning infrastructure and detailed 3D datasets) will be available in the future.

The 3D spatial data sets describing pedestrian locations were constructed based on aerial photos and existing map of these locations. This way of constructing 3D spatial datasets for pedestrian routing is time-consuming and prone to errors. More efficient ways of capturing and collecting spatial information of 3D data to support pedestrian routing in public transport advices on a national scale must be found. Tools to automatically create pedestrian surfaces from existing indoor building models (IFC or CityGML LOD4) could be beneficial in the creation of these datasets. Agreements on collection, exchange and maintenance of these 3D spatial data are important between the involved parties.

A spatial data infrastructure (SDI) supporting pedestrians routing around public transport is essential for this. The proposed conceptual model could be used as basis for a data exchange format in such a SDI. In this respect a research on the establishment of a spatial data infrastructure (SDI) for this would be a step towards a 3D geo-data model solution to support pedestrian routing in multimodal public transport travel advices on a national scale.

ACKNOWLEDGEMENTS

The authors of this paper want to express their gratitude to the anonymous UDMS reviewers for the constructive comments. This research was conducted in the context of the MSc GIMA thesis project of the first author (Schaap 2010).

REFERENCES

Benner, J., Geiger, A. and Leinemann, K. 2005. Flexible generation of semantic 3D building models, Proc of the 1st Intern. Workshop on Next Generation 3D City Models.

Choi, J. and Lee, J. 2010. 3D Geo-Network for Agent-based Building Evacuation Simulation. In: Showalter and Lu (eds.), Geospatial Techniques in Urban Hazard and Disaster Analysis Geotechnologies and the Environment, 2010, Springer Volume 2, Part 5, 415–429.

Gröger, G., Kolbe, T.H., Czerwinski, A. and Nagel, C. 2008. OpenGIS City Geography Markup Language (CityGML) Encoding Standard, Version 1.0. 0, International OGC Standard. Open Geospatial Consortium.

Grotenhouse, J.W., Wiegemans, B.W. and Rietveld, P. 2007. The desired quality of integrated multimodal travel information in public transport: Customer needs for time and effort savings, Transport Policy, 14(1), 27–38.

Hochmair, H. 2004. Towards a classification of route selection criteria for route planning tools, Developments in Spatial Data Handling, Springer, Berlin, 481–492.

Hagedorn, B., Trapp, M. Glander, T. and Dollner, J. 2009. IEEE Computer Society, pp. 692–697.

IFOPT, 2007. Road traffic and transport telematics Public transport Identification of Fixed Objects in Public Transport, Technical Specification, CEN/TC278, University Zurich. p. 144.

Isikdag, U. and Zlatanova, S. 2009. A SWOT analysis on the implementation of BIM within geospatial environment, in Krek, Rumor, Zlatanova & Fendel (eds.) Urban and Regional data Management, UDMS Annuals 2009, CRC press, Boca Raton, pp. 15–30.

ISO PAS 16739, 2008 Industry Foundation Classes Release 2x, http://www.iso.org/iso/iso_catalogue/catalogue_tc/catalogue_detail.htm?csnumber=38056

Lee, J. (2004). A Spatial Access-Oriented Implementation of a 3-D GIS Topological Data Model for Urban Entities. *Geoinformatica* 8(3): 237–264.

Meijers, M., Zlatanova, S. and Preifer, N. (2005). 3D geoinformation indoors: structuring for evacuation, in: *Proceedings of Next generation 3D city models*, 21–22 June, Bonn, Germany, p. 6

Miller, L.E. (2006). Indoor Navigation for First Responders-A feasibility Study, NIST Report, Available online at: http://www.antd.nist.gov/wctg/RFID/Report_indoornav_060210.pdf.

Murray, C. 2010. Oracle® SpatialDeveloper's Guide11 g Release 2. March 2010 edn. Redwood City: Oracle Corporation.

Raubal, M. and Worboys, M. 1999. A Formal Model of the Process of Wayfinding in Built Environments, in: Freksa & Mark (Eds.): COSIT'99, LNCS 1661, Springer-Verlag Berlin Heidelberg, pp. 381–400.

Rüetschi, U.J. 2007. Wayfinding in Scene Space Modelling Transfers in Public Transport, Dissertation, University Zurich. p. 137.

Rüetschi, U.J. and Timpf, S. 2005. Modelling wayfinding in public transport: network space and scene space. Spatial Cognition IV, pp. 24–41.

Schaap, J. (2010). Towards a 3D geo-data model to support pedestrian routing in multimodal public transport travel advices Master's thesis GIMA (Utrecht Univ., TU Delft, Wageningen Univ., ITC), 2010, p. 101.

Slingsby, A. and Raper, J. (2008). Navigable Space in 3D City Models for Pedestrians, in: Van Oosterom, Zlatanova, Penninga&Fendel (eds) *Advances in 3D Geoinformation Systems*, Springer, Berlin, Heidelberg, pp. 49–64.

Stoffel, E., Lorenz, B. and Ohlbach, H. (2007). Towards a Semantic Spatial Model for Pedestrian Indoor Navigation. In Hainaut, J., Rundensteiner, E., eds.: *ER '07 Workshops / Advances in Conceptual Modeling*, Springer, 328–337.

Stevens M. and Choi, J. 2006. CAD Data Conversion to a Node-Relation Structure for 3D Sub-Unit Topological Representation. Journal of the Korean Geographical Society, 41(2), pp. 188–194.

Yuan, W. and Schneider, M. 2010. Supporting 3D route planning in indoor space based on the LEGO representation, ACM, ISA'10, 2 November 2010, San Jose, CA, USA, p. 8.

Urban and Regional Data Management – Zlatanova, Ledoux, Fendel & Rumor (eds)
© 2012 Taylor & Francis Group, London, ISBN 978-0-415-67491-1

Towards a 3D network model for indoor navigation

L. Liu & S. Zlatanova
Delft University of Technology, GIS Technology Section, Delft, The Netherlands

ABSTRACT: Indoor navigation is a fast developing area, but researchers are facing various challenges. One of them is the 3D model that could support the navigation. After reviewing current development of models for indoor navigation, in this paper we generalize essential emergency phases of buildings and clarify requirements for indoor navigation models. Then typical indoor navigation models are compared and their pros and cons are given. Based on these work, we propose a 3D network called "Indoor Visibility Structure", which makes use of 3D geometry & semantics of buildings and allow for automatic generation of navigation network. The proposed approach is designed to resolve navigation in complex buildings. Ultimately, by relying on the IVS, its updating mechanism, and corresponding routing strategy considering indoor real-time changes, the 'door-to-door' navigation within buildings could be achieved.

1 INTRODUCTION

Indoor navigation approaches are actively developed in the last years for rescue and evacuation from buildings. Fire alert systems and light- & sound-based evacuation systems are commonly available in all large buildings. Many of the public buildings are even equipped with modern systems allowing for tracking people. Yet such ways are insufficient for evacuation or rescue processes. A successful indoor navigation relies on: an up-to-date navigation model of buildings, an accurate indoor positioning and adjustable navigation routes concerning real-time changes of buildings and pedestrians. One of the most significant factors in this process is the accurate and up-to-date navigation model.

At present there are various approaches to model buildings for the purpose of navigation (Stoffel et al., 2007, Lamarche & Donikian 2004, Li & He 2008, Li et al., 2010). Most of these methods are based on pre-defined 2D floor plan or pre-defined simple 3D models of buildings. Those models merely represent the static state of buildings and can not take changes within indoor environment into account. For instance, when a part of a building is unavailable during a disaster (e.g., fire), the pre-defined navigation plans would be ineffective and the rescue process may be disturbed.

3D building represented as a collection of 2D floor plans has also insufficiencies. Multi-story buildings may have intermediate levels within certain floors. It is impossible to use 2D plans to exactly represent this complexity. How to exactly represent these indoor structures becomes a tough problem. Use of 2D plans may fail to represent correctly vertical indoor environments (such as configuration of stairs, elevators and escalators).

The currently proposed 3D navigation models are demonstrated on simple (regular-structured) buildings and fail to provide automatic approaches for routing in complex interior structures. Generally, automatic derivation of the navigation model from complex 3D building models is one of the most critical obstacles presently. The changes (unavailable corridors, exits, etc.) in the environment due to smoke, water, or other types of damages are represented only on the network for navigation. Thus, the link with the geometry of the building is disturbed and may lead to a loss of important information. For example, windows on the first and second floors, which are not normally included in the network for navigation, might be omitted as options for escape.

This paper presents a new concept to navigation modeling, which makes use of 3D geometry & semantics of a building and allows for automatic generation of navigation network. Depending on the damages on the building different networks can be derived, i.e., taking into account different construction elements. The proposed approach aims at resolving navigation in complex buildings, by relaying on an Indoor Visibility Structure and a 'door-to-door' navigation within buildings.

Following this introduction, section 2 summarizes the essential emergency phases of buildings, navigation requirements and evaluation criteria of indoor navigation models. Section 3 introduces several typical indoor navigation models and assesses these models. Section 4 provides a possible better solution for indoor navigation models, including navigation model design and routing strategy. Section 5 presents the conclusion of this paper.

2 REQUIREMENTS FOR INDOOR NAVIGATION MODELS

While good for an emergency situation (e.g., fire, air pollution) a model and the corresponding system for navigation might be inappropriate for others (e.g., earthquake & flood). In case of flood usually the first floors are mostly damaged (i.e., escape to the upper floors is recommended), while in case of fire the upper floors might be more dangerous. It is never known which parts of the building are more secure, but estimates can be made on which construction elements (i.e., the semantics that is needed) might play a role in the navigation within buildings.

2.1 *Buildings status in emergency situations*

Accurate information about the status of the building is one of the most critical aspects in successful navigation and evacuation. The conditions in the building change during an incident: stairs might be not accessible due to rubble or fire, exits might be unavailable due to high water level, rooms and corridors might be not recommendable due to smoke or gas leakage. Therefore, to make sure navigation models could be applied in all kinds of cases, it is necessary to classify the status of buildings in an emergency. We propose following classification:

- Status 0. Normal state, no change within indoor environments (this is the everyday status). In this stage, elevators could be used, yet emergency exits of buildings cannot be open. Pedestrians can walk freely within the buildings.
- Status 1. Some accidents occurs in buildings (e.g., on fire or gas leak), yet it causes no damage of main structure in this status. In this phase, normal and emergency exits may be opened, but elevators may not be used, etc. Evacuees could escape from the incident building according to evacuation instructions, following sound and light instructions.
- Status 2. A certain detriment of indoor environments is obvious. Because the structure of buildings would be deformed or not accessible (e.g., smoke, water, etc.), some parts of the buildings will be shut-down.
- Status 3. If it is the case, the disaster building is seriously damaged (might be collapsed floors). That means it is difficult to find an escape way for people in some isolated indoor areas. It is also barely possible for rescuers to enter the building.

A scenario could be used to describe these Statuses: fire hazard happens in the 12th floor of a building with 30 stories. At first, the fire spread endangers several rooms in the 12th floor yet the staircases still can be used. Smoke from the fire has not disseminated too far. This scene could be considered as Status 1. In this case, elevators would be stopped while emergency exits would open. People can utilize passageways and safe exits for evacuation according to predefined evacuation plan.

After a while, if the fire extends to the adjacent floors, the stairs connecting the 12th floor with other floors would be destroyed and some parts of the floors may collapse. Then there would be isolated areas within the building. Since the structure of the building is changed and the vertical connections around the 12th floor are cut off, the original evacuation plan becomes invalid. There may be some other hazards as well, such as pervasive smoke which would greatly impede people's movements. Now the building emergency is in the Status 2. It is possible for people above the 12th floor to get to the roof platform and wait for rescue. While pedestrians below the 12th floor could escape by staircases, outside fire escapes and some temporary exits (e.g., windows could be broken and treated as exits below the 3rd floor).

If the whole building is almost enveloped by the fire and smoke and its structure is seriously impaired, then the situation is in Status 3. In this case, each story of the building suffers various degrees of damage. Besides, indoor space is split up severely. Not only vertical connections but also most connectors within every floor of the building are unavailable. Rescuers might break the ceiling and jump into the next less-damaged story for searching new outlets. Remaining available windows could be considered as exits if there is help from outside rescue equipments (such as fire fighting trucks). Nevertheless, it is hardly possible to find escape ways for people within indoor isolated regions by themselves under this condition.

2.2 *Requirements to indoor model for navigation*

To be able to provide navigation in all the different response phases (building statuses), we propose the model has to have the following properties:

- Semantically-rich;
- Automatic generation of network from original geometry;
- Dynamic routing capacity (re-computing alternatives and prediction);
- Considering door-to-door movement.

"Semantic-rich" is related to keeping useful navigation information. Since in an emergency situation, the most interesting question to evacuees is "where are the exits?" By means of affluent semantics, required knowledge could be readily abstracted (such as "what are doors?"), which facilitates automation of navigation model generation.

"Automatic generation of network" means that the network for navigation should be automatically derived from any 3D geometry representation of a building. Under the emergency condition, the indoor environment is more and more difficult for evacuation with constant changes within building. Therefore, it should be possible to derive a new network, which will incorporate new appropriate building elements (such as windows, doors, or other parts that can be used as exits) for specific users. Apparently this has to be fast and real-time computation.

"Dynamic routing capacity" means that the evacuation system should consider changing environment (smoke, blockages, crowds blocking areas and exits). Two kinds of routing are to be considered: 1) simple re-routing when user cannot continue due to blockages; and 2) re-routing by considering the movement of the treat (plume, flood).

"Door-to-door navigation" implies that the path-finding should represent the most natural movement in emergencies. For instance, people would like to head directly to the door they need to pass through, instead of walking along the wall (e.g., in case of a power outage and poor visibility of exits). We name this direct movement towards exits as door-to-door.

There are differences between indoor navigation models for individual persons and crowds. In the context of this research we consider individuals or a small group of people that move together. The navigation model will then treat crowds as "moving obstacles" and seek a feasible route around them. To be able to navigate crowds, they could be partitioned into separated crowd flows. Then evacuation for different groups of crowd becomes a dispatching problem at the macro-level, which will make use of all possible exits of the building and attempt to avert indoor congestion.

3 EXISTING INDOOR NAVIGATION MODELS

In this section, we will discuss several typical indoor navigation models, which are dual graph, navigable space, regular-grid and sub-division. According to aforementioned criteria in section 2.2, their pros and cons will be given in the following subsections.

3.1 *Dual graph*

There are several types of derived graph model of buildings. The first group of methods is the conventional Dual Graph approach proposed by Lee (2001), which uses a Node-Relation structure (NRS) to represent the connectivity of buildings based on Poincaré Duality theory. Poincaré Duality converts "Room—Door" relations in primal space to "Node—Edge" relations in dual space (Munkres 1984, Corbett 1985). In order to represent indoor environments more accurate, Lee (2004) extends the NRS to Geometric Network Model (GNM), which introduces geometrical metric. Lee (2004) also mentions a skeleton-abstraction algorithm to help construct 3D GNM. That is the Straight-Medial Axis Transformation (S-MAT) modelling method (Eppstein & Erickson 1999, Choi & Lee 2009), which can abstract linear features from simple polygons (such as corridors).

Another method named Multilayered Space-event model provides a multilayer representation for different spatial models, such as topographic space for 3D buildings and sensor space for sensor range partition (Becker et al., 2009). In the model, each space layer is divided into primal and dual spaces on one hand, and also divided into topology and geometry spaces on the other hand. Various layers of the topological space models are connected by so called "joint-state" edges, which represent there is space overlap of two nodes from different space models. At a time only one "joint state" edge and related nodes are active. Ultimately, "space-event" (people's movement) is expressed by topological space representation. In the dual space of topographic space layer, a dual graph could be derived by the same way as Lee (2004) mentioned.

Conventional Dual Graph model is accepted by many researchers (Lee 2004, Meijers et al., 2005, Li & He 2008, Becker et al., 2009, Boguslawski and Gold 2010). It indeed has benefits: 1) it contains geometry and explicit topology of buildings; 2) it can be generated automatically; 3) indoor route calculation (e.g., shortest path) can be conducted readily and fast (Kwan & Lee 2005). Musliman (2008) proves that conventional Dual Graph could support dynamic changes (insert or delete edges quickly) as well.

Except for the apparent merits of conventional Dual Graph (DG) model, according to the criteria in section 2.2, we summarize some deficiencies of this kind of navigation model. DG model may ignore some semantics (e.g., pedestrian-dependent factors) of indoor environments. Then it may not represent accessibility within buildings accurately. For instance, it could not reflect a door only opened for pedestrians who have the key.

Furthermore, S-MAT, which is mostly used for approximation of complex spaces in conventional DG model, fails to provide "door to door" routes for people. As shown in Figure 1, the straight medial axis of a corridor is *D1-M1-M2-M3-M4-D4*. Even if the person could not see the door *D3* from *D1*, the "door to door" route could be *D1-S1-D3* (the dash line), which means the person will see *D3* until he reaches *S1*. Yet the DG route would be *D1-M1-M2-M3-D3* in this case.

Another inconvenience of the conventional DG model is the localization of an individual person on a DG model. If the indoor location of a person could be acquired, the location

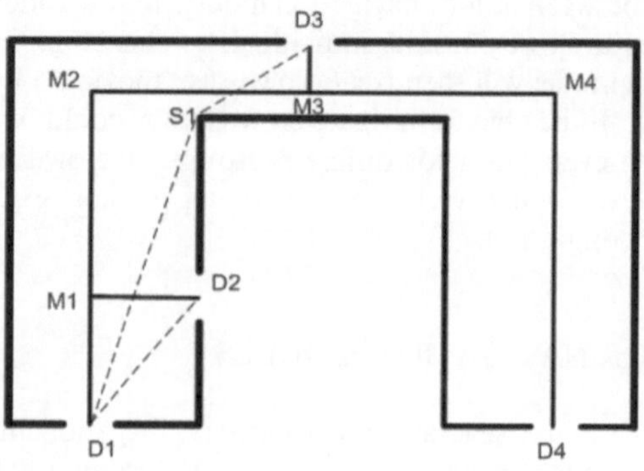

Figure 1. An example of "door to door" routes.

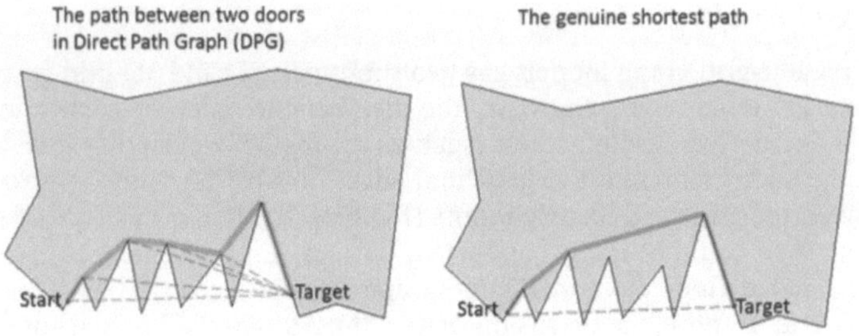

Figure 2. A comparison of the generated route in DPG with the genuine shortest path.

should be represented by its projection on the medial-axis of connections (e.g., corridors and stairs) or nodes (closed space) of the DG model, which might be quite different from the real position of the person (especially in complex shaped buildings).

Yuan & Schneider (2010) propose the Direct Path Graph (DPG). It considers three types of elements: cells, path segments and accessibility. Cells are the different units (spaces) in indoor environment. Path segment is referred to as a part of the shortest path between certain two access points, where access points (e.g., doors) are attached to cells. Concave vertices of cells are incorporated into paths between doors (in a single cell) as well. Accessibility of cells is determined by attributes of corresponding access points and path segments. Nodes of DPG consist of access points and concave points of cells, and each edge denotes the shortest route between any two nodes. Besides, DPG records distance information for every edge. Apparently, DPG is dedicated to improving the S-MAT approach. Theoretically, after employing the shortest path algorithm on DPG, the shortest route between two given nodes could be acquired. Here the "direct path" and "door-to-door" have the same meaning, that is, to look for the shortest way between any two doors (or exits).

Nevertheless, in the work of Yuan & Schneider (2010), the way to obtain the shortest path segments between two nodes could not always derive the real shortest path. Figure 2 illustrates a "direct path" between two doors in a strange shaped room. The bold line in left image is the "direct path", which is derived by dividing one straight line into two lines when the line intersects concave boundary. While the bold line in the right image is the genuine shortest path between the two doors. It is apparent that the path-generation approach used in DPG may not acquire the shortest path in some cases (e.g., some irregular-shaped room).

Moreover, the DPG may not be suitable for navigation in complex indoor environments. It is possible that there are many doors and lots of concave points. Then the amount of nodes (including doors and concave points) and edges of DPG will soar. The computational complexity of the routing algorithm may be very high and the performance will decrease (Goetz & Zipf 2011).

3.2 *Navigable space*

Navigable space model (Slingsby & Raper 2008, Schaap 2010) intends to construct topologically-connected and navigable spaces (surfaces) to point out where pedestrians are able to move. The approach for navigable space model could reconstruct a 2.5 D model from 2D plan with limited height and surface constraints. The group of models can provide some specific semantics (e.g., time and pedestrian-dependent properties within indoor environments) for navigation. Besides, it is relatively easy to locate pedestrians in this model, because the position could be pointed out in one piece of surface of the model.

Navigable space model could deal with all statuses of buildings. It is semantically-rich but the derivation of the network might be challenging for complex surfaces (Shaap 2010). It could provide the accessible space for diverse users, but there is no "door to door" style route in it. The possibilities to automatically generate navigable spaces of various levels might be another problem.

3.3 *Regular-grid model*

The group of regular-grid graph models has two sub-types: 2D and 3D grid graphs. By overlapping discrete 2D grids with a 2D plan, the membership value of each grid is assigned according to the underlying cellular unit (such as rooms). Then a grid-based graph could be generated: the nodes represent the grids and edges of a certain node represent link relationships between the node and its neighbours (Li et al., 2009). On this type of graph, static shortest path analysis could be easily conducted. Besides, it is convenient to conduct 2D diffusion analysis (smoke, flood etc.) on 2D grid graph models.

Basically 3D grid graph is the extension of the 2D grid graph. The purpose of 3D grid-based (voxel) graph is to represent the 3D structure of indoor space. Then in a floor, intermediate levels (e.g., ramps and obstacles in air) could be measured. Yuan & Schneider (2010) propose a model called LEGO graph based on 3D voxels. LEGO-graph only provides all feasible or possible paths with different accessible widths and heights. Yet they have not proposed an algorithm to compute optimal navigation routes. Bandi & Thalmann (1998) give a more specific solution. They discretize building elements into 3D voxel and compute an obstacle-free feasible route with consideration of surmountable and insurmountable obstacles and "hole" areas in space.

Generally, regular-gird graph can readily point out the membership (e.g., corridors, rooms and stairways) of a grid. Besides, 3D grid graph can concern height constraints for indoor navigation. Current 2D and 3D grid models could be derived from 2D plans or 3D geometry automatically. The path-finding computational efficiency of grid model depends on required granularities and different algorithms. Fine-grained grid graph might provide the "door to door" navigation routes for pedestrians.

3.4 *Sub-division*

This type of method concerns subdividing 2D plans into cells according to certain criteria (e.g., distances between walls, visibility etc.). There are different criteria to sub-divide indoor space in literature. With consideration of passing bottlenecks in a 2D plan, Lamarche & Donikian (2004) apply constrained Delaunay triangulation algorithm and Convex Cell optimisation to subdivide a 2D plan. The subdivision locally conserves bottleneck information.

A different approach is given by Lorenz et al. (2006). For buildings with simple rooms and corridors, indoor space (2D plan) is decomposed into cells to build a graph structure. As shown in Figure 3, cell centres are connected with doors. However, the graph may result in some unnecessary tortuous paths. For instance, if there could be a direct connection between two doors, it is unnecessary to pass through a cell centre (e.g., a node of the corridor). Besides, there is no completely automation of the cell decomposition.

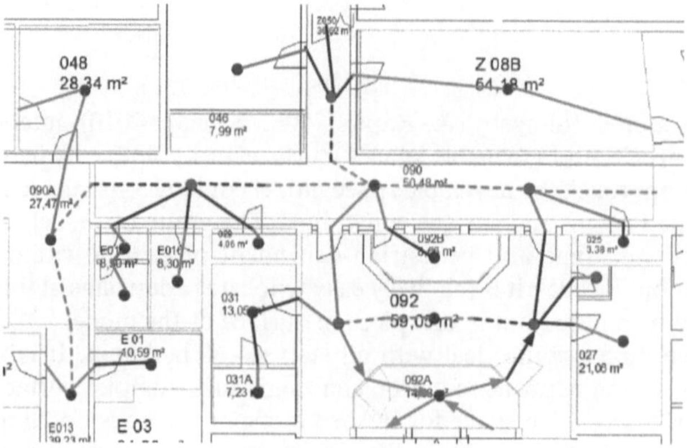

Figure 3. Cell centres and paths overlaid with a floor plan (Lorenz et al., 2006).

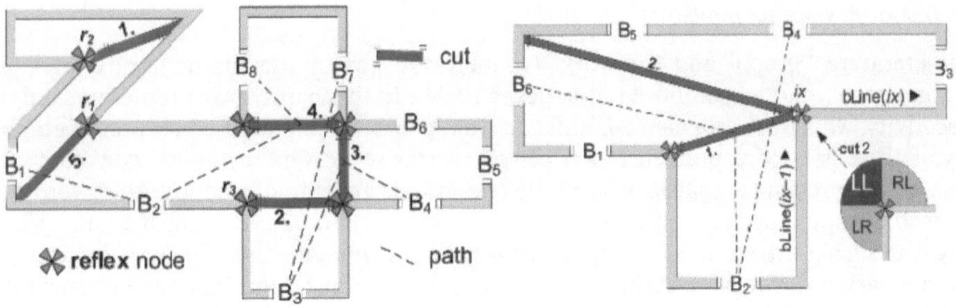

Figure 4. Visibility partitioning result (Stoffel et al., 2007).

Table 1. Comparison of different indoor navigation models.

	Models			
Properties	Dual graph	Navigable space	Regular-grid models	Sub-division
Semantic-rich	Partly	Yes	Yes	Partly
Automatic generation from geometry	Conditional	Semi-automation	Yes	Yes
Dynamically routing capacity	No prediction	No	Yes	No
'Door-to-door'	No	No	Conditional	Yes

Another subdivision method is proposed by Stoffel et al. (2007). This method partition a plan into non-overlapping convex sub-regions (cells) according to visibility criterion. Therefore, boundary nodes, which represent the openings on the shared surface of two room units, are mutually visible if they are in a same sub-region (Fig. 4). Accordingly, a navigation graph could be generated, which indicates the possible paths between boundary nodes.

The 2D sub-dividing methods are interesting because they can consider visibility and/or width constraints, people could be guide directly to openings. Based on appropriate subdivision of indoor space, it is helpful to generate "door to door" routes.

According to the properties of a favourable indoor navigation model mentioned before, there is a comparison among the four kinds of models listed below.

From Table 1, it is obvious that a DG model is difficult to provide "door to door" style routes for indoor pedestrians. Therefore, DG model is not taken into consideration in this paper. The only one fully supporting "door to door" navigation is the visibility subdivision method. In this way, people could obtain the most natural movement guiding in emergencies. Yet current subdivisions in literature are based on 2D plan, it could not take real-time changes and 3D indoor space into account. A 3D sub-dividing method has more potential to offer accurate route calculation. Therefore, we concern that a 3D network on the basis of visibility partition is more suitable for navigating indoor pedestrians (Fig. 1).

4 A POSSIBLE SOLUTION

Before we describe our approach for indoor navigation, we make the following assumptions: 1) there are 3D semantic-rich geometry models of buildings such as CityGML or IFC (Nagel et al., 2009, Isikdag & Zlatanova 2009); 2) sensor data (smoke detector & monitor etc.) could be used for real-time update; 3) as mentioned previously the navigation model aims to serve small groups of users (first responders and/or people who need help).

In this section, we will introduce a new indoor navigation structure making use of 3D geometry and semantics of buildings and provide corresponding routing strategy.

4.1 *Indoor navigation model*

In the literature, "graph" and "network" (or metric graph) are used in the same context, i.e., as a connectivity model composed of edges in nodes. In this paper, *graph* represents only the connectivity, while the term *network* indicates that geometry and metrics are preserved. Generally, if there is a correspondence between the vertex set of Graph A and the vertex set of Graph B that preserves adjacency, then the two graphs are isomorphic. In other words, two isomorphic graphs may have different configurations, yet they are essentially the same (Fig. 5). Many routing algorithms are developed for graphs, e.g., breadth-first, Dijskra, A*, etc.

In contrary a network contains geometrical information of the indoor environment. It could be two-dimensional (2D) (Fig. 5) or three-dimensional (3D). 3D network represents also the vertical geometrical information of buildings (e.g., staircases).

3D networks are useful in indoor navigation, because they:

- Can be employed for 3D visualisation. Although schematic this visualisation gives a good overview of the path to be followed
- Contain spatial information, which can be readily used in computations of optimal routes, i.e., there is no need to store coefficients. For example, the slope of a ramp can be computed from the coordinates of the nodes.
- Can be used to perform spatial operations with the geometrical model of the building, e.g., computing the distance from a node of the network to a door.
- Are convenient for tracking of user, i.e., the current position of the user can be compared directly with the nodes and edges, which are spatial abstractions of building elements.

In our approach we aim at automatic extraction and management of both the connectivity information and the metrics. We consider this model a network navigation model. The visibility criterion is deployed to define the navigation structure of buildings. Finally, for the purpose of indoor navigation, we concentrate on door-to-door navigation as looking for feasible exits have to be the top priority.

The concept for a navigation model named "Indoor Visibility Structure" (IVS) will be shortly presented for a building with a complex structure. As shown in the left image of Figure 6, there is a floor plan (on the 5th floor of a building) including inner rooms. Possible outlets encompass inner room doors (denoted by "*I*"), room doors (denoted by "*D*"), stairs (denoted by "*S*") and windows (denoted by "*W*"). Each opening is abstracted as a node of a network model (see in the bottom-left part of Fig. 6). If two openings are mutually visible to each other, then there is an edge of the network between the two nodes. The dash-line edges in Figure 6 indicate windows will be used for escape only in some specific conditions (e.g., broken in an emergency situation).

If we have to derive conventional DG model of the same floor plan, the indoor space should be sub-divided at first (see in the top-right image of Fig. 6). After that, a dual graph of the storey could be built. In contrast, the network we intend to construct will be more straightforward and effective to search exits for indoor pedestrians.

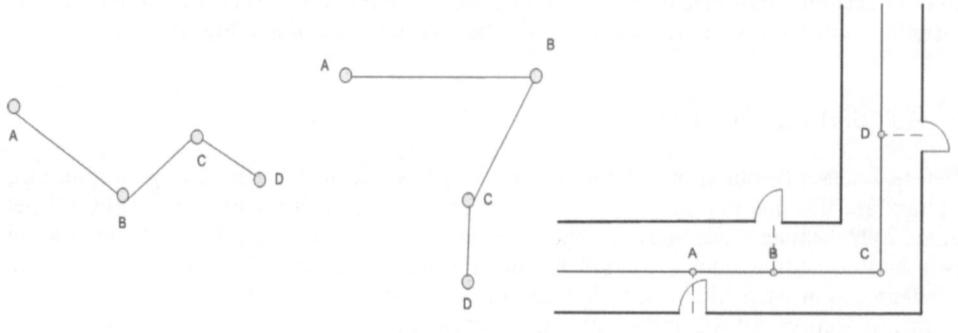

Figure 5. Two isomorphic graphs and the corresponding network.

A typical example of vertical elements is stairs. For instance, a staircase may consist of several landings (intermediate platforms). Thus, a staircase between two floors could be divided into several parts according to the visibility criterion. In Figure 7, the left image shows landings are treated as nodes in a regular staircase. They could be connected by "visible" edges. While the right image demonstrates a spiral staircase is partitioned into three nodes. Although *M1* is not a landing, it is the visible far-end point viewed from *L1*. So *M1* is inserted as an intermediate node between *L1* and *L2*.

Although there is certain work related to IVS, such as DPG, IVS has advantages over the counterpart. For a single room, they both follow the same principle (to find out the shortest route between two specific doors). As mentioned before (section 3.1), the path-derivation approach used in DPG may not always derive the shortest path in real sense between any two doors. Thus, we have developed a new algorithm to derive the shortest route between two openings in a single room, which is shown as follows:

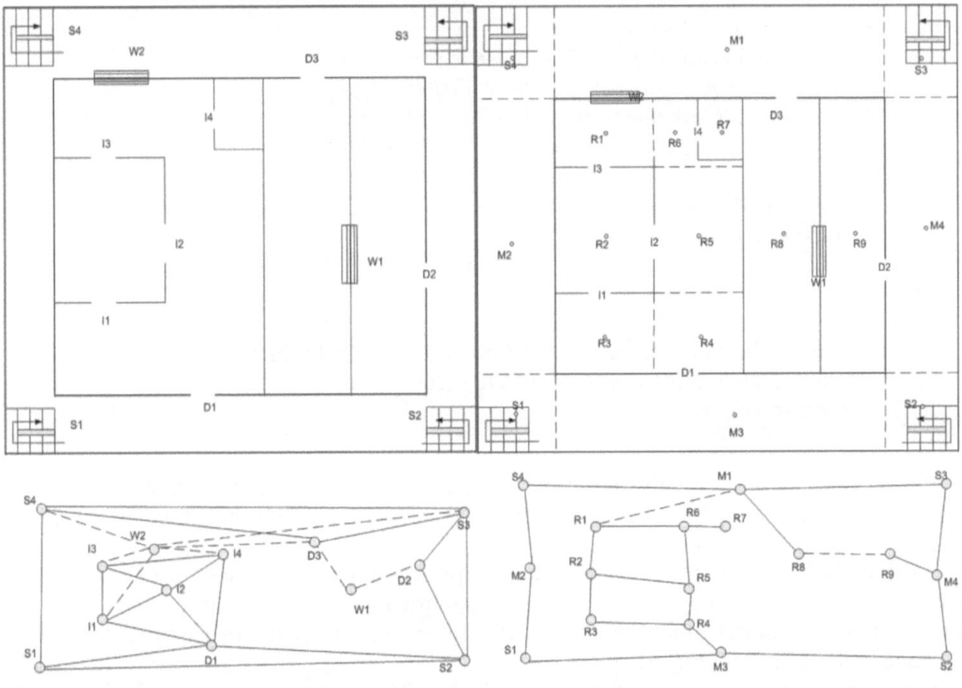

Figure 6. Indoor visibility structure & conventional DG model.

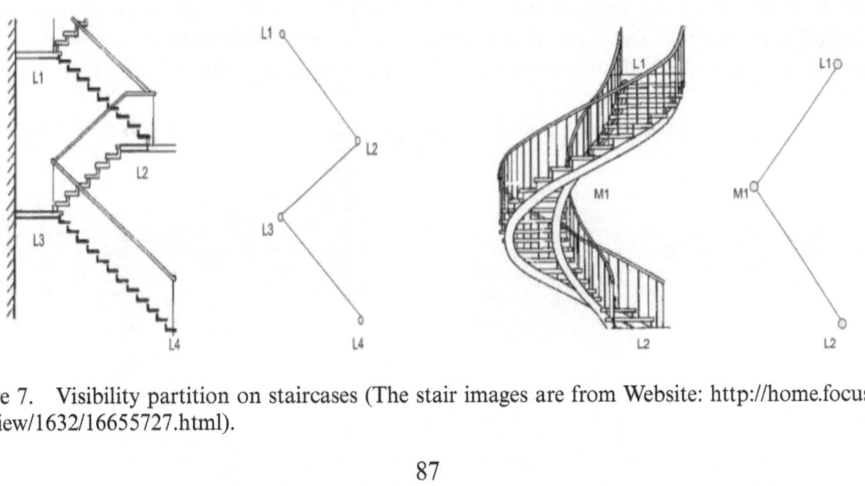

Figure 7. Visibility partition on staircases (The stair images are from Website: http://home.focus.cn/msgview/1632/16655727.html).

Algorithm Door-to-Door_Route (*startDoor, targetDoor, Pt*)

Input: *startDoor*, *targetDoor*, and a set *Pt* including start and target doors and all concave vertices in a single room.

Output: The sequence of vertices forming the door-to-door path, which is stored in a vector *Route*.

```
// VisibleCP denotes a vector storing visible concave vertices from current location.
// AdjMatrix represents the adjacency matrix of Pt.

if (IsNotVisible (startDoor, targetDoor) )
then
{
        Determine Pt in the room;
        AdjMatrix ← Φ;
        for each node Pt[i]
        {
                VisibleCP ← Φ;
                FindVisibleCP(Pt[i], VisibleCP);
                for each visible concave point VisibleCP[j]
                        Corresponding weight in AdjMatrix =
                                Distance (Pt[i], VisibleCP[j]);
        }
        ShortestPath (AdjMatrix, Route);
        return Route;
}
else
{
        Connect the startDoor and targetDoor with a straight line;
        Add startDoor and targetDoor to the vector Route;
        return Route;
}
```

With this door-to-door algorithm, we can acquire the genuine shortest path between two doors (i.e., as shown in Fig. 2, right). More information regarding the algorithm could be found in Liu & Zlatanova, 2011.

As discussed in section 3.1, door-to-door approach applied for the entire building may result in a very complex network. In this case, all mutually visible doors have to be connected with each other. For example, a building with 2 floors, 22 offices and two stairs will result in 50 nodes and 462 edges (Fig. 8).

Furthermore, to reflect real-time changes in buildings a dynamic way for IVS network generation is required. Therefore we propose a two-level approach for indoor navigation:

- The *first level* takes care about the room sequence. A room sequence to be traversed is selected according to the type of rooms (e.g., to give a preference to connector space/corridor), which is based on current situation. Then openings (doors) connecting any two

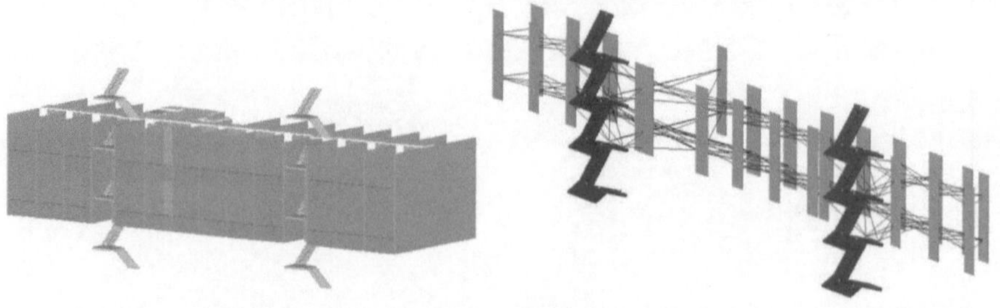

Figure 8. The geometry and IVS of a part of OTB building.

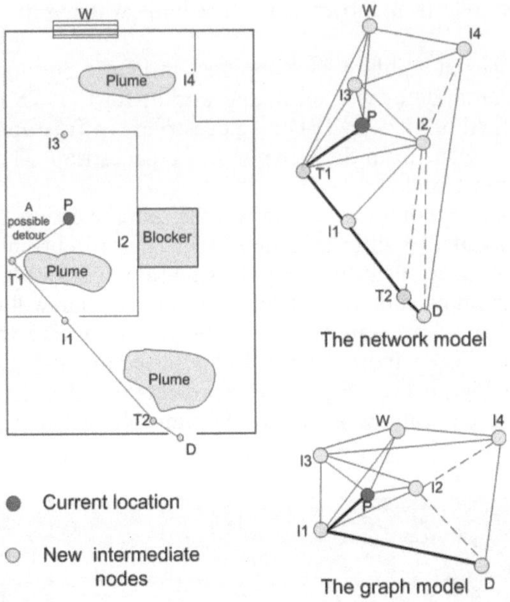

Figure 9. Floor plan (left), graph model (bottom right) and the corresponding network (top right).

rooms in the sequence could be determined according to semantics of the building model. After that, a graph is constructed, whose nodes represent these selected openings and edges denote the accessibility between them.

- The *second level* concentrates on the navigation in a single room. In each room (or cell), the door-to-door path between specified two openings is computed according to the "Door-to-Door_Route" algorithm as explained above. Here the generated network is still an IVS, since any two connected nodes (openings or concave vertices) in the network are mutually visible (when there are no obstacles).

With this two-level approach we expect to reduce significantly the computational complexity and allow for real-time modifications of the route with respect to the level of emergency. Compared to DPG, we don't attempt to utilize all concave points and all doors (or openings) of a building and we don't construct the entire detailed graph of the building at once.

If accidents occur in halfway (e.g., the target room is changed or the current room cannot be passed through), then re-calculation for the network is implemented. In such cases, probably it is necessary to re-compute the traversing sequence of rooms at first, and then adjust the graph model and the network. As shown in Figure 9, due to smoke pervasion two intermediate nodes are added in the network to build a new route for pedestrians at location *P*. Whereas the corresponding graph model will not be changed, since the accessibility between openings is not altered.

One thing that must be stressed at this point is that obstacles in emergency situations are not discussed until now. At this stage, only a network representing accessibility of a disaster building is built. In the next section, the routing strategy based on this type of indoor network will be considered.

4.2 *Indoor routing strategy*

Based on the Indoor Visibility Structure (IVS) proposed in the section 4.1, the optimal path for one or a group of people (5 or 6 persons) could be generated. An IVS of a building is a 3D network including connectivity relationships between indoor openings or concave vertices of rooms. For the purpose of providing effective rescue routes for the first responders, the network serving for pedestrian movement will be built, which is the base of optimal path-finding.

Since real-time changes in building have apparent influence on indoor path-finding, it is necessary to work out an update mechanism appropriate for the IVS network at first. Here it is assumed that the field of view (FOV) of a pedestrian is a frustum. Figure 10 illustrates a snapshot of an indoor scene from the viewpoint of a pedestrian. The grey parts are doors and the dark ones are inner walls.

As real-time changes in buildings would impact the route selection, local structure of IVS should be modified (updated) within a certain range around current pedestrian location. In an established IVS of an accident building, the situation of surrounding nodes in the current visible range (a certain neighbourhood) of a given pedestrian will be updated according to indoor real-time information. It means the openings (or concave vertices of cells) in the scope of the pedestrian's FOV will be checked and adjusted for current navigation.

As demonstrated in Figure 11, the point *P* represents a pedestrian and the person can access doors *I1*, *I2*, *I3*, and a window *W2* in normal state. Nevertheless, when the floor of the

Figure 10. Field of view of a pedestrian.

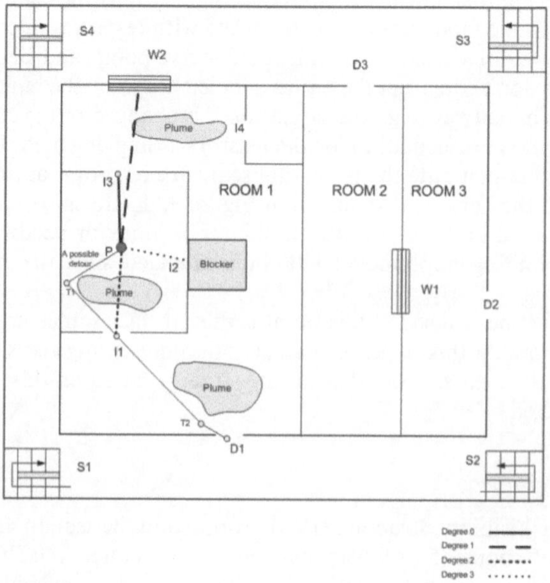

Figure 11. Various degrees of "visible" edges.

building is in an emergency situation (e.g., on fire), the "visible" connections between *P* and the four openings are changed. The "visible" connection between two nodes can be classified into four degrees:

- Degree 0, there are no obstacles between the nodes and they are completely connected.
- Degree 1, two nodes are partly blocked (the blockage ratio < = 50%), yet there are some detours between them.
- Degree 2, the connection is mostly blocked (the blockage ratio >50%). Yet there are still a few detours between them. For instance, there is a possible detour from *P* to *D1* in Figure 11.
- Degree 3, the connection between two nodes is completely blocked. One of the two nodes can not be used.

In this case, the graph model will be updated. At first, the "visible" edges around the person *P* in Figure 11 would be updated as follows: *P-W2* (Degree 1), *P-I1* (Degree 2), *P-I2* (Degree 3) and *P-I3* (Degree 0). Besides, the graph also would be updated (Fig. 12): *I4-I2* (Degree 3); *D1-I2* (Degree 3); *I1-I2* (Degree 3); *I3-I2* (Degree 3); *I3-I4* (Degree 1); *I3-W2* (Degree 1); *I3-I1* (Degree 2); *I1-W2* (Degree 2); *I1-D1* (Degree 1); *I4-W2* (Degree 1); *I4-D1* (Degree 2). Based on the local update, the next step is to look for optimal routes on the corresponding IVS for a specific user (here we assume the user is a rescuer). When the person moves to the next position, the graph model and the IVS would be updated in a similar way. This update mechanism lays a foundation for real-time indoor routing.

There are four critical factors in the dynamic routing problem: a prediction, a cost function, a path-finding algorithm and realistic-style routes. Firstly, it is necessary to predict how event factors (such as fire/smoke/gas) spread. Besides, crowds in a specific indoor environment could also be considered as "moving" obstacles to a rescuer. Based on real-time information, the prediction for crowd flows and event factors can provide the estimation of traffic cost function for path-finding. The cost of any graph edge comprises current accessibility and estimation of future changes. For example, in Figure 11, the person at *P* could break the indoor window *W2* to get out of *ROOM 1*. He also can get through it by the detour *P-T1-I1-T2-D1*. From the viewpoint of safety, the route selection relies on diffusion rate of plume, the convenience to break the window, the distance of the two routes and the possible required time.

After the traffic cost function is determined, the corresponding routing algorithm will be devised. It depends upon which kind of "optimal" routes will be derived. It is assumed the "optimal" means the "safest" in this paper. The final goal of path-finding is to seek a safe and collision-free path for a specific pedestrian.

For the purpose of exactly navigating rescuers, environment, event and human factors should be considered for simulating a realistic indoor environment (Zlatanova & Baharin 2008). Prediction of the accident development and the crowd flow is also significant.

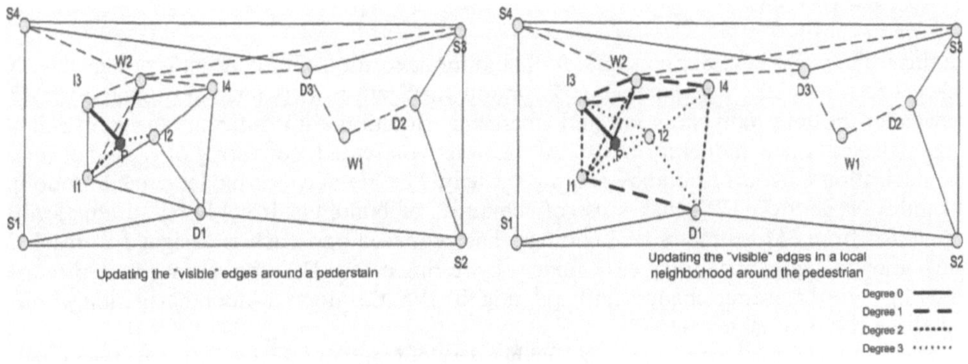

Figure 12. Update of the graph model around a pedestrian.

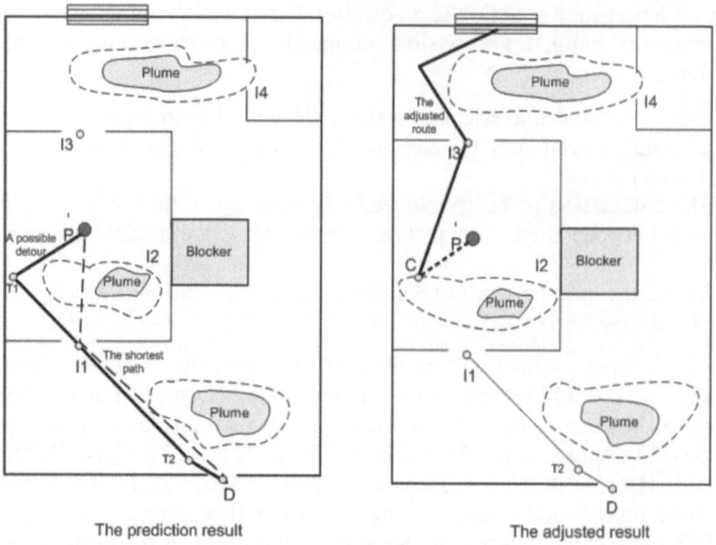

Figure 13. Prediction and re-computation in path-finding.

All of the obstructive factors, i.e., real and possible obstacles (e.g., indoor furniture, walls, predicable smoke or fire in a halfway), would be taken into account for determining an "optimal" path in fact (to stay or go somewhere and wait for help, or go along another route). As shown in the left image of Figure 13, according to the prediction result, a person may choose a detour for safety to evade possible fire or smoke in the halfway of another optional shorter path.

Although prediction will help pedestrian to select optimal routes and save computation time, it still inclines to generate some inaccurate results. Accordingly, dynamic adjustment also should be taken into account when pedestrians are blocked on half way (see in the right image of Fig. 13). That means a new route from current barrier location to a certain outlet should be re-calculated. As shown in Figure 13, pedestrians at location *P* were supposed to go along the detour (in the left image) based on prediction. Yet in the real situation, people will find the plume has extended and blocked their way forward when they move to location *C* (in the right image). In this case, a new route from *C* to a nearby outlet would be computed immediately. Also, the re-computation process should make prediction for accident development and decide which path is optimal at present. Therefore, the two methods (prediction & re-computation) to response indoor "dynamic changes" are complementary and should be used together.

5 CONCLUSIONS

In this paper we present our concept for 3D indoor navigation. The concept is designed with respect to a number of requirements for indoor navigation. It is intended to support 3D navigation in both daily situation and emergency situation with different stages of building damages. A new model for indoor navigation is presented, consisting of a 3D network for navigation and corresponding routing strategy. The 3D network model named "Indoor Visibility Structure" (IVS) makes use of semantics of buildings. It could be automatically generated from 3D geometry of buildings. The proposed approach is designed to resolve navigation in complex buildings. Ultimately, by resting on the IVS, the routing strategy considering indoor real-time changes and updating the IVS, the 'door-to-door' navigation within buildings could be achieved.

The future work will concentrate on further development and formalization of the model, implementation and performing tests.

REFERENCES

Bandi, S. & Thalmann, D. 1998. Space discretization for efficient human navigation. *Computer Graphics Forum* 17(3): 195–206.

Becker, T., Nagel, C. & Kolbe, T.H. 2009. A Multilayered Space-Event Model for Navigation in Indoor Spaces. In Lee, J. & Zlatanova, S. (eds), *3D Geoinformation Sciences*: 61–77. Berlin: Springer.

Boguslawski, P. & Gold, C. 2010. Construction Operators for Modelling 3D Objects and Dual Navigation Structures, in Jiyeong Lee and Sisi Zlatanova (eds.), 3D Geo-Information Sciences (Lecture Notes in Geoinformation and Cartography; Springer Berlin Heidelberg), pp. 47–59.

Choi, J. & Lee, J. 2009. 3D Geo-Network for Agent-based Building Evacuation Simulation. In Lee, J. & Zlatanova, S. (eds), *3D Geo-Information Sciences*: 283–299. Berlin: Springer.

Corbett, J.P. 1979. Topological Principles in Cartography. In *Technical Paper 48*: 61–65. Washington DC: US Bureau of the Census.

Corbett, J.P. 1985. A general topological model for spatial reference. In J.P. van Est (ed.), *Report of Spatially Oriented Referencing Systems Association (SORSA) Workshop*: 9–24.

Eppstein, D. & Erickson, J. 1999. Raising Roofs, Crashing Cycles & Playing Pool: Applications of a Data Structure for Finding Pairwise Interactions. In *14th Annual ACM Symposium on Computational Geometry*: 58–67. Minneapolis.

Franz, G., Mallot, H. & Wiener, J. 2005. Graph-based Models of Space in Architecture and Cognitive Science—a Comparative Analysis. In Leong, Y.-T. & Lasker, G.E. (eds), *Proceedings of the 17th International Conference on Systems Research, Informatics and Cybernetics*: 30–38.

Goetz, M. & Zipf, A. 2011. Formal Definition of an User-adaptive and Length-optimal Routing Graph for Complex Indoor Environments. In *Joint ISPRS Workshop on 3D City Modelling & Applications and the 6th 3D GeoInfo Conference, Wuhan, China*. Heidelberg: Springer.

Isikdag, U. & Zlatanova, S. 2009. A SWOT analysis on the implementation of BIM within geospatial environment, in Krek, Rumor, Zlatanova & Fendel (eds.), In: *Urban and Regional data Management, UDMS Annuals 2009*, 15–30 Boca Raton: CRC press, Taylor & Francis.

Kwak, S., Nam, H. & Jun, C. 2010. An enhanced indoor pedestrian model supporting spatial DBMSs. In *Proceedings of the 2nd ACM SIGSPATIAL International Workshop on Indoor Spatial Awareness (ISA'10)*: 25–32. New York: ACM.

Kwan, M.P. & Lee, J. 2005. Emergency response after 9/11: the potential of real-time 3D GIS for quick emergency response in micro-spatial environments. *Computers, Environment and Urban Systems* 29: 93–113.

Lamarche, F. & Donikian, S. 2004. Crowd of virtual humans: a new approach for real time navigation in complex and structured environments. *Computer Graphics Forum* 23(3): 509–518.

Lee, J. 2001a. 3D Data Model for Representing Topological Relations of Urban Features. In *Proceedings of the 21st Annual ESRI International User Conference*. San Diego, USA.

Lee, J. 2001b. A spatial access oriented implementation of a topological data model for 3D urban entities. In *University Consortium for Geographic Information Science (UCGIS) Summer Assembly*. Buffalo, New York.

Lee, J. 2004. A Spatial Access-Oriented Implementation of a 3-D GIS Topological Data Model for Urban Entities. *Geoinformatica* 8(3): 237–264.

Li, X. Claramuntb, C. & Rayb, C. 2010. A grid graph-based model for the analysis of 2D indoor spaces. *Computers, Environment and Urban Systems* 34(6): 532–540.

Li, Y. & He, Z. 2008. 3D Indoor Navigation: a Framework of Combining BIM with 3D GIS. In *proceedings of 44th ISOCARP Congress*.

Liu, L. & Zlatanova, S. 2011. A "door-to-door" Path-finding Approach for Indoor Navigation. In *Proceedings of GeoInformation For Disaster Management Conference 2011, Antalya, 3–8 May 2011*.

Lorenz, B., Ohlbach, H.J. & Stoffel, E.P. 2006. A Hybrid Spatial Model for Representing Indoor Environments. In *Proceedings of W2GIS (LNCS 4295)*: 102–112. Hong Kong, China.

Meijers, M., Zlatanova, S. & Preifer, N. 2005. 3D geoinformation indoors: structuring for evacuation. In *Proceedings of Next generation 3D city models*. Bonn, Germany.

Munkres, J.R. 1984. *Elements of Algebraic Topology*. Menlo Park, CA: Addison-Wesley.

Musliman, I.A., Rahman, A.A. & Coors, V. 2008. Implementing 3D network analysis in 3D-GIS. In *The International Archives of the Photogrammetry, Remote Sensing and Spatial Information Sciences* 37 (Part B2): 913–918. Beijing.

Nagel, C., Stadler, A. & Kolbe, T.H. 2009, Conceptual requirements for the automatic reconstruction of building information models from uninterpreted 3d models, in: Kolbe, Zhang & Zlatanova (eds), *International Archives of Photogrammetry, Remote Sensing and Spatial Information Sciences*, Vol. XXXVIII-3-4/c3, 46–53, Berlin, Germany.

Schaap, J. 2010. Towards a 3D geo-data model to support pedestrian routing in multimodal public transport travel advices. *MSc Thesis*. GIMA MSc. programme, Utrecht University, Netherlands.
Slingsby, A. & Raper, J. 2008. Navigable Space in 3D City Models for Pedestrians. In Van Oosterom P. Zlatanova, S. Penninga, F. & Fendel, E. (eds), *Advances in 3D Geoinformation Systems*: 49–64. Berlin: Springer.
Stoffel, E., Lorenz, B. & Ohlbach, H. 2007. Towards a Semantic Spatial Model for Pedestrian Indoor Navigation. In Hainaut, J. & Rundensteiner, E. (eds), *ER' 07 Workshops / Advances in Conceptual Modeling*: 328–337. Berlin: Springer.
Yuan, W. & Schneider, M. 2010. iNav: An indoor navigation model supporting length-dependent optimal routing. In *The 13th AGILE International Conference on Geographic Information Science*. Heidelberg: Springer.
Yuan, W. & Schneide, M. 2010. Supporting 3D Route Planning in Indoor Space Based on the LEGO Representation. In *proceedings of ISA'10*. New York: ACM.
Zlatanova, S. & Baharin, S. 2008. Optimal Navigation for First Responders. In Van der Walle, Song, Zlatanova & Li (eds.), *Information systems for crisis response and management*, 529–542. Harbin, China.

Part II: Data management for local governments

Urban and Regional Data Management – Zlatanova, Ledoux, Fendel & Rumor (eds)
© 2012 Taylor & Francis Group, London, ISBN 978-0-415-67491-1

Data interoperability for spatial planning: A tentative common description of European datasets concerning land use

F. Camerata & S. Ombuen
Dipartimento di Studi Urbani, Università degli Studi Roma Tre, Roma, Italy

F. Vico
Dipartimento Interateneo territorio, Politecnico e Università di Torino, Torino, Italy

T. Mildorf
University of West Bohemia, Pilsen, Czech Republic

ABSTRACT: In the framework of the INSPIRE Directive, the Plan4all project aims at establishing a European network of public and private organisations dealing with spatial planning and regional development, in order to build up a consensus about interoperability of data supporting planning activities. One of the key activities of the project has been the definition of a data model for the INSPIRE theme "Land Use". Working on such a data model has entailed a number of considerations that can be of interest for a common vision about planning in Europe. The result is a first attempt, shared by a number of partners working in the planning field and coming from different European countries, to describe the datasets relating to spatial plans, in such a way to be as comprehensive as possible, but also as generic as to be adaptable to different contexts.

1 INTEROPERABILITY OF SPATIAL PLANNING DATA

1.1 *Spatial data interoperability and harmonisation: Some definitions*

The practices of spatial planning are deeply rooted in each country's specific culture, traditions and legislation. Moreover, there are many cases in Europe where different pieces of planning legislation coexist within the same nation. This makes it particularly difficult to exchange planning-related data among nations or regions for various purposes, including researches by investors and developers, cross border policy programming, enhancement of transparency etc. Data are produced, stored, managed and distributed in different ways by each responsible actor, and this has an effect on how the same data can be exchanged and circulated around Europe.

"Interoperability" means "*the possibility for spatial data sets to be combined, and for services to interact, without repetitive manual intervention, in such a way that the result is coherent and the added value of the data sets and services is enhanced*" (INSPIRE Directive 2007).

"*Beyond the ability of two or more computer systems to exchange information, semantic interoperability is the ability to automatically interpret the information exchanged meaningfully and accurately in order to produce useful results as defined by the end users of both systems. To achieve semantic interoperability, both sides must defer to a common information exchange reference model. The content of the information exchange requests are unambiguously defined: what is sent is the same as what is understood*" (http://en.wikipedia.org/wiki/Interoperability).

"*Automatically*" means that data are always associated with their semantics, which have to be included in the system. For a complex theme like spatial planning, semantic interoperability is an absolute need but a difficult objective; and when you deal with existing datasets, like in this case, semantic interoperability is achievable mainly by means of data harmonisation.

"*Harmonizing data help to get a better consistency and comparability of data across information systems, usually within a particular program area. […] The harmonization process makes information available for the integration of data systems and improves meaning and format across different systems. […] The harmonization process therefore applies transformation rules and definitions to the existing heterogeneous data elements in order to have a common representation of the same elements with an improved quality and consistency*" (Viganò & Millot 2007, p. 11).

"Data harmonisation" in terms of INSPIRE means "*providing access to spatial data sets […] through network services in a representation that allows for combining them with other such spatial data sets in a coherent way*" (INSPIRE Drafting Team "Data Specifications" 2010a).

1.2 *Why spatial planning data interoperability is needed*

There are reasons related to planning theory and practice on the one hand, and communication and participatory reasons on the other.

1.2.1 *Reasons related to planning theory and practice*

In all European countries there are various planning levels (municipal, province or department level, regional level, national level); and various sector plans (e.g., commercial plans, transportation plans, etc.) or special plans (e.g., park plans, river basin plans, etc.).

In planning processes at a certain level, planning data referring to other levels have to be taken into account. A plan covering a large area (e.g., landscape plan or river basin plan) is most often designed on the base of a small scale map, i.e., a map with low spatial resolution (1:25,000 or less). But zonings and constraints introduced by such plan often have to be superimposed on municipal plans having a higher resolution and being drawn at larger scales (1:1000–1:5000) (Laurini & Murgante 2008), and this can give birth to some inconsistencies or misunderstandings. Where planning processes are still based on paper maps, these inconsistencies, although existing, are somehow hidden or less evident; but when topological vector maps are used, they are evident and generate errors, and they tend to weaken the planning process itself. The issue of looking at the same territory and the same problems in different ways and from different points of view is often called "transcalarity".

Another example related to this need for interoperability is given by the case of neighbouring planning areas. Planning areas necessarily share some boundaries with other planning areas. The context in which each plan is included has to be carefully considered. Plan boundaries could be boundaries between countries, regions (or states in Germany and Austria), municipalities, etc. Not only different countries can have different planning laws, but also in some countries (like Germany, Italy, Spain …) different regions can have different laws and norms. Different plans in adjacent municipalities can be designed by different consultants or professionals; similar concepts and goals can be expressed in slightly different manners, different planning styles and/or graphical codes can be used. Inconsistencies can arise from improper integration of datasets created in different spatial reference systems and scales, or from the use of different planning criteria.

A typical example is given by the representation of a river. In different contexts, the same river can be conceptualised in different ways (as a line or as a polygon, including or not banks or yearly flooding areas) and, consequently, its area can be delimited differently (Figure 1). Also, there can be inconsistencies caused by the use of different categorisations of the same areas into different risk classes (Figure 2).

1.2.2 *Communication and participatory reasons*

Planners themselves can have difficulties in understanding planning regulations of neighbouring countries and regions. Also for investors and decision makers, it is almost impossible

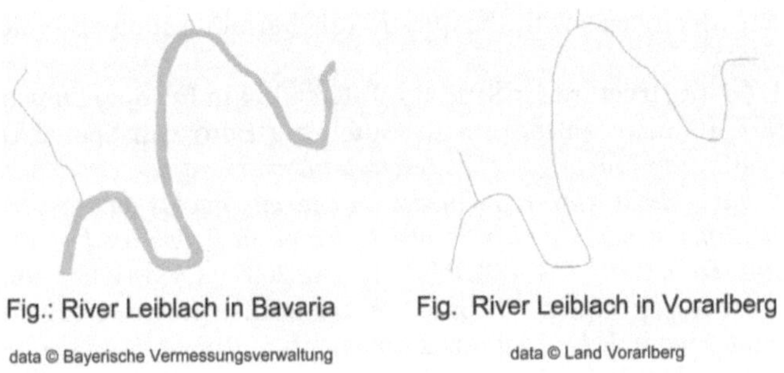

Figure 1. River Leiblach (Austrian-German border): different river data models across the border (source: HUMBOLDT Consortium 2010).

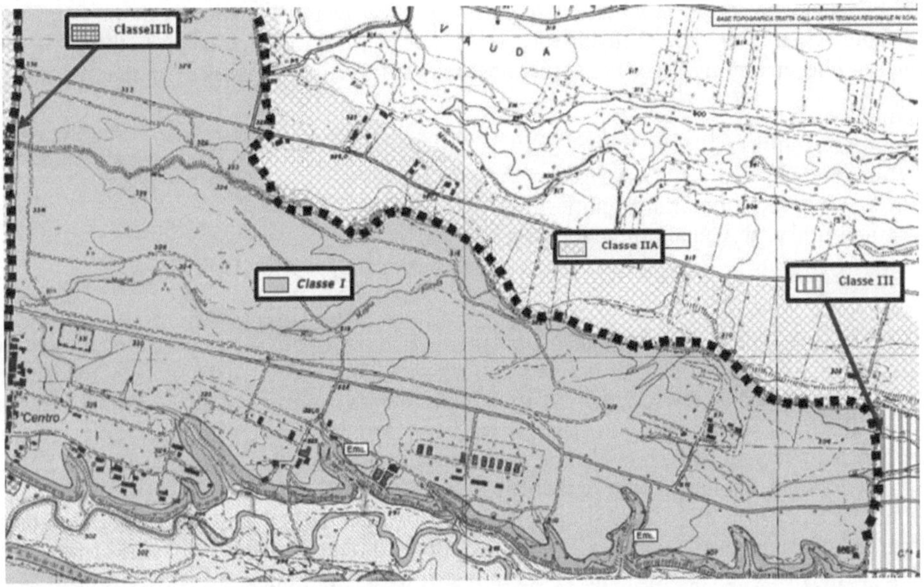

Figure 2. Buffer zones of a stream in three adjacent municipalities categorised into different risk classes. Dotted line depicts municipal borders (elaboration based on CSI-Piemonte's data 2010).

to compare planning regulations across Europe and to single out territorial development opportunities. Even citizens involved in planning participatory processes are interested in abundant, readable, consistent and easily accessible spatial planning information.

The need for harmonised, interoperable and easily accessible data is also entailed in the so-called "planning conferences". Planning conferences have been introduced in some countries or regions, in order to promote less hierarchical planning frameworks, to practice subsidiarity and to speed up planning processes. Planning conferences gather all institutional stakeholders, involved at different levels and with different competences. Planning conferences need to rely on robust, consistent and shared information (Giudice 2008) and are another example of a field where spatial planning data harmonisation and interoperability are required.

2 INSPIRE DIRECTIVE AND THE PLAN4ALL PROJECT

Among the actions aimed at supporting the process of implementation of the INSPIRE Directive, the European Commission Programme "eContent-plus" has financed the

"Plan4all" project, which deals with the question of harmonisation and interoperability of spatial planning data.

The INSPIRE (INfrastructure for SPatial InfoRmation in Europe) Directive entered into force in May 2007. Its main objective is to establish a European Spatial Data Infrastructure (SDI), defined in the Directive as "*… metadata, spatial data sets, spatial data services; network services and technologies; agreements on sharing, access and use; coordination and monitoring mechanisms, process and procedures, established, operated or made available in accordance with this Directive …*" (INSPIRE Directive 2007). INSPIRE is not aimed at establishing new infrastructures; rather, it is based on the infrastructures created by the Member States and made interoperable by common Implementing Rules (IRs) (JRC 2005). Although being specifically aimed at supporting the European environmental policy, INSPIRE is having a great impact on the European GI community. Its implementation could be a big step towards effective information sharing. INSPIRE represents a solid background on which to build interoperability of spatial planning in Europe, since it takes into consideration current standards and practices in the field of SDIs and the point of view of most stakeholders.

The Plan4all project is based on existing best practices from EU regions and municipalities and other research projects. It focuses on seven out of the 34 themes listed in the INSPIRE Annexes, chosen among those being the most relevant for the purposes of spatial planning:

- Land Cover;
- Land Use (subject of this paper);
- Utility and Government Services;
- Production and Industrial Facilities;
- Agricultural and Aquaculture facilities;
- Area Management/Restriction/Regulation Zones and Reporting Units;
- Natural Risk Zones.

After having explored the state of the art in the fields of interoperability, SDIs and related tools and standards, the project defined its own metadata profile, data models for each of the seven themes, and transformation tools able to make existing data interoperable. The existing data, chosen by each partner, are available through a specific geoportal after being transformed by using these models and tools. A validation process and extensive dissemination activities are also important parts of the project.

Plan4all started in May 2009 and will end in October 2011. The project consortium is composed of 24 partners from 15 European countries. It encompasses local authorities, technological partners, universities and research institutes, GI national and pan-European associations, and a planners' association. Its final objective is to build a network of local, regional and national actors dealing with planning issues and regional development, in order to create consensus on the harmonisation of SDIs for spatial planning, and to contribute to the standardisation of related spatial data themes in the framework of the INSPIRE Directive.

3 THE LAND USE DATA MODEL

3.1 *The meaning of a data model for land use datasets and the interest for planners*

Six of the themes chosen by Plan4all concern data *for* spatial planning, i.e., spatial data needed for building the overall "knowledge framework" lying at the basis of the planning process. On the other hand, data *of* spatial planning are mostly included in the "Land Use" theme, undoubtedly the most relevant for the planners' community. A shared European model for such data entails a series of considerations going beyond the mere "logical" data modelling solution: in fact, a debate was soon triggered among the project partners, especially those having planning competences, about what kind of elements and information such a model should include. The main points of this discussion are exposed here.

First of all, this theme appears to be considerably different from the other INSPIRE themes that Plan4all deals with. While other themes such as "Land Cover", "Natural Risk Zones" or "Agriculture Facilities" are more clearly and directly connected to the geographical data representing spatial objects, in this case what we talk about is a legal statement concerning the current and/or future function or purpose of a spatial object; in other words, the "geographical dimension" of the Land Use theme is somehow elusive. Therefore, besides containing information of a purely geographical kind, an appropriate land use data model has necessarily to be connected to plans and the planning process, and to the norms and regulations that they entail. Indeed, this kind of data must necessarily contain, in an explicit or implicit way, directly or indirectly, information about the planning process.

The same INSPIRE document *D2.3 Definition of Annex Themes and Scope* shows some confusion on this matter. It defines "Land Use" as the *"territory characterised according to its current and future planned functional dimension or socio-economic purpose (e.g., residential, industrial, commercial, agricultural, forestry, recreational)"*. Immediately after, it says that "*land regulation is the general spatial planning tool at regional and local levels. Land use may be characterised as ordinary mapping of existing functions as an objective picture of the use and functions of a territory, but may also be plans characterising how land may be utilised at present and in the future*"; and that "*there are two main land use definitions, a functional one and sequential one*". Choosing the functional definition ("*description of land in terms or its socio-economic purpose*"), the document recommends to stick to the ISIC classification (International Standard Classification of All Economic Activities) drawn up by the United Nations. This classification contains 17 main categories and a number of sub-categories. But, since it has been developed from a merely economic point of view, any planner would probably see it as incomplete. For example, it is questionable trying to frame a residential use into the ISIC category "*Real estate, renting and business activities*", which only considers the economic function of housing and not the social one; or a land use such as "protected area" or "public park" into any of the listed categories, because the environmental protection—not being immediately relatable to an economic value—is not even included among the categories. This seems to us a first important remark on the INSPIRE framework. The main objectives of land use planning are to take care of the public assets and to ensure and regulate the general public convenience in order to manage and protect those goods and activities—of all kinds—that combine to maintain the citizens' living environment. These functions are by no means to be considered only in terms of economic revenue, because such approach would exclude most of the core aspects of the public responsibilities related to planning. A land use data model based on the simple description of an economic function would definitely be limiting, because it would exclude a considerable amount of information which is essential to planners.

It has also to be considered that the planning process entails the coordination and involvement of different sectors. This means that much of the information that is essential to the definition of land use has to be collected from different actors, each having his own perception of reality, directly connected to his function and responsibilities. Interpretation of reality becomes a relevant matter here, and—talking in terms of data modelling—leads to give different attributes to the same objects. Going back to the example of the river, we can say that this object can be seen in different ways by different actors. A river basin authority—being responsible for the safety of human settlements—would see it in terms of flow intensity and flood recurrence interval, whereas an environmental protection agency—being responsible for nature conservation—would look at it in terms of biological quality and ecological functionality of the banks. All these data and interpretations converge towards the definition of the planned land use for the single geographical object. Each authority involved in the planning process has the power of introducing into the plan a certain land use "instruction", often deriving from an evaluation/interpretation of reality according to his functions and responsibilities. The planning process must definitely relate to all information and points of view that the different actors can supply.

There is also a question relating to which, among the many possible planning "levels", is to be described by the model. A restrictive interpretation of the INSPIRE definition leads to intend the term "land use" as "zoning", which has traditionally been the most "central" part of the planning activity in many European countries. Traditional zoning, i.e., assigning precise functions to each land parcel, is surely the first core information to be included in a land use data model; however, the recent evolutions in the field have made planning a more complex activity. At least in France, Italy, and the UK, there are plans of different levels in the municipal planning process (generally, at least two levels such as a framework plan or policy statement, and an implementation plan). Each of these plan levels contains provisions that can either have direct influence on the lower-level plan—which will acknowledge them, define them more precisely, and/or complete them with more detailed information –, or be directly binding for the land parcel. Within this "planning cascade" system, the same land parcel is progressively given further attributes, until obtaining the exact determination of the allowed transformations: and from the legal point of view, the exact definition of the allowed land use comes from all these provisions. Thus, a land use data model intending to be really useful for planners will necessarily have to include information relating to all the planning levels.

Furthermore, the ever more frequent cases where multiple land uses are present on the same parcel (such as a commercial mall built on a railway station, or a tunnel under a protected park area) highlight the need for describing in a more appropriate way the complexity of the modern planning system.

Another relationship to deal with refers to the upper level plans (provincial, regional, national plans) and to sector plans, which have influences on the municipal plans, but sometimes also directly on the land parcel. The data model should therefore describe also the connections among the municipal plans, the upper level plans and the sector plans. Moreover, certain kinds of constraints often derive from national/regional laws before being elaborated by plans, and this is another information to be taken into account.

Indeed, the "multiplicity of powers" governing the territory obviously has an influence on land use, but at the same time, powers are separated and distributed in very different ways depending on each country, and produce many different kinds of land use indications inside the plans. Therefore, the intention of the authors of this model has been to focus primarily on the municipal plan (since this is the most important for what concerns land use provisions), but also to keep it as generic enough as to allow to include the description of all possible territorial government systems.

3.2 *The methodology of the data model definition*

For the elaboration of the Plan4all data model, a preliminary survey has been carried out in order to collect relevant case studies across Europe. The Plan4all partners have been involved, and invited to propose one or more data models from their countries, which they considered representative and useful for the purpose of defining the Plan4all model.

The collected models, all related to planning, are the following:

- Austria: LISA (Land Information System Austria, www.landinformationsystem.at/). This data model is about both land use e land cover;
- Germany: XPlanung (www.iai.fzk.de/www-extern/index.php?id=679). Model on municipal, regional and landscape plans;
- Spain: SIOSE (www.siose.es). This model also is about both land use e land cover;
- Finland: national data model for the regional plans;
- Ireland: ePlan (www.donegal.ie/dcc/eplan/). This model concerns only the development permits and authorisations;
- Italy: basic data model provided by the Province of Bologna for the municipal framework plans (cst.provincia.bologna.it/ptcp/psc.htm);
- Italy: technical specifications provided by the Region of Tuscany for the digitising of geographic data (www.rete.toscana.it/sett/pta/cartografia_sit/sit/). It concerns municipal and regional plans;
- France: national prescriptions for the delivery of digitised planning documents. It concerns municipal plans (www.geomatique-aln.fr/article.php3?id_article=289);

- The Netherlands: IMRO 2008 (www.geonovum.nl/dossiers/rostandaarden/destandaarden). This model is about municipal and regional plans;
- Norway: SOSI (www.statkart.no/nor/SOSI/SOSI_in_English/). It is the national set of data models, comprising also those about spatial planning.

The first step has been to overlay all these models in order to highlight differences, similarities and possible convergences. This preliminary analysis showed that—in confirmation of the fact that defining a Land Use data model is not a task of immediate comprehension—the collected models differ a lot in terms of contents and approach. The first reason for these mismatches is obviously the difference among the national and regional planning laws and procedures that the models derive from; also, the differences concern the scope and purposes of the modelling activity itself.

The Austrian and Spanish models, for example, are about both land use and land cover. This approach is certainly motivated by adequate evaluations, but it turned out to be less useful for our purposes, given the considerations made at the beginning of the work and exposed above. In any case, also these models have been taken into consideration for the definition of the Plan4all model.

The Irish model also differs considerably from the others, because it is only about development permits and the related administrative procedures. This has resulted in a specific class in the final Plan4all model, but didn't make it possible to compare this model with the others in a more thorough way.

The Dutch, French, German and Italian models proved to be the most suitable for mutual comparison—being in line with the preliminary considerations made by the Plan4all consortium—, and their structures have been used to define the core of the Plan4all data model. The Dutch, French and German models are national standards, whereas the Italian models used for this research derive from regional and provincial specifications.

After having compared all the collected models, an attempt has been made to group their main classes into a number of "super classes" having the same characteristics. The result of this activity is not properly a data model but has proven useful for having a preliminary "map" of all the collected materials, and for starting a discussion among partners on the possible classes of the Plan4all model and their attributes.

Figure 3 above shows—yet in a simplified way—this preliminary classification. Each box is a "class" intended to group all the existing classes, so as to take into consideration, as far as possible, all the collected information. This exercise was intended to provide a solid basis for the final Plan4all model, meant to be as comprehensive as possible of all concerned planning systems and modelling approaches.

The analysis of the collected models led us to introduce four main "super-classes":

- Administrative and procedural information: comprises information related to the planning process/procedures, such as date of adoption of the plan and name of the responsible authority; and information related to the development applications (this part derives from the Irish model, see above);
- Textual information: the textual parts of the plans;
- Planning information: this is the core of the scheme and comprises all the specific information on land use: the classifications of the land, including dimensioning and construction prescriptions, and indications on infrastructure and services; the non-binding indications; and the conditions and constraints acting on urban development, both those coming from other plans, laws or regulations, and those generated by the plan itself;
- Graphical information: any information containing prescriptions on how to publish the plan in paper form.

This general structure has been later discussed among the project partners—especially those normally involved in planning activities—and refined according to their observations, and formed the basis from which the actual modelling started. The final model is the result of many further adjustments and has considerably evolved from this scheme,

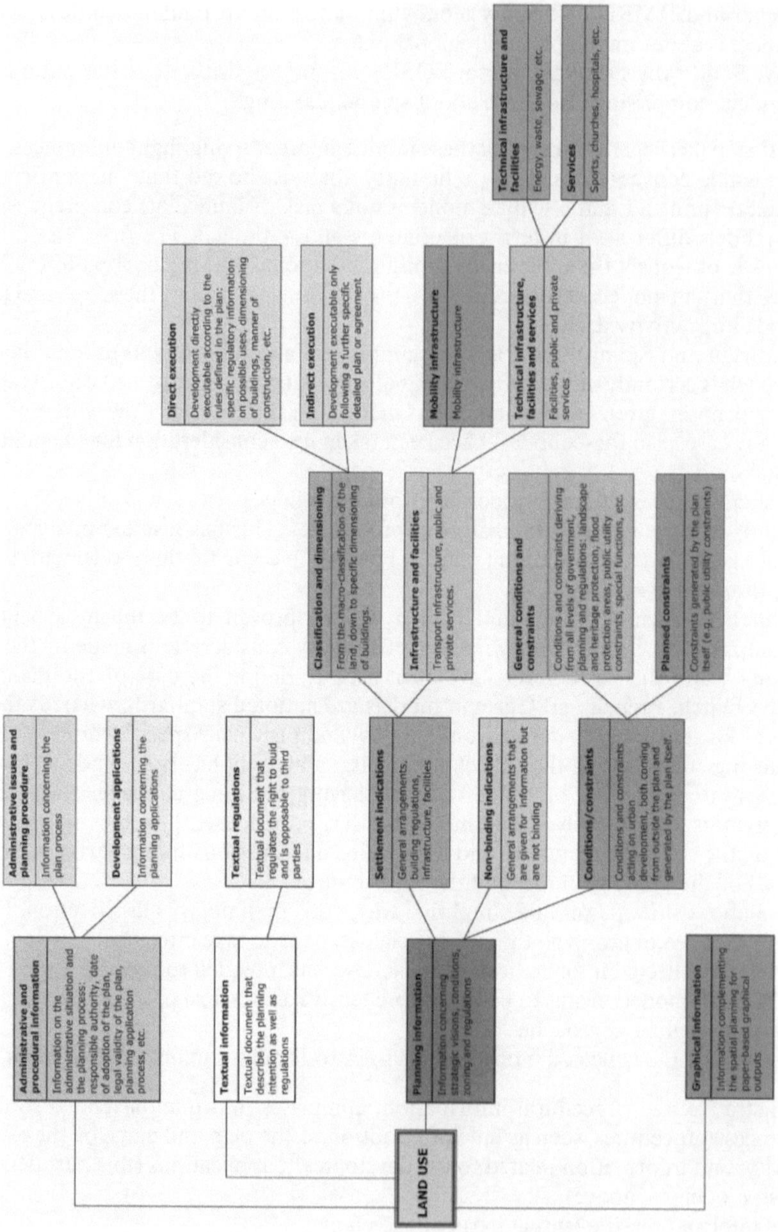

Figure 3. First tentative classification.

which has in any case proven very useful as a basis for discussion and as a guide for the modelling task.

A similar grouping exercise has been made on the attributes of all the collected models. Given the specificities of the different national planning systems, this task has been possible only to a certain extent, but the final result is reasonably comprehensive.

3.3 *Description of the Plan4all Land Use data model*

The construction of the actual data model, too, has been accompanied by continuous feedback from the involved partners. The final version is shown in Figures 4 and 5 as UML

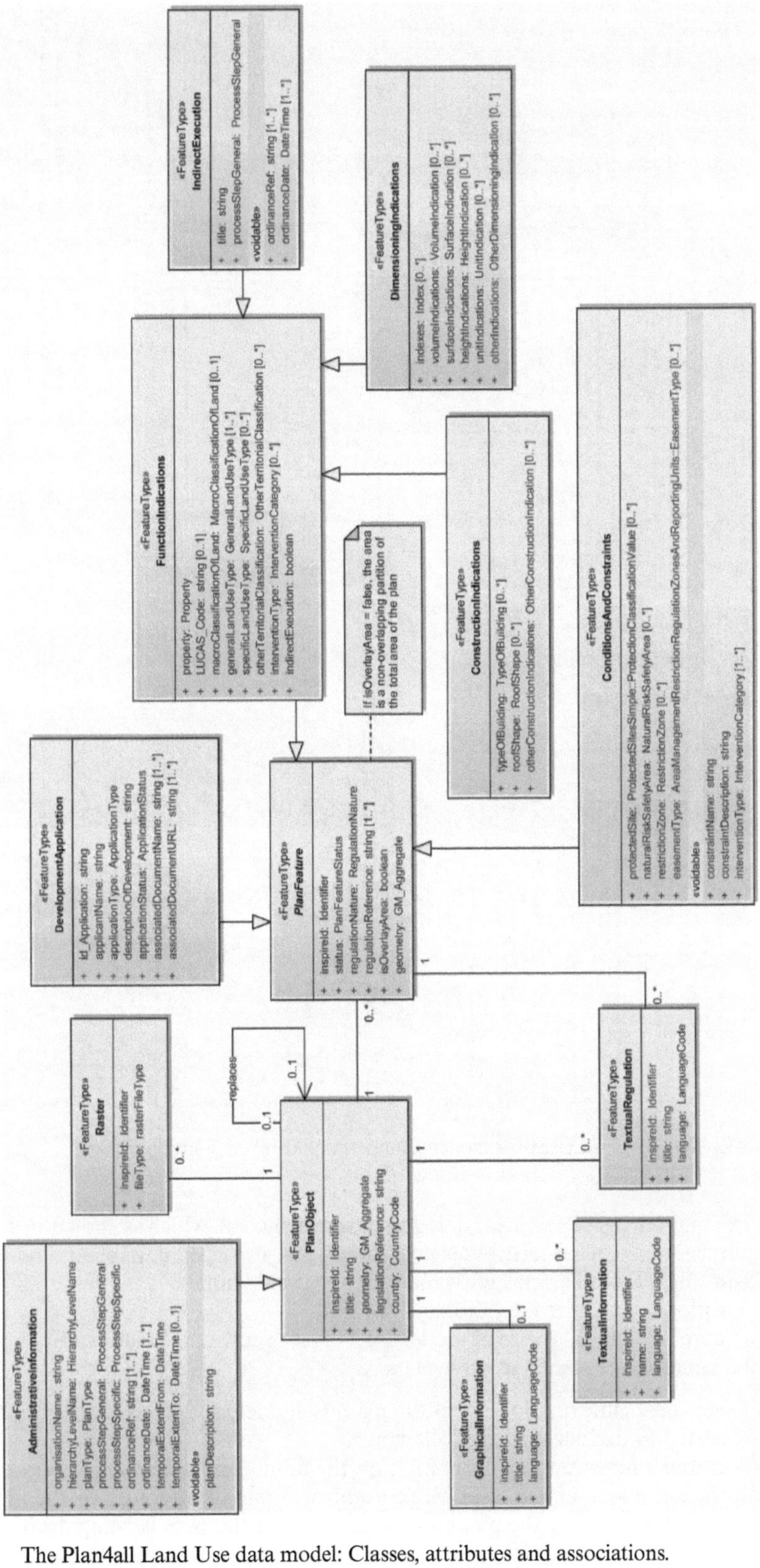

Figure 4. The Plan4all Land Use data model: Classes, attributes and associations.

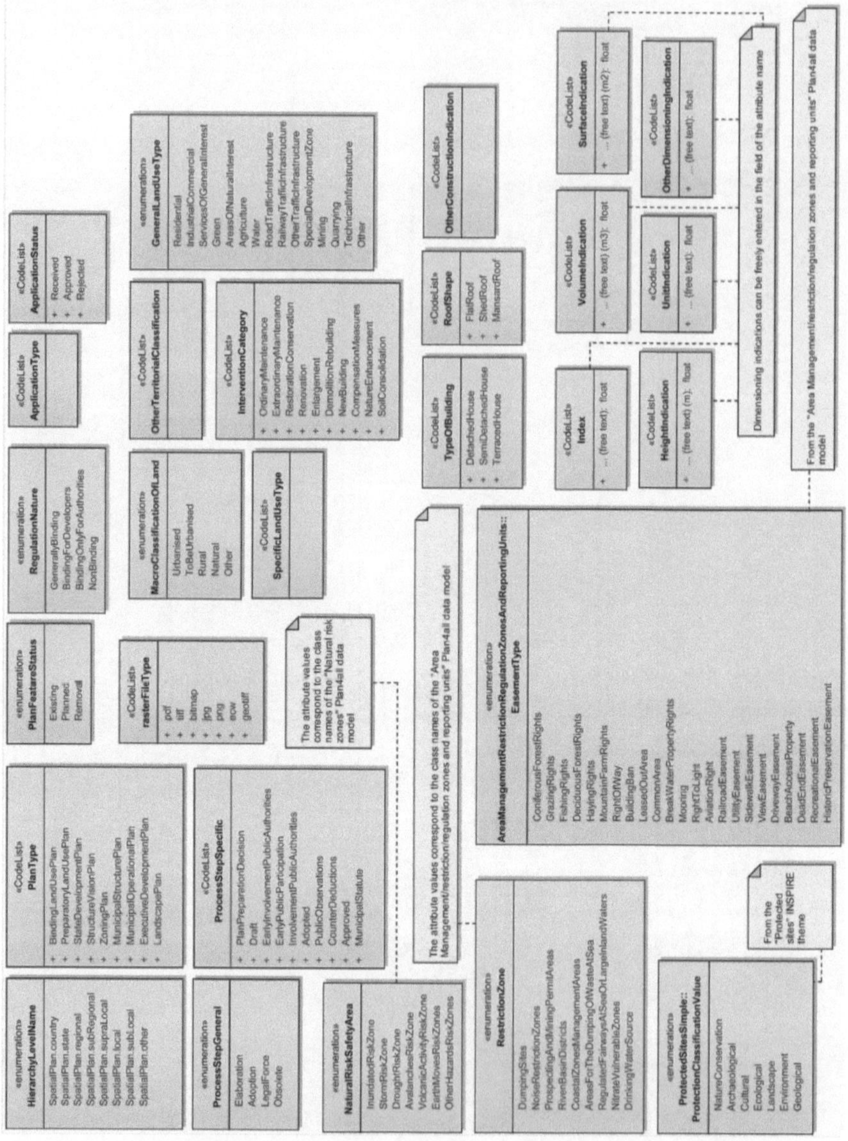

Figure 5. The Plan4all Land Use data model: Enumerations and code lists.

diagrams. A comprehensive feature catalogue (a screenshot of which is shown in Figure 6) describing in detail all classes, attributes and values, has also been developed, and is downloadable from the Plan4all portal www.plan4all.eu (the document containing all the data models is downloadable under the "Library" section).

The two "core" classes of the final model are "PlanObject" and "PlanFeature"; all other pieces of information are related to these classes.

- "PlanObject" bears information about the plan itself, including the geometry of the plan borders. Related to this class are the following:
- "AdministrativeInformation": information on the administrative situation and the planning process, e.g., name of the responsible authority, date of adoption of the plan, legal validity of the plan, step of the planning process (i.e., if the plan is being drafted, if it is adopted, approved, obsolete, etc.);

IndirectExecution	
Subtype of:	PlanFeature
Definition:	Information about a further plan, programme or agreement that is necessary for implementing the land use indications given in the plan.
Description:	NOTE This class gives information about the name of the further plan and its legal status. EXAMPLE 1 When a developer cannot start a development application according only to the general zoning plan, but has to make an executive plan first and get it approved. EXAMPLE 2 When an upper level plan (such as a regional landscape plan) doesn't give exact determinations about the land use, but is acknowledged and/or further defined by a municipal plan.
Stereotypes:	«featureType»
Attribute: title	
Value type:	String
Definition:	Name of plan.
Attribute: processStepGeneral	
Value type:	ProcessStepGeneral
Definition:	Information on the status of implementation of the plan.
Description:	NOTE The enumeration provides four values intended to be common to most planning systems.
Multiplicity:	1
Attribute: ordinanceRef	
Value type:	String
Definition:	Reference to relevant administrative ordinance, if any.
Description:	NOTE This attribute is multiple because, independently from the current legal status of the plan, there can be references to more than one ordinance, in relation to the different steps that the planning process has already undergone (e.g. ordinance for the preparation of a new plan, ordinance of adoption, ordinance of approval, etc.).
Multiplicity:	1..*
Stereotypes:	«voidable»
Attribute: ordinanceDate	
Value type:	DateTime
Definition:	Date of the relevant administrative ordinance, if any.
Description:	NOTE This attribute is multiple because, independently from the current legal status of the plan, there can be references to the dates of more than one ordinance, in relation to the different steps that the planning process has already undergone (e.g. ordinance for the preparation of a new plan, ordinance of adoption, ordinance of approval, etc.).
Multiplicity:	1..*
Stereotypes:	«voidable»

Figure 6. The Plan4all Land Use data model: Excerpt from the feature catalogue.

- graphical specifications for the paper-based outputs, if existing ("GraphicalInformation");
- files containing the textual parts of the plan ("TextualInformation" and "TextualRegulation");
- raster files referring to old plans in paper form ("Raster");
- "PlanFeature" contains the specific land use indications. It gives information such as the status of the land use indication (i.e., whether it refers to an already existing land use, or to a planned one), the type of regulation (i.e., for whom the land use indication is binding), the reference to any specific norm it refers to, etc. It contains also the geometry of the land use indication (the land area that the specific indication refers to). Specialisations of this class are the following:
- "FunctionIndications", comprising all kinds of land use indications, from the most general classification of the municipal land (e.g., urbanised/to be urbanised/rural/natural), down to the specific function for the single land parcel. These indications can be also about dimensions ("DimensioningIndications"; e.g., indexes, volume ratios, maximum heights), type of construction ("ConstructionIndications"; e.g., type of building or roof shape allowed),

and/or indirectly executable ("IndirectExecution", in the case that the task of specifying in detail the function of a certain area is entrusted to other plans);
- "ConditionsAndConstraints" acting on urban development, both coming from outside the plan and generated by the plan itself. In order to use code lists that have been already defined, some connections to other themes have been attempted: whether developed by INSPIRE Thematic Working Groups ("Protected Sites Simple"), or by other Plan4all partners ("Area Management, Restriction, Regulation Zones and Reporting Units" and "Natural Risk Zones");
- administrative information regarding the procedures for issuing building permits and other kinds of authorisations referring to the same plan ("DevelopmentApplications").

The data model has then undergone, within the framework of the Plan4all activities, a validation process involving, in addition to the Plan4all partners, some external experts and stakeholders. They contributed with useful observations, comments and modifications, which are deemed essential for improving this proposal before definitely submitting it to the INSPIRE community. Moreover, this model has been useful as a starting point for the work of the INSPIRE Working Group on Land Use, which is carrying out its activities during the same period, and which includes two of the Plan4all partners who have worked on this same project task.

4 CONCLUSIONS

The proposed model is a first attempt, shared by experts involved in spatial planning and coming from different European countries, to describe in a normalised yet complete way the datasets concerning spatial plans of very different kinds and origins. In order to make this possible, this tool, having gone through many stages of refinement and rationalisation, ends up to be rather simple and generic, but at the same time it could be suitable to bear a quite exhaustive core of planning information.

But planners love creativity, and many could dislike what might appear to them as an excessive degree of "normalisation". However, we want to stress the fact that normalisation is a necessary step in the path towards interoperability. And interoperability is intrinsically linked to the transcalar nature of many planning issues.

A traditional input to planning practices at the intermediate level (e.g., in Italy the provincial level) is the so called "mosaic" of the municipal plans. In some cases, this "mosaic" is built "ex post", through the normalisation of the existing municipal plans, with some loss of information. In other cases, the higher level authorities define "ex ante" a normalised legend or a glossary that municipalities must apply, with some loss of autonomy for the municipalities themselves.

The data model approach may differ from both the approaches mentioned above. "Telescopic" data models may be defined in order to deal consistently with geographic objects and rules at different scales and at different levels of detail. While the main aim is to make the data flow among various levels possible, the principle of preserving some of the municipalities' autonomy is applied, and data models are designed so that other pieces of information belonging to a particular level can be integrated later into the models themselves. The experience of the Province of Bologna, mentioned above, goes in this direction.

Another consideration regards the legal value of spatial plans published in digital form. The significance of spatial planning harmonisation and interoperability diminishes if digital plans do not have legal value. If not, you need to check paper maps and documents in order to have definitive information. As far as we know, The Netherlands is the only European country where the plan in digital form has legal value, on the basis of a law of 2010. This seems to be the culmination of the process towards planning data interoperability, but it does not seem to be fully agreed by the same local governments. The adoption of a plan in form of a database may result in ambiguous interpretations of the constraints and requirements, partly because different cartographic representations of the same data may occur. In addition to being a technical issue, moving towards a similar result also appears to be a cultural issue, because a digital plan which unambiguously defines a land use provided by the law can result in a feeling of loss of decisional power for the local authorities and civil servants.

Finally, we want to stress the fact that this exercise has highlighted a further issue that shows up when it comes to data modelling: the relation between the experts of the theme to be modelled, and the IT experts whose contribution is essential in the modelling task. The dialogue between planners and IT experts has not always proven smooth and easy, but is a necessary step—as many local authorities experience when, having the purpose of building their own SDI, they have to ensure a dialogue between two very different departments.

As a conclusion, we want to stress the fact that the Plan4all Land Use data model presented in the previous section is only a first proposal, matured within the strict time schedule of a project focusing on a wider set of objectives. It is intended as a first step towards the beginning of a dialogue among different approaches to planning and data modelling.

A future, and more exhaustive, research—following also the work of the related INSPIRE Land Use Thematic Working Group—could deepen this proposal towards the purpose of a smoother data exchange at all administrative levels; this can be useful, besides the establishment of a European SDI, for implementing European-level actions related to spatial planning and, eventually, for a progressive harmonisation of the different national and regional spatial planning practices.

REFERENCES

Camerata, F., De Marco, G., Magaudda, S., Ombuen, S. & Pellegrino, G. 2010. La pianificazione territoriale in Europa: una proposta di modello dati per il tema *land use* di INSPIRE. In ASITA, *14° Conferenza Nazionale ASITA, Brescia, 9–12 Novembre 2010.*

CETE Normandie Centre 2008. *Nomenclature pour la nouvelle base de données de l'occupation du sol du littoral 2000–2006.* France: Ministère de l'Ecologie, de l'Energie, du Développement durable et de l'Aménagement du territoire.

Commission de Validation de Données pour l'Information Spatialisée 2010. *Plan local d'urbanisme—Plan d'occupation des sols (PLU et POS)* (Version 0.9). France: COVADIS.

Conseil National de l'Information Geographique & CERTU 2009. *Plan Local d'Urbanisme—Prescriptions nationales pour la livraison des documents d'urbanisme dématérialisés.* France: CNIG.

De Vries, M., Di Donato, P. & Penninga, F. 2008. *Concept of application-specific harmonised data models.* HUMBOLDT project deliverable.

Di Donato, P., Salvemini, M. & Berardi, L. 2007. HUMBOLDT: armonizzazione dei dati a supporto delle Infrastrutture di Dati Territoriali. In ASITA, *11° Conferenza Nazionale ASITA. Torino, 6–9 Novembre 2007.*

Eremitaggio, A.M., Procaccini, P., Ombuen, S., Vico, F. & Zanetti, N. 2010. Plan4all: l'armonizzazione dei dati per la pianificazione territoriale in accordo con INSPIRE. In ASITA, *14° Conferenza Nazionale ASITA, Brescia, 9–12 Novembre 2010.*

GEONOVUM 2008. IMRO (*Information model for Spatial Planning*). The Netherlands.

GeoVille Information Systems GmbH 2010. *LISA—Land Information System Austria.* Innsbruck: GeoVille Information Systems GmbH.

Giudice, M. 2008. Carte e statuti. In Properzi P. (ed.). *Rapporto dal Territorio 2007*: 80–86. INU. http://www.rapportodalterritorioinu.it/Pagine/Carte_Statuti.pdf

Hopkins, D., Kaza, N. & Pallathucheril, V.G. 2005. Representing urban development plans and regulations as data: a planning data model. *Environment and Planning B: Planning and Design* 32: 597–615.

INSPIRE Directive 2007. *Directive 2007/2/EC of the European Parliament and of the Council of 14 March 2007 establishing an Infrastructure for Spatial Information in the European Community (INSPIRE).* http:// http://inspire.jrc.ec.europa.eu/

INSPIRE Drafting Theme "Data Specifications" 2008a. *D2.3 Definition of Annex Themes and Scope.* INSPIRE Drafting Theme "Data Specifications".

INSPIRE Drafting Team "Data Specifications" 2008b. *D2.6 Methodology for the development of data specifications.* INSPIRE Drafting Theme "Data Specifications".

INSPIRE Drafting Team "Data Specifications" 2010a. *D2.5 Generic Conceptual Model, Version 3.3.* INSPIRE Drafting Theme "Data Specifications".

INSPIRE Thematic Working Group "Hydrography" 2010b. *D2.8.I.8 INSPIRE Data Specification on Hydrography—Guidelines*. INSPIRE Thematic Working Group "Hydrography".

JRC 2005. *Draft Guidelines for the Development of the INSPIRE Implementing Rules*, http://inspire.jrc.ec.europa.eu/reports/dt/ir_draft_guidelines.pdf. JRC.

Karlsruhe Institute of Technology 2010. *XPlanGML Version 4.0—Objektartenkatalog*. Karlsruhe: Karlsruhe Institute of Technology.

Laurini, R. & Murgante, B. 2008. "Interoperabilità semantica e geometrica nelle basi di dati geografiche nella pianificazione urbana". In Murgante B. (ed), *L'informazione geografica a supporto della pianificazione territoriale*: 229–244. Milano: Franco Angeli.

Local Government Computer Services Board 2007. *iPlan Dasebase Table Layouts*. Ireland.

Norwegian Mapping Authority 2008. *SOSI standard*. http://www.statkart.no/nor/SOSI/SOSI_in_English/. Norwegian Mapping Authority.

Plan4All Consortium 2009. D.2.1—Cluster of leading regional in SDI for spatial planning. Plan4all Consortium.

Provincia di Bologna 2007. *Modello dati di base del PSC per il territorio provinciale bolognese*. Provincia di Bologna.

Regione Emilia-Romagna 2000. *L.R. Emilia-Romagna 20/2000—Disciplina generale sulla tutela e l'uso del territorio*. Regione Emilia-Romagna.

Regione Toscana 2003. *Specifiche tecniche per l'acquisizione in formato digitale dei dati geografici tematici*. Regione Toscana.

Regione Toscana 2005. *L.R. Toscana 1/2005—Norme per il governo del territorio*. Regione Toscana.

SIOSE National Technical Team 2005. *Land cover and use information system (SIOSE)—Technical document* (Version 2.0). Madrid: SIOSE National Technical Team.

Tóth, K. & Nunes de Lima, V. 2005. Data Quality and Scale in Context of European Spatial Data Harmonisation. EU JRC, *11th EC GIS&GI Workshop*. Alghero.

Viganò, S. & Millot, M. 2007. *Data State of Play. Compliance Testing and Interoperability Checking*. JRC.

Urban and Regional Data Management – Zlatanova, Ledoux, Fendel & Rumor (eds)
© 2012 Taylor & Francis Group, London, ISBN 978-0-415-67491-1

Fully automated generalisation of topographic data in current geo-information environments

J.E. Stoter
Delft University of Technmology, GIS Technology Section, Delft, The Netherlands
Kadaster, Apeldoorn, The Netherlands

J. van Smaalen
Esri NL, Rotterdam, The Netherlands

R. Nijhuis, A. Dortland, J. Bulder & B. Bruns
Kadaster, Apeldoorn, The Netherlands

ABSTRACT: This paper studies the feasibility of fully automated generalisation of topographic data in the context of nowadays use of geo-information and available technologies. The starting point is that the requirements with respect to up-to-dateness may get priority over cartographic principles, although the result should still be of acceptable quality. In addition the study acknowledges that the results obtained by an automated process may differ from results from interactive processes. A pilot was carried out that takes these perspectives and set up a fully automated generalisation workflow using commercial software. In the pilot the workflow has been refined and the optimal implementations and parameter settings have experimentally been determined by an expert team in five weeks time. The results show potentials to generalise a 1:50 k map from 1:10 k data in a fully automated manner. Issues for further research are identified.

1 INTRODUCTION

This paper presents a study on the feasibility of fully automated generalisation of topographic data in current geo-information environments, which also comprises present requirements of multiscale topographic data. These requirements may differ from the requirements underlying existing multiscale topographic databases at National Maping Agencies (NMAs) which often have their origin decades ago. In addition the study takes into account that results of newly introduced automated generalisation processes may differ from topographic data obtained by interactive generalisation processes. The paper reports on a pilot that sets up a fully automated generalisation workflow using these different perspectives. In the pilot generalisation techniques are tested that recently have become available in commercial software, such as displacement of infrastructure and buildings. The ultimate aim is to obtain small scale maps from large scale data in a fully automated manner.

The motivation of this study is the current production line of topographic maps at different scales of the Dutch Kadaster which is a legal task. Since 1 January 2010 Kadaster is obliged to produce both object oriented topographic vector data and raster maps at scale 1:10 k, 1:50 k, 1:100 k, 1:250 k, 1:500 k and 1:1.000 k. Other governmental organisations are obliged to use these data and they are no longer allowed to acquire their own multiscale data (apart from the scale 1:1 k). To meet this legal obligation Kadaster has started the conversion of their vectorised maps into object oriented databases since 2007 and these products are updated by cartographers in an interactive manner according to generalisation guidelines.

Although the conversion showed the best potentials to meet the law in time, the approach consolidated the history of maps at predefined scales of which the origin is often more than 60 years old. This has several drawbacks. At first the approach of taking the existing maps as starting point may overlook new requirements of multiscale topographic data. In addition it may limit the potentials for automated generalisation as the problem to automate a process which used to be interactively may be too complex. Another drawback is that the small scale databases start from the vectorised maps and therefore includes cartographic representations of topography that may have been enlarged or displaced to be able to obtain a readable map of the symbolised objects. These geometrical adjustments have not been controlled to meet the accuracy of that scale (see Figure 1). In contrast distinguishing between requirements for multiscale data and multiscale visualisations may offer better potentials to move forward in this domain. Finally, updates in the current multiscale data require considerable time, i.e., 500–2000 hour on average to update a 1:50 k map sheet of 500 square kilometres. Besides the costs involved, this lengthy process does no longer fulfil current requirements with respect to up-to-dateness of data and maps. Specifically since the step-wise process (i.e., every scale is generalised from the next larger scale) causes updates to be propagated even later in smaller scale maps.

In conclusion it is likely that the existing workflow, heavily relying on interactively generalised maps, cannot simply be replaced by automated generalisation. In addition it may be not the optimal workflow for meeting changing users' demand. This motivated a study to the optimal implementation for sustainable multiscale topographic data provision where existing maps are no longer the main drive. Instead current technologies and new requirements should play a role as well. An example of a new requirements is that a user may accept a cartographically lesser product in favour of more recent information (because generated fully automatically).

The Dutch Kadaster has no history in investing in automated generalisation in contrast to other NMAs such as IGN, France (Lecordix et al., 2007), KMS, Denmark (Foerster et al., 2010), ICC, Catalonia (Baella & Pla 2005) and Germany (AdV, 2007). Therefore a pilot study was carried out to generalise a map at scale 1:50 k from a 1:10 k topographic data (i.e., TOP10NL) in a fully automated manner. The aim of the pilot was to show the potentials of automated generalisation for Dutch landscape and data, and to compare the products of fully automated generalisation with the maps we are used to. In addition it would identify topics for further research.

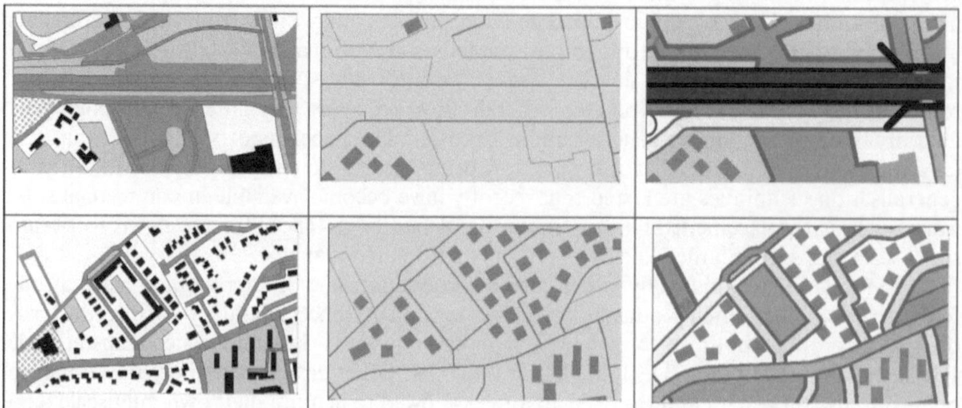

Figure 1. Displacement of roads in TOP50vector triggered by graphical conflicts of symbolised roads in 1:50 k map. Displacement is propagated to buildings to avoid buildings overlapping the roads.

This paper reports on the pilot. Section 2 further describes and justifies the scope of this research. The methodology that we used is presented in Section 3. Section 4 presents the results and Section 5 ends with conclusions and identifies topics for further research.

2 THE OBJECTIVES AND SCOPE OF THE RESEARCH

The research in this paper starts from two assumptions. At first one of the most important requirement of today's use of topographic data is highly up-to-date maps at different scales. Therefore the requirement of up-to-dateness may get priority over cartographic principles, although the results should still be of acceptable quality (see Section 5). Secondly, requirements for object oriented topographic data (e.g., for GIS analyses or to link to information of third parties) can be fulfilled by the largest scale (1:10 k in our case), where as topographic information at smaller scales is mainly used for visual applications. The consequence is that the smaller scale data are only cartographic representations and therefore disruptions of the geometry to meet cartographic requirements do not have to be controlled. With these assumptions in mind, we started a pilot on fully automated generalisation of small scale topographic maps from TOP10NL data using available technologies, starting from 1:50 k map. The following aspects further refine the scope of the research:

- Our aim is not to replicate the small scale maps and therefore we allow ourselves some freedom to distract from existing cartographic generalisation guidelines. We do use these guidelines to copy the important principles, but at the same time we may reconsider the guidelines in case our experiments show reasons for this. Therefore this research is different than the research reported in (Stoter, et al., 2009a), where requirements of current maps where the starting point.
- The topographic maps should serve orientation requirements as well as navigation requirements, initially of car drivers (to be adjusted for other users in the future).
- The ultimate aim of the generalisation is the map (on paper or screen), i.e., the aim of generalisation is to quickly obtain small scale maps of updated 1:10 k data. Model generalisation (i.e., generalising the data without considering the symbolisation) will be applied, but only to prepare the data for the optimal visualisation.
- The most straightforward way for updates in automated generalisation is to replace the old version of a whole partition with updated data. Therefore we did not attempt to maintain links between the objects at the two scale levels. If these links are required, this can be part of a subsequent study.
- The scope is limited by the static visualisations on the screen or on paper. Dynamic generalisation for Internet use (real time and not per se related to the predefined scales) has to meet different requirements but may have also different solutions and is therefore outside the scope of this research.
- Generalisation without any interaction is the best guarantee for efficiency and consistency. Therefore the results in our research are generalised fully automatically.
- Because this is our first investment in the topic, we wanted to make use of available commercial software on generalisation. For the same reason we wanted to build on previous research in which we studied the suitability of generalisation guidelines at scale 1:50 k for automated generalisation (see Stoter et al., 2009b). Algorithms and approaches that were developed in this research have been implemented in ArcGIS 10 (released in 2010). Therefore we use Esri ArcGIS tools for the implementation.

3 METHODOLOGY

This section presents the methodology of the research by first describing the global approach and secondly the detailed action plan of the automated generalisation workflow that we implemented to generate the 1:50 k map fully automatically.

3.1 *Global approach*

The research was carried out with a research method used in Kadaster to study the feasibility of innovative technologies. The method is based on an intensive collaboration of people with the necessary domain expertise in a limited timeframe (i.e., five weeks). The team established for the purpose of this pilot combined expertise of Dutch topographic data and the underlying architecture, expertise of cartography, expertise of (Geo-) IT tools and expertise of (automated) generalisation.

The pilot refined the generic action plan to generalise TOP50 map from TOP10NL which was the result of the research reported in (Stoter et al., 2009b; Smaalen en Stoter, 2008). This generic action plan was obtained by studying the generalisation guidelines for interactive generalisation (Kadaster, 2005) and by comparing intermediate results with existing maps. The obtained insights in how cartographers interpret the guidelines were used to improve the process. Also external data was used to enrich the process. The enrichment is done in two ways. Either external data sources are used to obtain the knowledge or the reuqired knowledge is made explicit by computation.

The pilot of this research refined the initial action plan by experimentally determining the optimal approach, algorithms and parameter values for each step in the action plan and also the best sequence of steps. If appropriate the steps were prototyped with different alternatives to identify the best results. At the end of the five weeks pilot, the optimised action plan was implemented as one workflow consisting of eleven steps.

3.2 *Detailed action plan*

This section details the eleven steps of the action plan. The steps can be divided in three parts:

i. data generalisation aiming at reducing the data that has to be visualised
ii. symbolisation of the data
iii. cartographic generalisation to solve cartographic conflicts of symbolised objects.

The success of the generalisation process depends also on the information available in the start data set. Therefore part of the process is enriching TOP10NL with relevant knowledge, preferably as automated as possible. Also this enrichment is included in the following steps.

I Data generalisation

1. Reclassify all TOP10NL codes into TOP50 codes.
2. Compute urban areas by calculating building density to distinguish between urban and rural area.
3. Convert areas with many buildings (i.e., coverage > 17%) to built-up areas. This step was implemented with a couple of refinements. Firstly, buildings are only converted to built-up area if they are located within urban area (computed in step 1) and if they are not part of an industrial estate (external data). In addition, if buildings are not equally distributed, building areas are only generated on the locations of the buildings as was done in the interactive process, shown in Figure 2. Unbalanced distribution of buildings is determined by analysing if 'holes' exist in the polygon being analyzed. Another refinement is that prominent and important buildings as schools, hospitals, churches are kept separately. This functional information is included in TOP10NL.
4. Convert TOP10NL road centrelines into TOP50 centrelines, since TOP50 centrelines differ from TOP10NL centrelines. In TOP10NL every lane has a centreline in the database (apart from polygon-geometry which is used for the map) and in TOP50 every road has one centreline (also used for the map), see Figure 3. In our pilot the TOP50 centrelines are generated with an algorithm that merge lanes of a single road into one centreline. Merging is also done in case of diverging roads for a refuge. Other approaches were also considered—for example generating centerlines from the road polygons—but these gave unsatisfying results, specifically at crossings. This is a known problem, see (Thom, 2007).

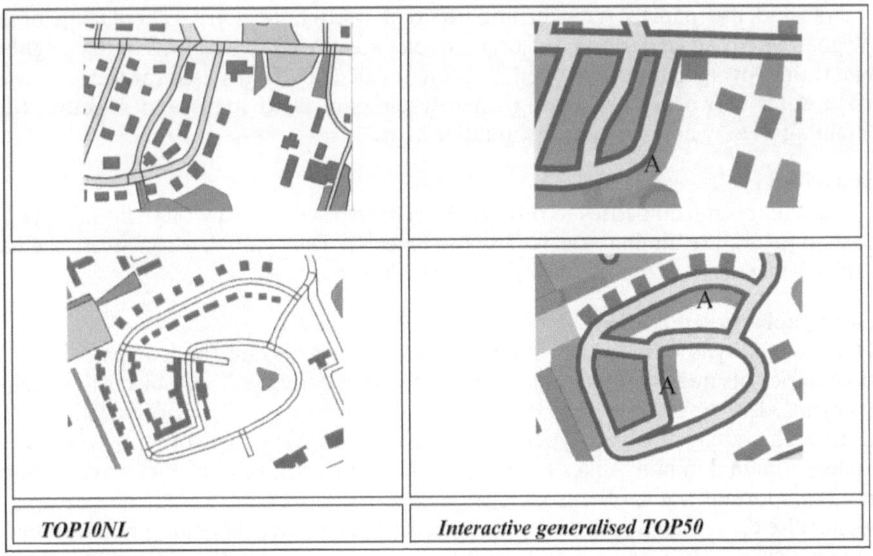

Figure 2. Conversion of TOP10NL buildings to TOP50 built-up areas obtained from interactive generalisation. 'A' shows locations where uneven distribution of buildings was taken into account.

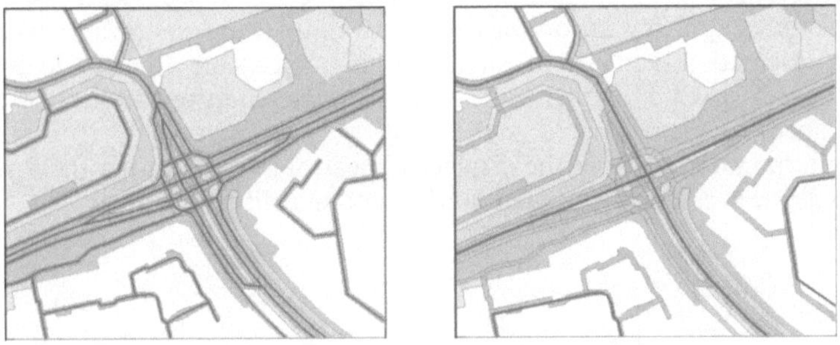

Figure 3. Centrelines in TOP10 NL (left) and TOP50 (right), both projected on TOP10NL polygons.

5. Prune the road network because of insufficient space to visualise all roads, see also Chaudry and Mackaness (2005) and Thompson and Richardson (1999). This is not straightforward, specifically because TOP10NL data does not contain many attributes for pruning. The pruning of the road network as implemented consists of several steps:
   - Free cycle paths are maintained while cycle paths parallel to roads are deleted. Both types are automatically detected with a buffering algorithm.
   - Dead ends shorter than 50 m in urban areas are deleted.
   - Non-paved dead-ends shorter than 1 km in rural areas are deleted but access roads to buildings are automatically detected and kept. Access roads are detected as such if buildings are located in the vicinity of the end of the road AND no other roads are located in the neighborhood of the building.
   - For pruning the remaining roads two methods were compared: a) a self-developed algorithm which calculates continuing roads (see Stoter et al., 2009b) and b) an algorithm that generates a simplified road network that retains connectivity and general character also using a hierarchy for the relative importance and the minimum length of a road. Based on experiments the latter algorithm was selected as the optimal. The results of this algorithm are a balanced compromise between the requirements posed by hierarchy, resolution (minimum length), and the morphology and connectivity of the road geometry.

6. Extend land use parcels (i.e., terrain) to road centrelines. Due to replacing of road polygons with road centerlines, the original road areas need to be added to the neighbouring terrain. An algorithm developed in (Stoter et al., 2009b) was used for this purpose.
7. Prune water network by removing small water parallel to roads and selection of the remaining water where free water should be kept.

II **Symbolisation**

8. Assign symbols to geometries to display the geometries in the way they should appear on the map (as we are aiming at cartographic objects). The symbolisation steers the cartographic generalisation process to improve visualisation.

III **Cartographic generalisation**

9. Displace the symbolised dikes, railways, roads, water, terrain (both linear objects and boundaries of areas) with an algorithm that displaces symbolised objects and reshapes them in order to avoid overlap. The algorithm is implemented via an optimization method using an hierarchy of object types (Punt and Watkins, 2010). The order used in the hierarchy is: main damming dikes (may hardly be displaced), railway, highways, important roads, important water, other roads.
10. Generalise the remaining buildings from step 3 to avoid overlap and to meet a minimum building size. At first small details of buildings are deleted with an algorithm that simplifies the boundary or footprint of building polygons while maintaining their essential shape and size. Secondly an algorithm is applied that displaces the buildings to meet the minimum distance between buildings and also between buildings and other surrounding objects. This algorithm resolves symbol conflicts of objects with respect to linear barrier features applying an optimisation technique (Punt and Watknins, 2010). The position, orientation, and size of the concerning objects are adjusted to avoid overlap or violation of spacing requirements, while the representative pattern and distribution is maintained.
11. Recreate terrain objects from the displaced linear object and assign the former terrain codes to the new terrain areas by an overlay with the old terrain areas.

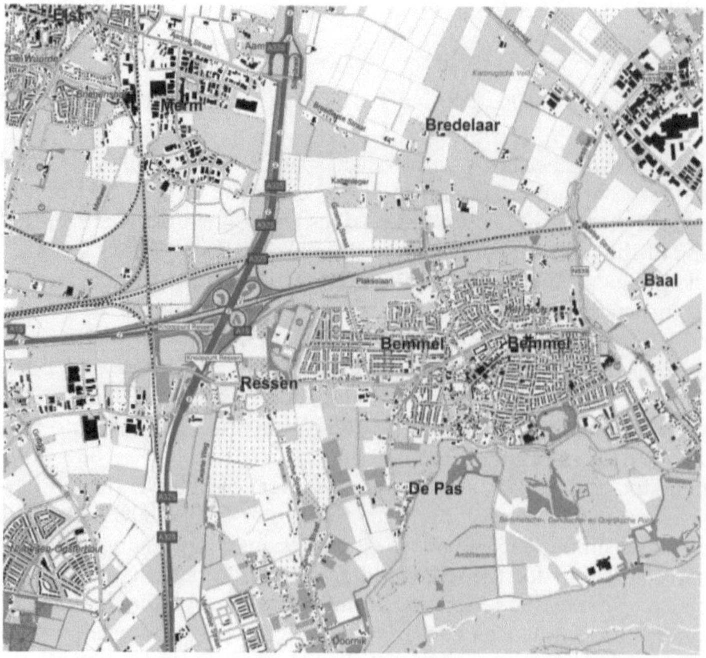

Figure 4. Source data of first test area (TOP10NL, displayed at smaller scale).

3.3 *Test data*

We selected test areas in such a way that we cover the main types of topography in The Netherlands. The four test areas that we selected are characterised by respectively:

1. Relief and dense road pattern (see Figure 4)
2. Complicated crossings and dense parcel boundaries
3. Dense water network
4. Urban and industrial areas.

4 RESULTS

This section reports on the results of the pilot.

We succeeded to implement all the steps in one workflow within four weeks of time and run it for one of the test areas. The results of the first test area look promising as shown in Figure 5a. It took 15 minutes to run this model for an area of 55 km^2. Figure 5b shows

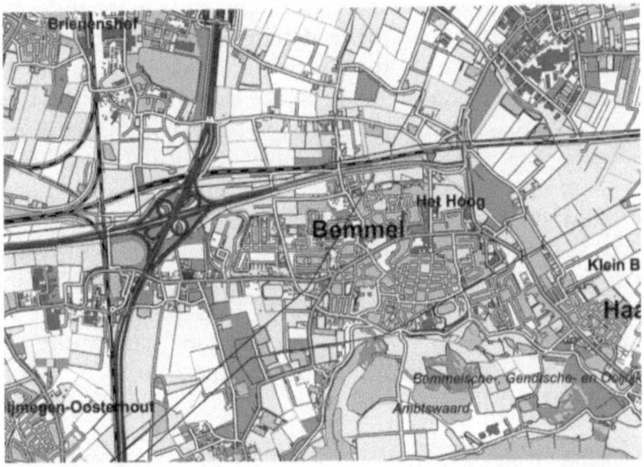

a: 1:50 k map, obtained fully automatically; displayed at smaller scale

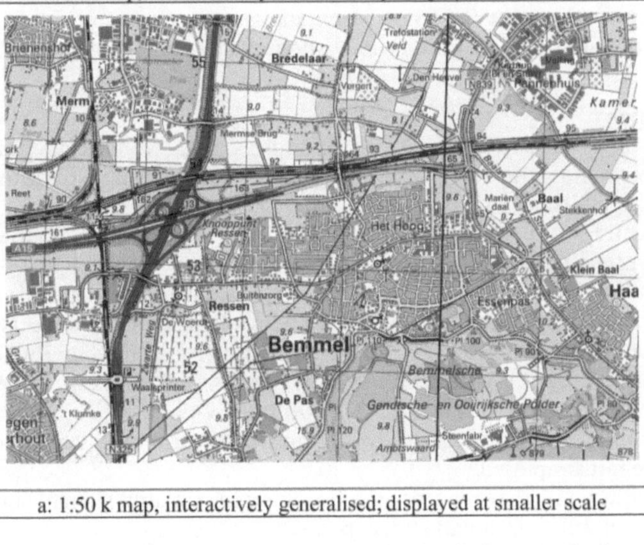

a: 1:50 k map, interactively generalised; displayed at smaller scale

Figure 5. Results of automated generalised map (a) and interactively generalised map (b).

the existing interactively generalised maps. However the two maps at scale 1:50 k should be compared with caution. At first the results of the automated generalisation process are not meant to replicate the results of the existing maps, as was explained in Section 2. Secondly, the existing map is based on older data. Also it should be noted that names are not generalised in Figure 5b.

This section further elaborates on situations where we decided to ignore the existing generalisation guidelines (Section 4.1) and possible refinements that could further improve the results which could not be studied in the limited time available for the pilot (Section 4.2).

4.1 *Ignoring the existing guidelines*

In three cases we decided to ignore the existing guidelines.

1. The generalisation guidelines prescribe that detached houses may never be converted into built-up areas. We see problems to always meet this rule, because detached houses are often enlarged to meet the minimum size (15 × 15 m). In most building blocks with detached houses both sides are covered with such houses. Consequently the widths of the building blocks should be at least 40 m to accommodate enlarged detached houses at a minimum distance of 10 m (required for readability). At the same time streets in TOP50 are symbolised with line widths of 20 m, which is wider than streets widths in reality. Because of this high competition of space, in the existing maps these houses are often converted into built-up areas, despite the guidelines. Because it is hard to automate a rule which has many exceptions and because the conversion of detached houses into built-up areas was not considered as problematic, we decided to always convert detached houses in built-up areas if building density threshold is exceeded.
2. Ditches are treated as ordinary terrain boundaries, in contrast to the traditional generalisation, because of their limited importance and because their visualisation is identical to terrain boundaries.
3. All water with linear geometries in TOP10NL (i.e., smaller than 2 meters) parallel to roads are eliminated to save space as we regard them as unimportant.

4.2 *Possible refinements of the pilot results*

The refinements listed in this section may improve the pilot results to a TOP50 product that is good enough to meet the main customers' demands as well as legal obligations. We identified these refinements from the observations and experiments of the pilot.

Data generalisation

- Prune the (artificial) water networks can be improved on two aspects: pruning based on drainage pattern and pruning the network with a road thinning algorithm. For the first aspect drainage information needs to be added to the source data (can be obtained from external source). For the second the source data need to form a complete water network, which is currently not always the case.
- Detect damming dikes as objects. Although very important for generalisation (i.e., they may never be displaced), damming dikes are not available as single objects in TOP10NL. Instead they can only be recognized by hatches on the map. If these objects cannot be detected automatically, it is worthwhile to copy them once from a different source dataset, as they are predominantly static objects not subject to change.
- Study if smoothing and weeding of roads brings further improvements as currently these geometries are not simplified.
- Better generalisation of fly-over crossings by taking the (relative) heights into account.
- Generalisation of relief like dikes and slopes.

Cartographic generalisation

- Guarantee that touching roads and bending roads are not pushed away from each other in the displacement process. This needs further study.
- Improve the proper placement of buildings in a limited amount of space with. For example by assigning lower hierarchy to small and unimportant buildings (information to be retrieved from external sources).
- Improve the generalisation of small areas of forest and water which are surrounded by roads (or water). Since these areas are often important for orientation on the road, they cannot always simply be eliminated. At first it should be evaluated if these are treated sufficiently (i.e., aggregated, enlarged or eliminated if appropriate) in step 9. Secondly more advanced solutions should be studied, for example aggregation of several small areas or by using a propagate displacement algorithm. With this algorithm, the location of input features are adjusted based on the vector displacements contained in the displacement features. Adjustments are a compromise of all displacements, such that large displacements that occurred quite near an input feature will have more influence than smaller displacements further away. Conceptually, this action is similar to a rubber-sheeting process that moves features in various directions by various amounts to fit them back to the spatial relationship that they originally had with the roads. It does not resolve graphic conflicts and in fact may introduce new conflicts.
- Generalisation of additional objects as isolated forest (tree lines) and power pylons.
- Generalisation of annotations.

5 CONCLUSIONS

This paper presents a feasibility study on automated generalisation of topographic data that is steered by available techniques and (new) requirements of multiscale topographic data.

A pilot was executed to show in a limited amount of time the potentials of automated generalisation and how well the results obtained fully automatically serve customers of topographic maps.

The results of this pilot, obtained within the limited timeframe of five weeks, show that it is feasible to obtain a 1:50 k map fully automatically from TOP10NL data. Because the whole generalisation process is implemented in one automated workflow, the workflow can be adjusted to serve other customers. For example cyclists, for which the cycle paths are more important to be kept.

Despite the promising results, the results are not ready to go into production. The topics that were identified in our pilot for further research are listed below.

5.1 *Fine tuning*

Because of the limited timeframe of the pilot, several aspects could not be studied in detail. These have been listed in Section 4.2. For some of these aspects experiments have been done but time was too short to implement the findings in the workflow. Other aspects were identified but not tested at all. For these latter aspects more experiments are needed to show their complexity (how difficult is it to implement a solution) and what are the consequences if it is ignored or done in a simpler way.

5.2 *Evaluating*

Besides optimising the separate steps, no structured evaluating has been carried out (see also Mackaness and Ruas, 2007). Questions that need to be addressed are: how are the results in relation to the complexity of the implementation? How can we quantify and qualify the results, i.e., how can we determine the merits of this approach compared to the existing approach? Which results are below expectation and what are the reasons (poor input data?)?

Where is it easy to gain improvements? How do specific situations behave, e.g., a railway on a dike?

Having the aim of our study in mind, i.e., study the potentials from fully automated generalisation within the context of existing techniques and (new) requirements, two levels of evaluation can be identified.

At first the target map should meet minimal conditions to be a "true" and "readable" representation of reality. This includes keeping consistency (i.e., objects should keep their relative positions) and preservation of the main characteristics of the initial data (i.e., the target data should still reflect reality). In addition the target map should be sufficiently readable to serve its purpose. For example a railway parallel to the road in the initial data should not be covered be the road in the target map. The evaluation should check whether the minimal conditions are completely met.

The second level of evaluation addresses the issue of how acceptable the results are for a user given the fact that (s)he needs to pay for the improvements. The starting point for this evaluation is that the product that is obtained fully automatically (meeting the first level of conditions) will be the main product of the Kadaster and also the product to meet Kadaster's legal obligations. If customers request other products for which (more) interaction is required, customers are asked to pay.

5.3 *Validating*

The model needs to be run in the other test areas and adjusted if needed to validate and improve the results. This will make the implementation more generic. For example it is not clear yet if the current implementation will sufficiently prune dense road- and water network as they appear in the other test areas. It may be needed to enrich the water and road data with more attributes to better thin these networks.

5.4 *Scaling and partioning*

A future study should look at the scalability of the implementation, i.e., how can the workflow be implemented for the whole of The Netherlands? Most probably automated process should not be done in one run for the whole of The Netherlands. Larger areas may contain exceptional situations which were not yet accounted for. In addition some steps (such as displacement) or computationally very intensive, which requires the generalisation process to run in parts.

The workflow can be made scalable by partitioning the whole area of The Netherlands, preferably using topography that never can be displaced such as dikes, large waterways, roads and railways also based on information density of the resulting partitions. Other objects such as secondary roads in different partitions should still fit after the generalisation. See also (Chaudry and Mackaness, 2008; Touya, 2010).

To get insight into partioning potentials, we partitioned the Netherlands using dikes, railways, main roads and highways. This resulted in 1500 areas, of which the single largest was 750 km^2 (western dune area), the rest was all under 500 km^2 with an average of 23 km^2. Having in mind that the test area was 55 km^2, the implementation may well be scalable based on such a partitioning. However more research is required to properly detect dikes and to guarantee that infrastructural objects properly connect at the partition boundaries.

5.5 *Generalising for smaller scales*

Although the research question is broader, in the pilot only the generalisation from TOP10NL to TOP50 was considered. A future study (currently carried out) will extend the results to the smaller scales, i.e., 1:100 k etc. This will identify the action plan for smaller scales and also if implementations realised for TOP50 can be reused. Important is the question on the best approach: the ladder approach where every scale is derived from one scale larger, or the star approach where every scale is derived from the same base data set, i.e., TOP10NL in our case (EuroGeographics, 2005).

5.6 *Generalisation of whole area or updates*

The pilot only focused on the generalisation of the whole area. For future production lines the generalisation of updates should be considered. Since we strive for fully automated solution and the focus is on multiscale maps and not multiscale data, generalisation of whole partitions would be a feasible alternative, given that the processing time is reasonable and the results are not improved by any interaction afterwards.

The result looks promising and therefore currently a follow up research is carried out to evaluate the results in more detail, to scale and validate the results by also applying the workflow on other test areas and to extend the results to the smaller scales.

REFERENCES

AdV, 2007. *Produktblaetter Arbeitskreis Geotopographie*, Arbeitsgemeinschaft der Vermessungsverwaltungen der Laender der Bundesrepublik Deutschland.

Baella, B. & Pla, M. 2005. Reorganizing the topographic databases of the Institut Cartographic de Catalunya applying generalization. In *Proceedings of the 8th ICA workshop on Generalisation and Multiple Representation*. A Coruña, Spain.

Chaudhry, O.Z. & Mackaness, W.A. (2005). Rural and Urban Road Network Generalization Deriving 1:250,000 From 1:1250. In *International Cartographic Conference* pp. 10–16 July Spain.

Chaudry, O.Z. & Mackaness, W.A. 2008. Partitioning Techniques to Make Manageable the Generalisation of National Spatial Datasets, In: *Proceedings of the 11th ICA workshop on generalisation and multiple representation*, 20–21 June 2008, Montpellier.

EuroGeographics. 2005. Generalisation Processes: a benchmark study of the expert group on quality, *EuroGeographics*. Available at: http://www.eurogeographics.org/eng/05_quality_reports.asp.

Foerster, T., Stoter, J.E. & Kraak, M. 2010. Challenges for Automated Generalisation at European Mapping Agencies: A Qualitative and Quantitative Analysis, In: *The Cartographic Journal*, Volume 47, 1, pp. 41–54.

Kadaster, 2005. Generalisatievoorschriften TOP50vector, (Generalisation Regulations), *Topografische Dienst*, Emmen.

Lecordix, F., Gallic, J.L., Gondol, L. & Braun, A. 2007. Development of a new generalization flowline for topographic maps. In *10th ICA workshop on Generalisation and Multiple Representation*. Moscow.

Mackaness, W.A. & Ruas, A. 2007. Evaluation in Map Generalisation Process, Chapter 5 in *Generalisation of Geographic information: cartographic modelling and applications*, edited by W.A. Mackaness, A. Ruas and L.T. Sarjakoski, pp. 89–112. Elsevier. ISBN 978-0-08-045374-3.

Punt, E. & Watkins, D. 2010. user-directed generalization of roads and buildings for multi-scale topography, *13th ICA Workshop on Generalisation and Multiple Representation*, 2010 Zurich.

Stoter, J.E., Burghardt, D., Duchêne, C., Baella, B., Bakker, N., Blok, C., Pla, M., Regnauld, N., Touya, G. & Schmid, S. 2009a. Methodology for evaluating automated map generalization in commercial software, Pages 311–324 In: *Computers, Environment and Urban Systems* Volume 33, Issue 5, September 2009.

Stoter, J.E., van Smaalen, J., Bakker, N. & Hardy, P. 2009b. Specifying map requirements for automated generalisation of topographic data, *The Cartographic Journal* Vol. 46 No. 3 pp. 214–227 August 2009.

Stoter, J.E., Baella, B., Blok, C., Burghardt, D., Duchêne, C., Pla, M., Regnauld, N. & Touya, G. 2010. State-of-the art of automated generalisation in commercial software, *EuroSDR*, p. 231 http://www.eurosdr.net/projects/generalisation/eurosdr_gen_final_report_mar2010.pdf

Thom, S. 2007. Automatic resolution of road network conflicts using displacement algorithms orchestrated by software agents, *10th ICA Workshop on Generalization and Multiple Representation*, Moscow.

Thomson, R.C. & Richardson, D.E. 1999. The Good Continuation Principle of Perceptual Organization applied to the Generalization of Road Networks, *Proceedings of the ICA 19th International Cartographic Conference* (Ottawa, 14–21 August), 1215–1223.

Touya, G. 2010. Relevant Space Partitioning for Collaborative Generalisation, *13th Workshop of the ICA commission on Generalisation and Multiple Representation*, 2010 Zurich.

Van Smaalen, J.W.N. & Stoter, J.E. 2008. Automatische generalisatie van TOP10 NL naar TOP50 NL haalbaar?. In: *Geo-info: tijdschrift voor geo-informatie Nederland* (GIN), 5 (2008) 2, pp. 48–53.

Urban and Regional Data Management – Zlatanova, Ledoux, Fendel & Rumor (eds)
© 2012 Taylor & Francis Group, London, ISBN 978-0-415-67491-1

Volunteered geographic information in the context of local Spatial Data Infrastructures

T.S. Miranda, J. Lisboa-Filho, W.D. de Souza & O.C. da Silva
Universidade Federal de Viçosa (UFV), Viçosa, MG, Brazil

C.A. Davis Jr.
Universidade Federal de Minas Gerais (UFMG), Belo Horizonte, MG, Brazil

ABSTRACT: Geographic information and geo-technologies are essential for the municipal administration to learn the reality of the city and its infrastructure. Spatial Data Infrastructures promote data sharing and access to data by the community in an easy and effective way. However, acquisition and update of spatial data has changed in recent years because it is a time-consuming activity and of costly implementation. In this context, volunteered contribution can complement the efforts for mapping geographic information. This paper proposes an architecture for information systems that receive volunteered geographic information in a SDI at the municipal level. Among the contributions of this work is the adaptation and integration of a SDI architecture to services that include new collaborative components.

1 INTRODUCTION

A municipal administration routinely makes policy decisions that directly affect people's lives (Carrera and Ferreira, 2007). Geographic information is crucial for such decisions, as it assists the government in its various activities, such as urban planning, environmental monitoring, crime prevention, among others. Geographic Information Systems (GIS) are crucial tools to succeed in achieving the desired results when performing these tasks.

In recent years, the production of geographic information has gone through significant changes, especially in urban scales. The growing need for details and the intensive dynamics of urban processes impose a heavy toll on governmental data-gathering initiatives, even more so when the wide variety of GIS data used by local governments is taken into consideration. Traditional techniques are still in use, but the frequency and diversity of updates required to keep the municipal GIS up-to-date can easily scale up towards becoming unfeasible, both technically and financially.

On the other hand, citizen access to geographic information has never been easier. Sources such as digital globes and Web-based mapping sites have become commonplace, along with other geotechnological resources, such as GPS-equipped smartphones and navigation devices. The governmental side of such advances is the emergence of Spatial Data Infrastructures (SDI), a set of technologies, policies and people dedicated to improve geographic information sharing (Goodchild, 2007b). SDIs often materialize partnerships, in which public- and private-sector institutions cooperate in the creation, maintenance and sharing of geographic information (Davis Jr. et al., 2009). Thus, geographic information is currently available for the common citizen from both commercial (Web mapping sites) and public (SDI) sources. Citizens, however, are often aware of the limitations of these data, especially within the geographic scope of their daily experience: the vicinity of their homes, the paths they take to work, frequently visited areas, public places, and so on. Therefore, citizens can be seen as a source of valuable geographic information that, through geotechnological tools and resources, can revise, correct and augment the contents of current geographic data sources.

Volunteered Geographic Information (VGI) (Goodchild, 2007a; McDougall, 2009) is the generic name given to citizen-supported creation and maintenance of geographic data sets. In this paper, we see VGI as a component of SDI, through which users can contribute to the advancement and enhancement of publicly-available geographic data. We are especially interested in VGI dedicated to urban mapping, since we consider that the most important and unique contributions citizens can make are related to their individual interests and first-hand knowledge, and these are mostly related to local GIS applications and data.

This article presents an overview of SDIs, highlighting the context in which they arise and their most important characteristics (Section 2). The phenomenon of Volunteered Geographic Information and its relationship with SDI are presented in Section 3. We then propose, in Section 4, an architecture that integrates volunteered geographic information and SDI. Section 5 presents a case study in which the proposed architecture is implemented and integrated, taking on the form of a Municipal Collaborative SDI. Finally, Section 6 presents conclusions and outlines future work.

2 SPATIAL DATA INFRASTRUCTURES

Spatial data are available from many public and private organizations. Exchange of geospatial data among these organizations requires agreements on standards and an appropriate organizational and technical structure, in an attempt to gain access to new datasets while avoiding duplication of costs associated with data generation and maintenance. The emergence of Spatial Data Infrastructures is a result of that need (Vogel et al., 2004).

An SDI is more than a simple database or dataset. In an SDI, data must be appropriate documented (metadata), there must be means to retrieve, visualize and analyze data (catalogs and Web mapping) and methods that provide data access (Web services) (Nebert, 2004). Moreover, to establish an SDI, Pereira et al., (2009) argue that organizational and institutional arrangements to coordinate and manage the acquisition and interchange of data, based on established standards and adequate politics, are necessary. In recent years,

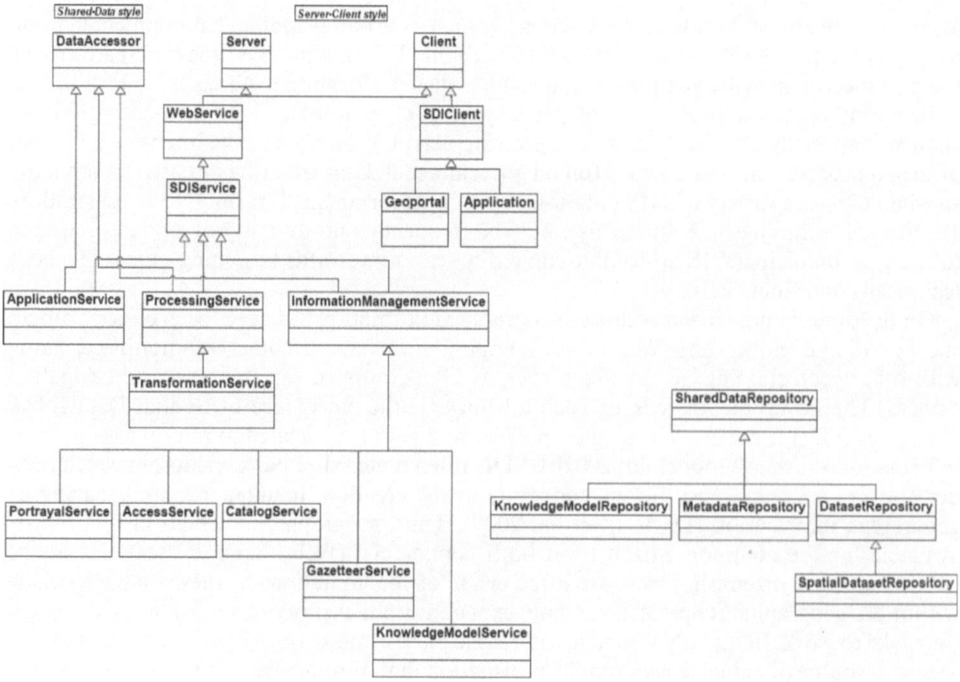

Figure 1. An architectural style for Spatial Data Infrastructures (Béjar et al., 2009, p. 6).

with the advancement of Web-based information systems, SDI have evolved and are being implemented with a clear preference for Service-Oriented Architectures (SOA) (Bernard and Craglia, 2005). Bernard and Craglia (2005) also point out that the adoption of geographic web services by SDI makes them interoperable and distributed from the ground up. There are in the literature some architectural SDI models such as ISO/TC 211 (ISO 19119, 2003), The Open Geospatial Consortium (Whiteside, 2005), The Canadian Geospatial Data Infrastructure Architecture (Geoconnections, 2005), The USA Federal Geographic Data Geospatial Committee (Evans, 2003) and the European Union Directive for the establishment of a European SDI (INSPIRE, 2007). However, these architectural models do not follow any kind of common structure or pattern, making it difficult to know their components, properties and relationships (Béjar et al., 2009). A pattern of architecture for SDI was proposed by Béjar et al. (2009), seeking to capture, unify and systematize the knowledge of existing SDI architectural models, and to clearly define the elements not considered in these models (constraints), or only implicitly considered (data repositories). Figure 1 illustrates this SDI architecture using a UML class diagram. For more information on UML notation, see Fowler (2003).

Davis Jr. et al., (2009) state that the type of sharing that motivates the creation of SDI can be extended to create "communities of practice" in which different actors (scientists, policymakers and citizens) can participate, interact, collaborate and contribute to solving real problems. This paper, therefore, proposes two additional elements to the SDI architecture introduced by Béjar et al., (2009) (described in Section 4), adapting it to the phenomenon of Volunteered Geographic Information.

3 VOLUNTEERED GEOGRAPHIC INFORMATION

According to Hudson-Smith et al., (2009), the world of geographic information is undergoing a transformation based on a connected digital environment, in which data are organized and created by citizens in an increasingly fine geographical scale. Along the same line, Craglia et al., (2008) argue that recent developments such as Web 2.0 and the use of Global Positioning System (GPS) extend the capture of spatial data beyond the exclusive domain of trained professionals, and open new possibilities for citizen involvement. In turn, Goodchild (2007a) believes that every human being is able to capture spatial information on social and environmental phenomena, and that the Internet provides a means for the generated observations to be uploaded and shared with other users.

The process of acquisition, generation, and dissemination of spatial information is expensive. Goodchild (2008) mentions that the United States Geological Survey (USGS) invests approximately $100,000 to produce a topographic map, plus other costs involved in the mapping process, such as satellite launches, leasing of aircraft, and others.

In this scenario, marked by such high investment in production processes and updating spatial data by mapping agencies, the idea of Volunteered Geographic Information (VGI) arises. This expression was proposed by Goodchild (2007a) to define user-generated geospatial content, combining elements of Web 2.0, Collective Intelligence and Neogeography, with the goal of meeting the needs of industry, government, communities and social networks. Coleman et al., (2009) also highlights that, in order to improve the process of updating geospatial data and their own change detection capabilities, these agencies should better exploit the existing potential of volunteer participation, as it can complement the current activities and practices.

Some examples of services and applications focused on VGI contribution using Web 2.0 technology are (1) Wikimapia,[1] a site in which individuals provide place names and descriptions over maps or images; (2) OpenStreetMap,[2] a tool to create urban maps collaboratively

[1]Wikimapia—http://wikimapia.org

[2]OpeenStreetMap—http://www.openstreetmap.org

Table 1. Differences between the characteristics of VGI and SDI based on SDI components.

	Differences	
SDI component	VGI	SDI
Policy	Community of registered users	Formal organizations
Network Access	Bidirectional	Focused on a direction
Patterns	Data	Services, data and metadata
Data	A theme or specific focus	Broad scope of data
People	Large base of non-professional users	Database limited to professional users

and online, using GPS, air photos, and the user's personal knowledge; and (3) WikiCrimes,[3] a Web system that allows accessing and recording criminal incidents directly into a digital map, thus informing people and authorities about crime hot zones, especially for types of events that usually are undernotified by the population.

Flanagin and Metzger (2008) argue that, although the volunteer information has expanded and improved spatial data, we need to be aware of their quality and reliability, since in most cases, data are provided as perceived by non-geographers. Furthermore, when data are volunteered in an anonymous system, vandalism and misinformation sometimes occur. Therefore, there must be some process of evaluating the quality of collaborative data, either through a validation conducted by trained professionals or by volunteer users themselves within the network, among other processes.

Advances in geospatial positioning, Web mapping, cellular communications and collaboration technologies based on Wiki surpassed the initial view of SDI architects, because they allow both experienced professionals and enthusiasts to create and share geographic information on a particular theme. Along these lines, the convergence between SDI and VGI has been explored by recent works (Craglia, 2007; Goodchild, 2007c; Budhathoki, 2008). The general observations indicate that infrastructures gain a broader audience and the possibility of receiving near-real-time data. SDI also facilitate the exchange of geographic information among individuals in a community, and allow the roles of the users to change from simple consumers (and critics) to producers of geographic information. In view of this, Castelein et al., (2010) characterize the VGI phenomenon and explore its relationship with SDI, pointing out differences between them (Table 1).

Finally, for Goodchild (2007c) the most important value of VGI may lie in what it can tell about local activities in various geographic locations that go unnoticed by the world's media, and about life at a local level. The author concludes by pointing out that these are the areas in which VGI may offer the most interesting, lasting and compelling value to geographers.

4 AN ARCHITECTURE FOR VOLUNTEERED GEOGRAPHIC INFORMATION IN SDI

Following the indications found in earlier work, we propose the integration of VGI resources to an SDI setting. The idea is to collect contributions from citizens on top of existing official (i.e., government-cureted) spatial data. Citizens would, through the use of VGI tools, contribute with information such as (i) new layers of popular interest, not usually covered by official cartography, (ii) indication of necessary corrections and updating, (iii) inclusion of additional details of local interest, and (iv) place name registration and geo-tagging. Voluntary contributors would not modify official SDI layers; either citizen contributions are recorded in classes and layers of their own, or contributions serve a source for corrections and updates

[3] WikiCrimes—http://www.wikicrimes.org

by the cartographic officials, which would then check the accuracy of the information before making changes to SDI data. Naturally, metadata for VGI layers should adequately reflect the possible limitations and shortcomings of the volunteered data, informing the user of the nature of the data collection process.

This section presents an architecture to develop such VGI systems integrated with SDI. It aims at ensuring the collection and presentation of information associated with a location, by means of interaction and volunteer participation. Moreover, given the difficulties in keeping the geographic content of an SDI up-to-date, we hope to minimize parts of these problems and expand data collection with collaborative action.

The proposed architecture takes into account the typical infrastructure needs of a local SDI. However, it was developed so that it can be applied in other types of SDI, because in its core it provides the basic general SDI functionality. The architecture can be used in various scenarios for achieving different objectives, within the scope of various applications, such as:

- Providing social assistance by means of mapping disease outbreaks, locations of shelters, schools, and so on;
- Recording complaints or exposing problems of a community;
- Identifying and disseminating information about community services such as police, post office, transportation terminals, and others.

The systems supported by the proposed architecture are characterized by a set of principles originated from the Web: user-generated content, easy content publishing, information sharing, validation, interaction and collaboration. The presence of these concepts seeks to ensure public participation through the Internet in the process of data expansion in a SDI.

UML diagrams are used to describe the functionalities of these systems, particularly use case, sequence and components diagrams. Figure 2 illustrates the use case that addresses services of integration and management of collaborative database contributions.

The use case *Register Collaboration* should be able to register contributions, complement the content of the existing contributions, persist comments and evaluate the contributions published in the collaborative base. Figure 3 illustrates the sequence diagram related to this use case.

The use case *Manage Collaboration* should be able to view the pending contributions in the collaborative base, validate content of contributions, request additional or complementary collaborated information and publish the recorded content. Figure 4 shows the sequence diagram of this use case.

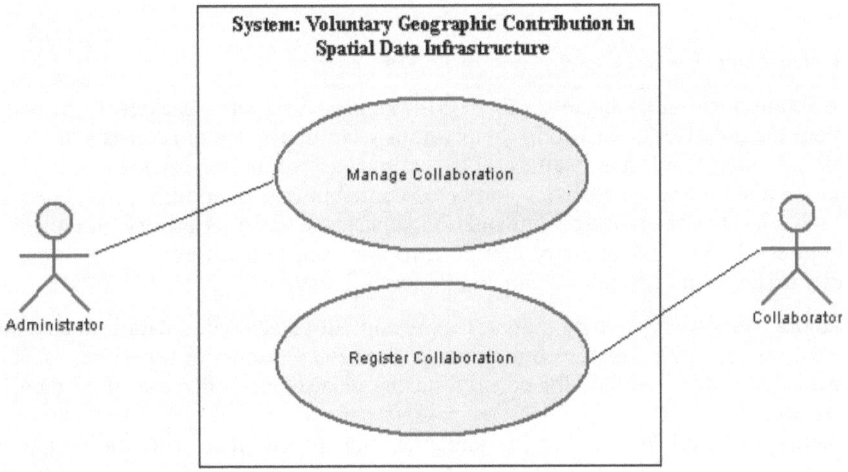

Figure 2. Use case for VGI systems.

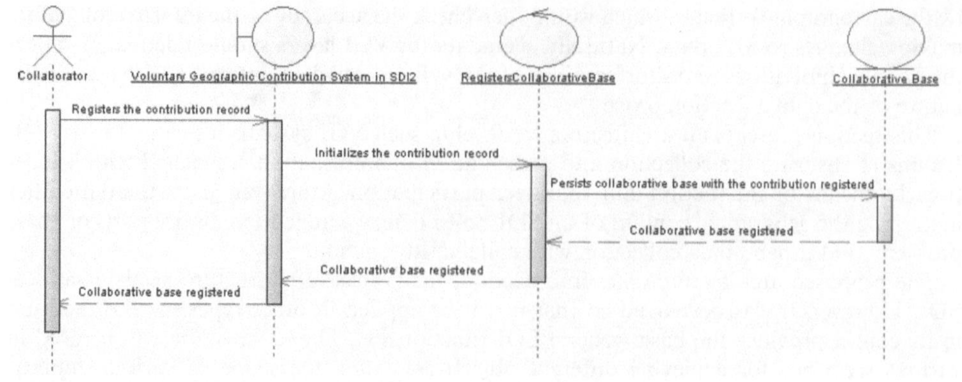

Figure 3. Sequence diagram of use case *Register Collaboration.*

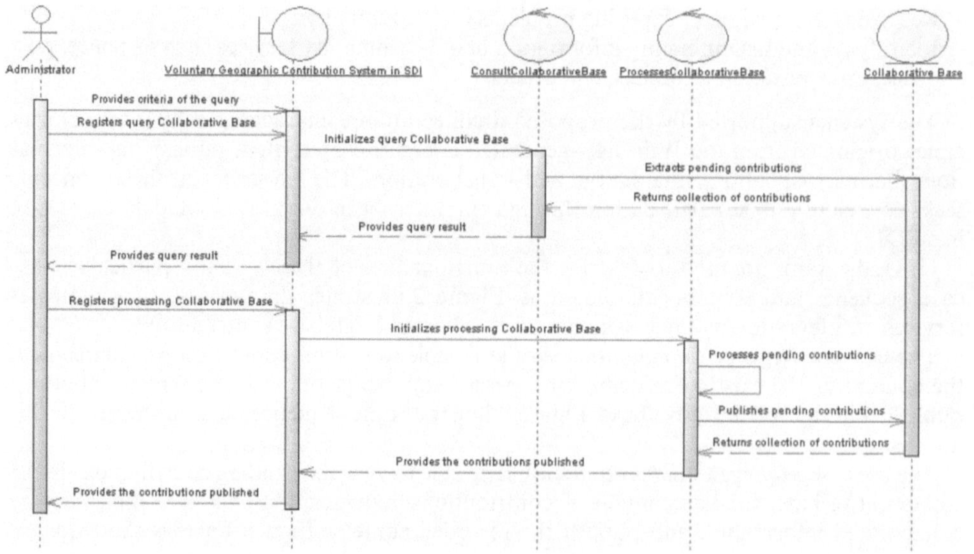

Figure 4. Sequence diagram of use case *Manage Collaboration.*

4.1 *System components*

A set of components that make up the VGI system in SDI was specified to provide and implement the use cases described in the previous section. The system consists of two modules that provide their functionalities to two applications, one for managing collaborated information and another to receive volunteered contributions. In addition, the system architecture has a module that controls information persistence and a repository for collaborative data. Figure 5 shows the diagram of components and their relationships.

A brief description of each diagram component follows:

- Collaborative Base: repository that stores the content of generated data in the collaborative environment, either by the community of user collaborators or user administrators;
- Persistence Module: module for controlling the persistence and retrieval of data in the repository;
- Collaboration Module: module for implementing an interface with the collaboration application and manage the process of contribution generated by the collaborative community;

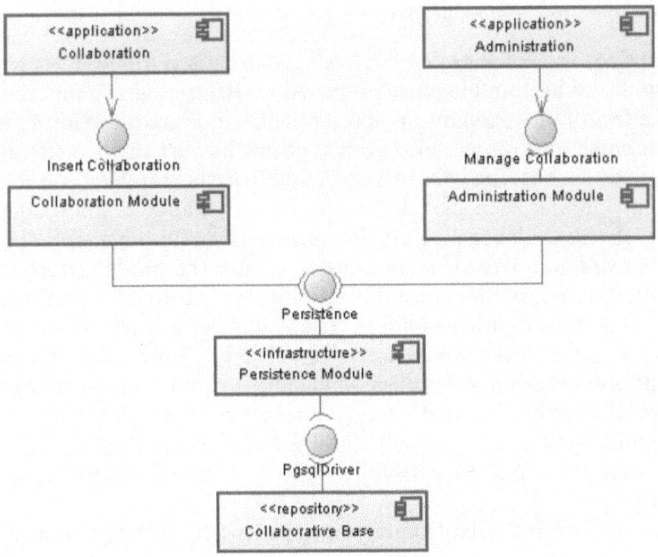

Figure 5. Component diagram of collaborative architecture.

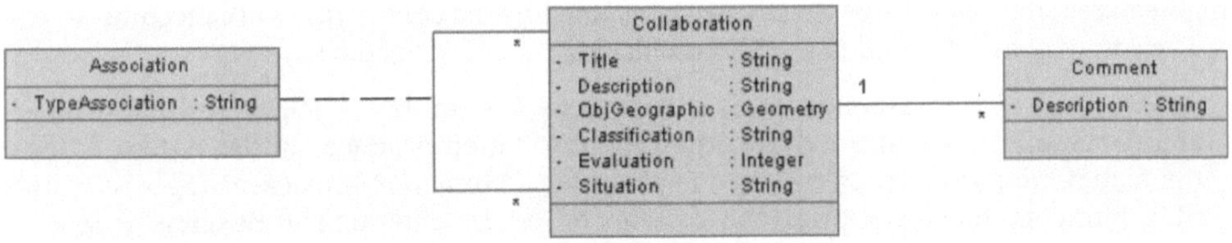

Figure 6. Conceptual schema of collaborative repository.

- Administration Module: module for implementing an interface with the administration application and manage the process of data validation and evaluation carried out by the administrator users;
- Collaboration Application: application for interpreting the functionalities of the collaboration module so that the user can understand and interact with the system. This application displays the geographic information and allows the user to browse, add and interact with the content existing in the environment;
- Administration Application: for interpreting the functionalities of the administration module so that the user can understand and interact with the system. This application displays the information generated by the collaborator and allows the administrator to approve and publish the contributions registered in the system.

A repository for collaborative data was designed for the system. Figure 6 shows its conceptual schema which is based on the set of information required for the functions described earlier. This conceptual schema was later mapped to tables in a relational database.

The function of each object in this conceptual schema is described below:

- Collaboration: stores the contents of the community-generated collaboration about a theme of interest. In addition to its description, it also has a classification, assessment and a situation for availability. The information complement of a collaboration is characterized by an object "Association";
- Comment: stores community-generated content related to collaborations of interest, for example, more information, reviews, and opinions about the contribution.

4.2 *SDI architecture extended to VGI*

This section describes the collaborative architecture proposed for VGI systems and present its integration with the SDI architecture proposed by Béjar et al., (2009), described in Section 2. The objective of this integration is to enhance and adapt the infrastructure so that it can provide support and receive VGI contributions as part of its contents. In addition, this adaptation complements the vocabulary offered to system architects when designing the infrastructure that contains such specificities, ensuring exchange of knowledge about them.

The structure of this collaborative architecture consists of three tiers (Figure 7): *Data*, *Business* and *Presentation*. This division aims to ensure the modularized development of the system, facilitating its maintenance, extensibility by means of new functionalities and reusability of its functions in other modules. The Data Tier is made up of the collaborative data repository produced by the volunteer community. The Business Tier is made up of the *Collaboration* and *Administration* Modules, which implement business classes, as well as Web services that allow the application customers to interact with the system. The business classes contain the system functionalities. Finally, the Presentation Tier interprets the functionalities of business classes and Web Services for a more user-friendly interface, so that users can interact with the system.

The two new collaborative architecture components (the collaboration module and the administration module) are coupled and added to the SDI architecture. Figure 8 illustrates the integration between the architectures. The relationships directed from the collaborative architecture to the SDI architecture correspond to the new components inserted in the infrastructure. Conversely, the relationship directed from the SDI architecture to the collaborative architecture indicates services that can be consumed by the collaborative architecture.

The components included in the SDI architecture are described below:

- *Application VGI*: corresponds to the collaborative system that is able to register, manage and communicate volunteered geographical information produced in the system. It uses the map service *PortrayalService*, available in the *InformationManagementService* of the SDI. This component comprises the elements of the Presentation and Business tiers;
- *Collaborative Repository*: corresponds to the repository of volunteer data produced in the collaborative system. It stores the type of spatial object and its coordinates, as well as the descriptive information provided to the object of interest. This component includes the element present in the Data Tier.

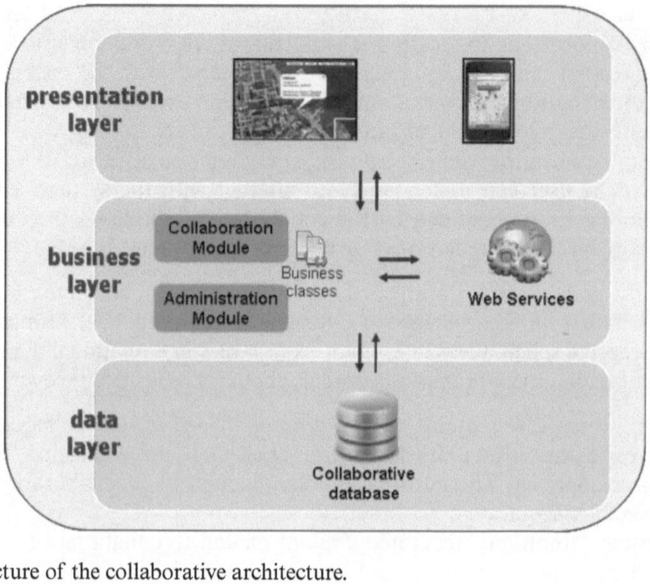

Figure 7. Structure of the collaborative architecture.

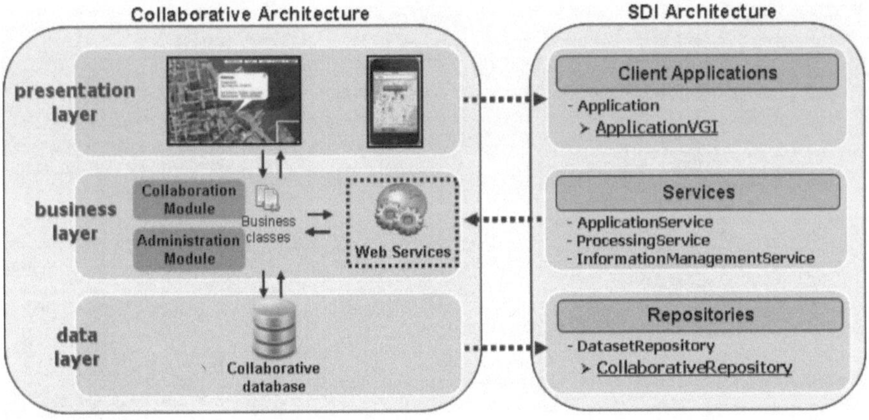

Figure 8. Integration between collaborative architecture and SDI architecture.

5 CASE STUDY

This section presents a case study in which this work was applied, describes the development of the Collaborative SDI and its computing artifacts, and details the collaborative process associated with the VGI system linked to SDI.

This case study is related to a project of the municipality of Viçosa (Minas Gerais state, Brazil), named as *Viçosa-Digital—Databases and Municipal Information*. The goal of Viçosa-Digital is to build and share georeferenced data and municipal information so that the design, planning, implementation and management of local and regional projects may take place with quality and efficiency.

The project portfolio consists of three themes: *cartographic base* (data on the physical-territorial aspects such as streets, buildings, rivers, etc.), *social economics* (information on population distribution, living conditions, employment and income, etc.), and *environment* (information on vegetation and conservation areas, climate, water quality, soil and air, etc). All this content is registered and stored in various formats, including vector and raster spatial data, theses, articles, photos, spreadsheets, and so on. The task of cataloging and making this dataset available was also part of the development and implementation of the Collaborative SDI for Viçosa-Digital. The next section presents the Collaborative SDI and its computing artifacts.

5.1 *Collaborative SDI Viçosa-Digital*

The Collaborative SDI was developed to support the sharing of geospatial data produced in Viçosa, Minas Gerais State. It was necessary to establish a structure to make its contents available and accessible. Interoperability is the highlight of this proposed SDI, because technological compatibility is a concern when it comes to the exchange of spatial data. The standards and specifications proposed by the Open Geospatial Consortium (OGC) were adopted to meet this requirement, as they enable and simplify the interoperability of geospatial content through Web services. Figure 9 shows the structure of the Collaborative SDI Viçosa-Digital including the volunteer collaboration system and its repository, which were described in the previous section.

The dissemination of geospatial data by an SDI is also very much related with the proper documentation of its information. Thus, the pattern of geospatial metadata that was chosen for this infrastructure is the MGB profile (a Brazilian standard for geospatial metadata), which is based on ISO 19115:2003 and has been adopted by INDE,[4] the Brazilian National Spatial Data Infrastructure.

[4]INDE—http://www.inde.gov.br/

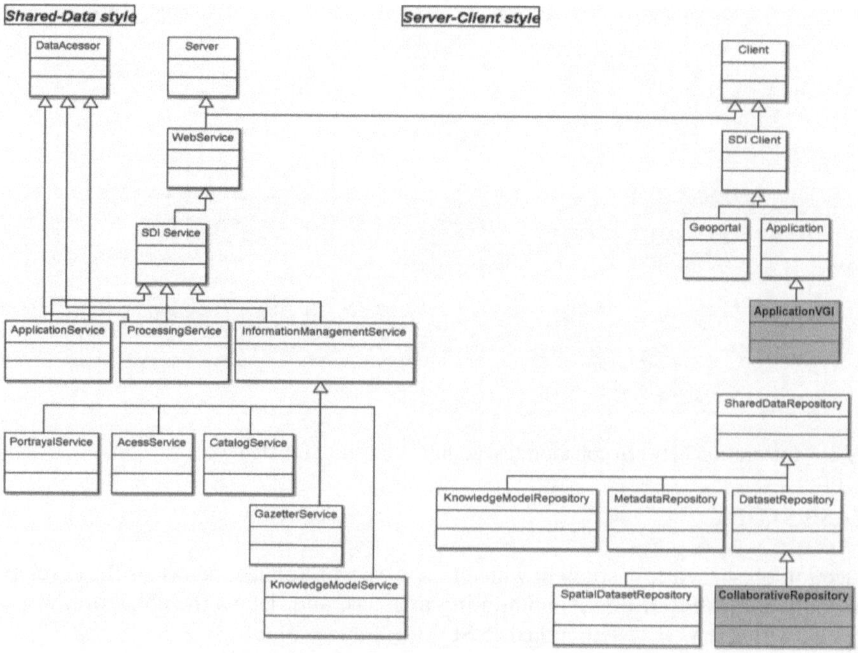

Figure 9. Collaborative SDI Viçosa-Digital (adapted from Béjar et al. (2009)).

Table 2. Collaborative SDI components and its computing artifacts.

Component of SDI collaborative	Computational artifact
Geoportal	Website
Application	I3Geo, Openlayers
ApplicationVGI	Collaborative system
Portrayal service	MapServer
Access service	
Catalog service	Geonetwork
Metadata repository	PostgreSQL
Dataset repository	
SpatialDataset repository	
Collaborative repository	

The collaborative infrastructure comprises a collection of computing artifacts that facilitated and optimized its development and implementation. All fall under the category of free software, which include: I3Geo[5], MapServer,[6] PostgreSQL[7] and Geonetwork.[8] Access and metadata description are provided by Geonetwork. The metadata repository is maintained by PostgreSQL. MapServer ensures the implementation of OGC Web services, while the visualization and analysis of spatial data is provided by I3Geo's viewer.

Table 2 shows the relationship between each Collaborative SDI component and its respective computing artifact.

[5]I3Geo—http://softwarepublico.gov.br/
[6]MapServer—http://mapserver.org/
[7]PostgreSQL—http://www.postgresql.org/
[8]Geonetwork—http://geonetwork-opensource.org/

Figure 10. Initial interface of the Viçosa-Digital Geoportal.

Maguire and Longley (2005) define a Geoportal as a key component in an SDI, because it promotes the organization and provision of geospatial services and content of the infrastructure. Zevenbergen et al., (2009) argue that, besides being a thematically organized environment, a Geoportal must include well described and complete metadata, in order to facilitate the location, evaluation and use of geospatial data sets. Thus, a Geoportal for Collaborative SDI Viçosa-Digital was developed, providing the community with tools for searching, accessing, online collaboration and visualization of geographic data, along with the description of such data through metadata. Figure 10 illustrates the interface of the Viçosa-Digital Geoportal.

5.2 *The volunteered geographical information module*

In the context of the Viçosa-Digital project, a collaborative system prototype was implemented to assess the usefulness of the proposed collaborative architecture. Its development employed open source, freeware and multiplatform components. Its functionalities were implemented using the programming languages PHP, AJAX and jQuery, while the CSS language was used in the interface design. Web access to the prototype is provided by the Apache Web server, and geographic data display is accomplished by OpenLayers. The collaborative data repository is stored in the PostgreSQL RDBMS. Table 2 lists these computational artifacts.

A key feature of the prototype is the spatial data viewer, because the functions for registration and management of contributions are based on a geographic reference. Thus, the spatial data displayed in the prototype are provided by MapServer using WMS (Web Map Service), which corresponds to the *PortrayalService* of the Collaborative SDI. This service confirms the interoperability aspect of the collaborative architecture, since it allows access to a set of spatial data using an XML (eXtensible Markup Language) specification. This architecture also allows the use of data from a spatial infrastructure provided by other organizations. This can be accomplished through their API. Examples of some organizations that provide access to their data are: Google Maps, Yahoo Maps, Microsoft Bing Maps, and others.

SDI allow access to data sources, metadata and geoservices of a given domain. Among data provided in a Municipal SDI are those of the Cartographic Base, which correspond to the physical-territorial aspects of a municipality. These aspects change and evolve over time; hence it is necessary to update them.

Thus, a collaborative geographic system invites and allows citizens to act as volunteer sensors (Goodchild, 2007c), i.e., to use their knowledge allied to aspects of collaboration and participation promoted by the Web and contribute to the improvement and expansion of geographic data of the municipality.

The developed prototype operates on spatial data of Viçosa, available in its Geoportal. Then, a map with layers for "Municipal Limit", "Blocks", "Streets" and "Districts" was used to support the collaboration process.

5.3 *Flow of collaboration mechanism*

The set of functionalities available in the prototype characterizes the flow of collaboration around the VGI system in an SDI. Figure 11 shows a diagram to elucidate the flow of execution of activities from beginning to end of cooperation, and explain the profiles of individuals involved in the mechanism.

The execution of activities depicted in the flow of collaboration is related to the level of knowledge and responsibility of the roles involved in the collaborative process. The profiles that make up the mechanism for collaboration are:

- Collaborator: users that feed the system by registering their contributions. Their level of activity is limited to the registration or to supplementing the information in a contribution;
- Administrator: users that manage the contributions recorded in the system. Their level of activity requires a better knowledge about the geographic information of the domain they administer;
- Producer: users that produce geographic data. These users receive communication from the administrators of the collaborative system about a new relevant input (e.g., new point of interest, indication of errors, among others).

The flow of activities within the collaboration mechanism is divided into three stages: *Collaboration*, *Management* and *Communication*. Each of these stages has one or more specific activities that are the responsibility of each role. These activities include:

- Contribution: volunteered geographic information created or supplemented in the system;

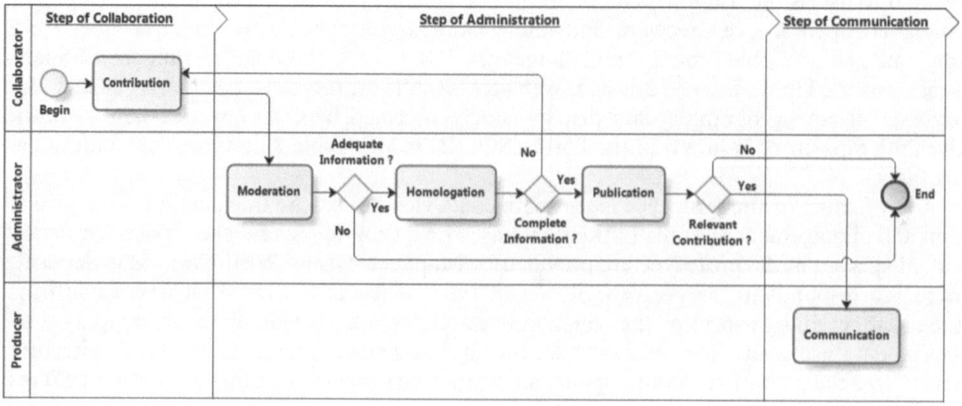

Figure 11. Flow of collaboration mechanism.

- Moderation: the first stage of assessment by the Administrator of a new contribution registered in the system. In this activity, for example, the administrator verifies whether the content of the provided information is inadequate or inappropriate;
- Homologation: the second stage of the assessment by the Administrator of a contribution registered in the system, but now he verifies whether the information content is sufficient to characterize it. The contributor may be required to supplement the information on the registered contribution;
- Publication: publishing the newly accepted contribution in the system, so it can be viewed by the community;
- Communication: informing the Producer of the geographic data about new contributions provided by the community.

5.4 *The collaborative prototype*

The prototype has two interfaces, one for data administration and another for registering contributions. This separation characterizes the different environments according to system functions and their respective user profiles. Figure 12 shows the interface of the Collaboration Module.

In this interface, volunteers register their contributions (point, line or polygon) on a spatial data viewer, i.e., the contribution is associated with a location. The collaboration interface also lists some indicators generated by the community involved in the collaborative process. These indicators improve the content existing in the system, because the community itself performs a self-management on the negative contributions in the system. Examples of indicators include:

- Most active contributors: users with the largest number of accepted contributions, i.e., those who collaborate best achieve recognition by the system;
- Best assessed contributions: contributions whose content is considered interesting or important by the community, and therefore deserve mention; for example, a contribution about a street whose traffic direction was recently changed;

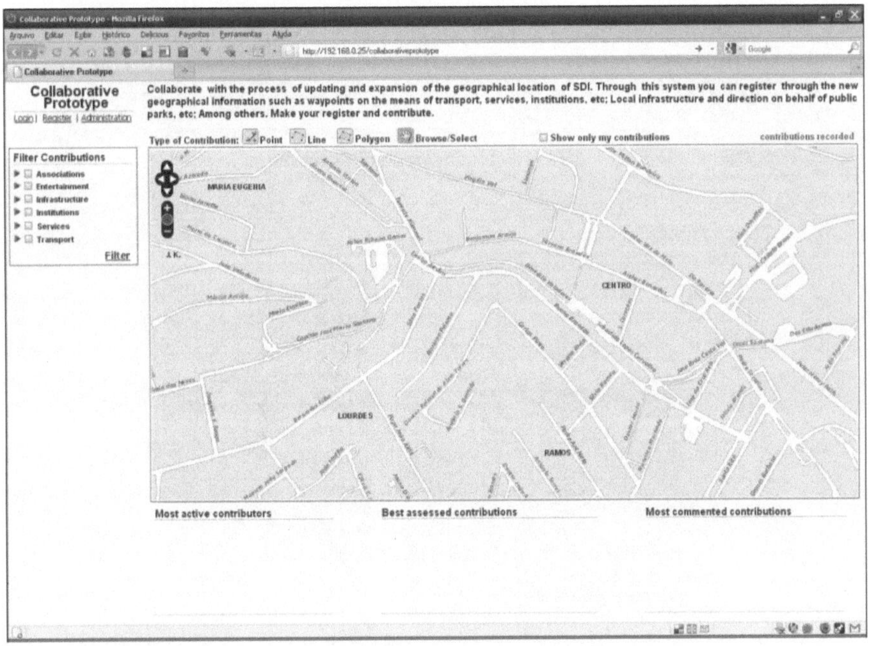

Figure 12. Interface of the collaboration module.

- Most commented contributions: similar to the previous section, they are contributions about which the community commented the most, regardless of whether comments are praising, criticizing, and even suggesting more details on the data. Comments are important because they can help to refute a false contribution.

To register a contribution to the collaborative system, the user must complete a set of descriptive fields arranged in three forms, with each form related to a specific theme: "Identity", "Multimedia" and "Comments". In addition to organizing the information to be recorded, this division seeks to guide the user through the correct sequence for filling out their contribution, because the last two forms are dependent on the first. Figure 13 illustrates the interfaces of "Identity" and "Multimedia" forms.

The Administration Module's interface is illustrated in Figure 14. In this interface the administrator queries and evaluates the content of the contributions registered by the community in the system. As described in section 5.3, using the module of activities of the cooperation flow, the administrator can: check whether the information provided by contributors is appropriate, i.e., if improper or inadequate terms have been used; check whether the description of the contribution is complete, i.e., if provided data are enough to understand

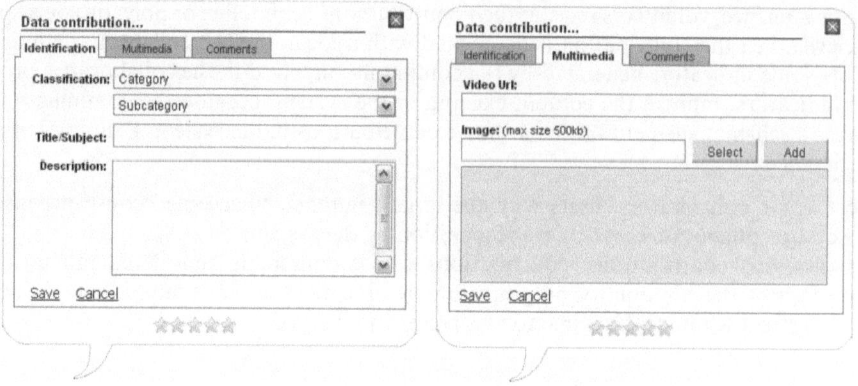

Figure 13. Interfaces of collaboration forms: Identification and Multimedia.

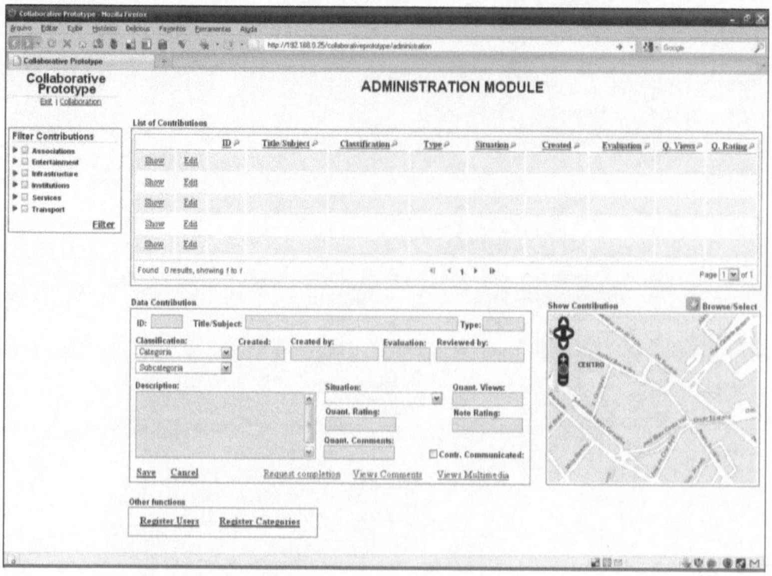

Figure 14. Interface of the prototype related to the Administration Module.

and locate the contribution geographically; request supplement to information of incomplete contributions, publish or refute the contribution in the collaborative environment, making it accessible to the community; and communicate a new important contribution identified by the community to the data producer.

6 CONCLUSION

This paper described an architecture for information systems that receive volunteered geographic information in a service-oriented Spatial Data Infrastructure at the municipal level. Through volunteered contributions, the original SDI content can be enhanced with new data layers, indications of errors and points that need updating, and more detailed local coverage.

The main contributions of this work are: (1) the integration and adaptation of a SDI architecture with components of collaborative architecture; (2) the specification and implementation of a municipal VGI-enhanced SDI; (3) the specification of a collaborative system for the registration and administration of volunteered geographic information on SDI data; (4) the implementation of a collaborative prototype that shows the feasibility of the proposed system for the expansion of a geographic database.

The prototype implementation of this collaborative architecture use simple functional resources known in the interactive environments on the Web, in order to make them more user-friendly. All contributions take place on a geospatial data viewer that uses SDI data and mapping services. In addition, the storage of the contributions is ensured by the proposed collaborative data repository.

Finally, the extension of the SDI architecture, proposed by Béjar et al., (2009), with the new collaborative components, was implemented and validated in the Viçosa-Digital project. Therefore, this extension allows the use of the Collaborative SDI architecture of in other levels such as regional and global.

As future work, we intend to investigate the feasibility of implementing all collaborative data gathering using OGC Web services such as WFS-T, thus improving on the interoperability aspects of the prototype.

ACKNOWLEDGEMENTS

This project was partially funded by the Fundação de Amparo à Pesquisa do Estado de Minas Gerais (FAPEMIG) and the Conselho Nacional de Desenvolvimento Científico e Tecnológico (CNPq / MCT / CT-Info).

REFERENCES

Béjar, R., Latre, M.Á., Nogueras-Iso, J., Muro-Medrano, P.R. & Zarazaga-Soria, F.J. 2009. An architectural style for spatial data infrastructures. *International Journal of Geographical Information Science* 23(3):271–294.

Bernard, L. & Craglia, M. 2005. SDI: from spatial data infrastructure to service driven infrastructure. In: 2nd Research Workshop on Cross-Learning between Spatial Data Infrastructures (SDI) and Information Infrastructures, *Proceedings* ..., Enschede: Holanda.

Budhathoki, N., Bruce, B. & Nedovic-Budic, N. 2008. Reconceptualizing the role of the user of spatial data infrastructures. *GeoJournal* 72(3–4):149–60.

Carrera, F. & Ferreira, Jr., J. 2007. The future of spatial data infrastructures: capacity-building for the emergence of municipal SDIs. *International Journal of Spatial Data Infrastructures Research* 2:54–73.

Castelein, W.T., Grus, L., Compvoets, J.W.H.C. & Bregt, A.K. 2010. A characterization of Volunteered Geographic Information. In: 13th Agile International Conference on Geographic Information Science, *Proceedings* ..., 10–14 May, Guimarães: Portugal.

Coleman, D.J., Georgiadou, Y. & Labonte, J. 2009. Volunteered geographic information: the nature and motivation of produsers. *International Journal of Spatial Data Infrastructures Research* 4:332–58.

Craglia, M. 2007. Volunteered geographic information and spatial data infrastructures: when do parallel lines converge? Available in http://www.ncgia.ucsb.edu/projects/vgi/ docs/position/Craglia_paper.pdf. Accessed in December 2010.

Craglia, M., Goodchild, M., Annoni, A., Câmara, G., Gould, M., Kuhn, W., Mark, D. & Parsons, E. 2008. Next-Generation Digital Earth. *International Journal of Spatial Data Infrastructures Research* 3:146–167.

Davis Jr., C.A., Fonseca, F.T. & Câmara, G. 2009. Infraestruturas de dados espaciais na integração entre ciência e comunidades para promover a sustentabilidade ambiental. In: I Workshop de Computação Aplicada à Gestão do Meio Ambiente e Recursos Naturais, Bento Gonçalves, *Proceedings*..., Porto Alegre: Sociedade Brasileira da Computação.

Evans, J.D. 2003. *A geospatial interoperability reference model, 1.1*. FGDC Geospatial applications and interoperability working group. Available in http://www.nonmonotonic.net/ ISOTEIA/metadata/ girm1.1.pdf. Accessed in November 2010.

Federal Geographic Data Committee—FGDC. 1998. *Content standard for digital geospatial metadata*. Washington: Federal Geographic Data Committee. p. 78.

Flanagin, A.J. & Metzger, M.J. 2008. The credibility of volunteered geographic information. *GeoJournal* 72(3–4):137–48.

Fowler, M. (3 ed.) 2003. *UML Distilled:* A Brief Guide to the Standard Object Modeling Language. Addison Wesley Longman Publishing Co.

Geoconnections. 2005. *The Canadian geospatial data infrastructure*. Architecture description version 2.0, Canada. Available in: <http://www.geoconnections.org/publications/ tvip/arch_E/ CGDI_Architecture_final_E.pdf>. Accessed in November 2010.

Goodchild, M.F. 2007a. Citizens as voluntary sensors: spatial data infrastructure in the world of Web 2.0. *International Journal of Spatial Data Infrastructures Research* 2:24–32.

Goodchild, M.F. 2007b. Citizens as sensors: Web 2.0 and the volunteering of geographic information. *Geofocus* 7:8–10.

Goodchild, M. 2007c. Citizens as sensors: the world of volunteered geography. *GeoJournal* 69(4): 211–221.

Goodchild, M.F. 2008. Commentary: whither VGI? *GeoJournal*, 72(3–4):239–344.

Hudson-Smith, A., Crooks, A., Gibin, M., Milton, R. & Batty, M. 2009. NeoGeography and Web 2.0: concepts, tools and applications. *Journal of Location Based Services* 3(2):118–45.

Inspire. 2007. Infrastructure for spatial information in the European community. European parliament and the European council. *Directive of the European Parliament and of the Council establishing an Infrastructure for Spatial Information in the European Community*. Available in. http://register. consilium.europa.eu/pdf/en/06/st03/st03685.en06.pdf. Accessed in December 2010.

International Organization for Standardization—ISO. 2003. *ISO 19115*. Genève, Switzerland: International Organization for Standardization.

International Organization for Standardization—ISO. 2003. *ISO 19119*. Genève, Switzerland: International Organization for Standardization.

Maguire, D.J. & Longley, P.A. 2005. The emergence of geoportals and their role in spatial data infrastructures. *Computers, Environment and Urban Systems* 29:3–14.

McDougall, K. 2009. The potential of citizen volunteered spatial information for building SDI. GSDI 11, *Proceedings*..., Rotterdam, Netherlands.

Nebert, D.D. 2004. *Developing Spatial Data Infrastructures*: the SDI Cookbook, Version 2.0. GSDI-Technical Working Group.

Pereira, G.C., Davis Jr., C.A. & Rocha, M.C.F. 2009. Establishing a sub-national SDI in Bahia State (Brazil)—its limits and possibilities. In: 27th Urban Data Management Symposium. Ljubljana: Slovenia.

Rajabifard, A. & Williamson, I.P. 2001. Spatial data infrastructures: concept, SDI hierarchy and future directions. In: GEOMATICS'80, *Proceedings*..., Tehran: Iran.

Vogele, T. & Spittel, R. 2004. Enhancing spatial data infrastructures with semantic web technologies. In: 7th Agile International Conference on Geographic Information Science, *Proceedings* ..., Heraklion: Greece.

Whiteside, A. 2005. *OpenGIS best practices paper*. [S.l.]: Open Geospatial Consortium Inc. (No. OGC 05–042r2).

Zevenbergen, J., Koerten, H., Donker, F.W., Bulens, J., Kuyper, M. & Jellema, M. 2009. Geopor-tal network—more process catalyst than project. In: 27th Urban Data Management Symposium. Ljublijana: Slovenia.

Urban and Regional Data Management – Zlatanova, Ledoux, Fendel & Rumor (eds)
© 2012 Taylor & Francis Group, London, ISBN 978-0-415-67491-1

Collaboration and participation: Exploratory research of web interaction

G. Corso Pereira
Federal University of Bahia, Salvador, Brazil

M.C.F. Rocha
PRODEB, Salvador, Brazil

ABSTRACT: The paper presents some considerations on how to improve the dialog between Public Administration and citizens based on the research project "Internet and Interactivity for Public Participation" (2i2p project) coordinated by PRODEB—Data Processing Company of Bahia State—and conducted in partnership with Federal University of Bahia. The project called for reflection and innovation in public websites design to convey spatial information and encourage interaction between citizens and Public Administration. The 2i2p project used social media as experimentation for different purposes. An experimental site was implemented iteratively, using Google Maps APIs. Interface evolved to increase the level of user interaction with geographic information. The final product was a guide that offers recommendations for Public Administration Websites. It works towards greater transparency, participation and collaboration for innovation in Public Administration, broadening the perspective of e-Government to incorporate the citizen in the policies definitions for the territory.

1 INTRODUCTION

States are mostly still organized according to typical models of industrial era. Authorities continue to view with suspicion citizens, keeping them at distance and limiting its participation to voting or participating in administrative decisions. As Belisario (2010) noticed, the introduction of information and communication technologies (ICT) in those processes retains the old bureaucratic model. The Open Government movement represents a break in this trend.

The policy of Open Government from Obama administration established in 2009 a normative-organizational model for Public Administration based on three pillars: a) transparency: it must provide information about the Public Administration work, b) participation: it should encourage citizens to participate in decision making since the collective intelligence can improve the institution's decisions, c) collaboration: it must revise organizational models to ensure continued collaboration between public entities, nonprofits organizations, individuals. The possibility of using Internet to allow citizens to talk to each other, help each other and cooperate with the government is in the basis of this revolution (Belisario, 2010).

Considering that governance activities derive and are built on shared interests and goals, the State must coordinates the interests involved and take charge of defining the context involving the participatory process (Debernardi & Rosso, 2007). It should also stimulate the action monitoring and promote the resolution of any conflicts between stakeholders (Debernardi & Rosso, 2007). The use by citizens of interactive Web 2.0 platforms for reporting problems would be one of the most effective monitoring systems with higher capillarity, allowing the Administration to act more quickly and timely (Belisario, 2010).

The site FixMyStreet (http://www.fixmystreet.com/) was one of the first to provide something like that. Belisario (2010) cites similar experiences in Italy: the e-Part (http://www.epart.it/) and the Iris Project (http://iris.comune.venezia.it/) from City of Venice.

Earlier in this century, Laurini (2001) mentioned that many countries proposed the use of computerized systems to support communication and interaction between actors of the urban planning process—whether technical, political or citizens, either alone or through the their representative groups. Beyond merely exercising the right to information and giving opinion when asked, the public could participate in decision making in diverse degrees: defining issues of interest, analyzing impact of their decisions, recommending solutions, and finally even being able to avoid implementation of the proposed intervention (Laurini, 2001).

To facilitate participation in territorial issues, systems would be used to visualize the area under discussion through different media (maps, photos, thematic data) in an attempt to build a collective understanding of their characteristics. Geographic Information Systems (GIS) could be used as tools to support decision processes involving multiple participants with different points of view. The definition of such usage as participatory GIS or PPGIS it is from 1998 (Nembrini & Joerin, 2002).

Today, after the first decade of this century, one can say that more than instruments and tools that provide digital representation of space, what actually becomes really effective to discover, assimilate, incorporate local knowledge and facilitate public participation on urban planning process is the availability of these resources in an interactive and an open network. Internet and Web 2.0 applications make it feasible at low cost.

In 2009 and 2010, the Data Processing Company of Bahia State—PRODEB adressed these questions by conducting a research project named "Internet and Interactivity for Public Participation" (2i2p) in partnership with the School of Architecture at UFBA—Federal University of Bahia, with support from Brazilian public organizations for research funding (CNPq and FAPESB). It intended to establish principles and recommendations for websites design focused in to expand the dialogue between Public Administration and citizens about actions on the territory. The use of geographic information in the dialogue between government and citizen was addressed by the research project. The project identified and used free platforms, applications and services for interaction through Internet (Web 2.0) and emphasized the user experience as a way to improve the usability of the sites.

This article will address some aspects of designing the dialogue between Public Administration and citizens in a specific usage: the Internet as a resource to access, create and disseminate collective intelligence and local knowledge.

2 INTERNET, INTERACTIVITY AND COLLABORATIVE MAPPING

The project has taken the idea of network (Web) as a dominant cultural interface capable of mediating sources of culture, information and knowledge. It considers Internet as a cultural creation where users play an increasingly important role in creation of new management methods and knowledge sharing.

Nowadays the network experience deepens, multiplies the number of users that self-taught, through it, experiment, collaborate, develop relationships and social practices of design and construction of knowledge. For this they make use of both, applications and services affiliated with Web 2.0, among them are blogging, social networking and wikis.

According to Monaci:

> Within these environments the prevailing idea is knowledge as dialogue, exchange and shared rewriting. Heirs of the virtual communities typical of origins of web, web 2.0 tools are inspired by open source philosophy. The transparency and the logic of exchange and sharing knowledge are used as regulative principle of participation and experience for on-line environments (Monaci, 2008, authors' translation).

In the context of establishment of the Net as the dominant cultural interface, governments are starting use these interactive elements to enhance their communication, to inform, to provide services or to develop a dialogue with citizens.

Examples of what it is now called Government 2.0 are: in early June 2009, the mayor of San Francisco (California) announces that the citizens can use Twitter to request services or make complaints, for example, garbage, street cleaning and walls, including shipping photo via cell phone. The City of Turin participates in the SanpaBlog project (http://www.sanpablog.it), an urban blog (or local blog, metro blog, regional blog, to handle matters of strictly local interest) that supports and encourages participation for those who live and enliven the neighborhood of San Paolo. Turin, incidentally, uses very well the capabilities of Facebook to provide useful information and interact with their large number of fans.

Projects of platforms for more interactive use of geospatial information for planning over the territory was launched in 2008 in Salzburg, Austria. This city was elected to be the first pilot project conducted by Autodesk Digital City, intended to provide a collaborative environment for visualization, analysis and simulation impacts in urban planning.

Still far from projects that involves more processing capacity, use and integration of technologies, a number of applications has emerged in the context of Web 2.0 to allow users to provide data and contribute to the description of geographic issues. This is the case of MapTube (http://www.maptube.org/map.aspx?mapid=195) site for sharing thematic maps without costs proposed by the Centre for Advanced Spatial Analysis at University College London (Hudson-Smith, 2009), and Wikimapia, which provides editing capabilities for collaborative mapping. A significant example of the possibilities of this phenomenon is the OpenStreetMap project, which made possible to build a cartographic database updated by volunteers in a few days (http://haiti.openstreetmap.nl/) in the case of Haiti Earthquake in January 2010.

In Brazil, an interesting example is the Citix (http://citix.terra.com.br) application developed by CESAR—Centre for Advanced Studies and Systems of Recife (PE), with support from the Federal Public Ministry (MPF) that allows geocoding of events and inclusion of reports by users of what is happening in the city of Recife. Another example, which uses Google Maps, is Urbanias (http://www.urbanias.com.br/), a non-governmental site that receives and publishes complaints about public services in São Paulo.

The phenomenon of user generated content on the Web was named volunteered geographic information (VGI) by Michael F. Goodchild in an article published in 2007 about the growing engagement of large numbers of individuals in the creation of geographic information, regardless of their formal qualification.

Google's contribution to the process of "democratization of GIS" is undeniable, as it turn geographical information available, and provide a number of features that previously could only be found in sophisticated GIS software and it was not easily accessible, either by it's cost or by lack of necessary skills. As Goodchild (2008) said, someone using Google tools do not need to make decisions about sources, symbolizations, projections or other issues related to cartographic practice, since the decisions were made when the software was designed.

This fact, according to him (2007), allowed the growth of VGI, along with:

a. the possibility offered by Web 2.0 that allow users access and edit content of databases created by others;
b. the possibility of knowing the exact geographic location of objects and events, by obtaining coordinates in standard systems with tools increasingly accessible, as is the case of GPS (Global Positioning System);
c. the diffusion of the Internet with ability to provide dynamic visualization of three-dimensional objects.

Common users can now contribute voluntarily to collect geographic information, often without training or expertise, and in these way to assemble a list of geographical names (gazetteer) in Wikimapia and to describe entire cities, towns, buildings, sights, everything that can be defined in space by geographical coordinates.

OpenStreetMap (http://www.openstreetmap.info/) is an example of successful experience in spatial data voluntarily provided to the collaborative construction of cartographic database. The OpenStreeMap proposal is to build a world map that can be used by everyone, and especially to aggregate data from people all over the world in a single geographic database.

According to Goodchild (2007), the phenomenon of low-cost production of geographic information stated by its creator, without any mention or reference to another authority, has the potential to redefine the role of traditional mapping companies and agencies, and even the role traditionally assigned to the specialist responsible for creating cartographic data.

To acquire, compile, print and disseminate geographic information is costly, which favors inevitably to the production by those agencies of the kind of information with less need to update and that may have multiple applications. The existence of gaps in the coverage of geographic information provided in this way is inevitable, according to the author (Goodchild, 2008).

Brazil is certainly a good example of this: with the time taken to structure national and sub-national SDIs, with projects that started more than a decade ago and not yet operational, (Pereira, Davis & Rocha, 2009) difficulties to access official data, mostly outdated, grew.

3 THE USE OF INTERNET TOOLS

During the 2i2p project an experimental site was implemented iteratively using Google Maps APIs. Beta versions were evaluated by the team and collaborators. Initially supported by a geovisualization typology formulated by Crampton (2002), the interface evolved from the plan proposed by UFBA to deepen the level of user interaction with geographic information:

a. for the interaction with the context: the choice of layers,
b. to interact with data: selection of solid color or transparency,
c. for interaction with the representation of the data: use of Google Maps web-mapping resources (zoom, scale choices).

Figures 1 shows the experimental site interface.

Besides these features it offers application or services to include comments on the site, comments on maps and provide vectorial layers and thematic maps simultaneous visualization over Google Maps basic layers.

For the layer with collaborative input of user objects, it were used the Ushahidi open source platform (Figure 2).

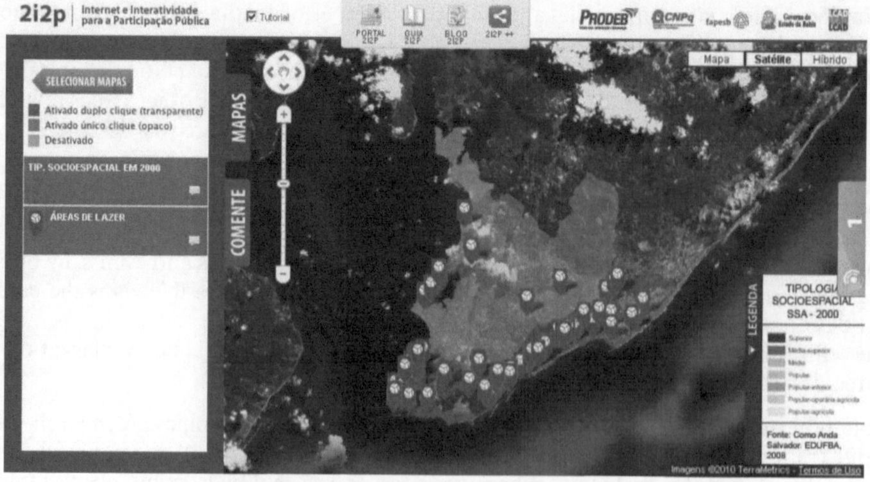

Figure 1. Experimental site (2i2p +).
Source: www.2i2p.ufba.br/portal.

Figure 2. Collaborative layer (2i2p ++).
Source: www.2i2p.ufba.br/ushahidi

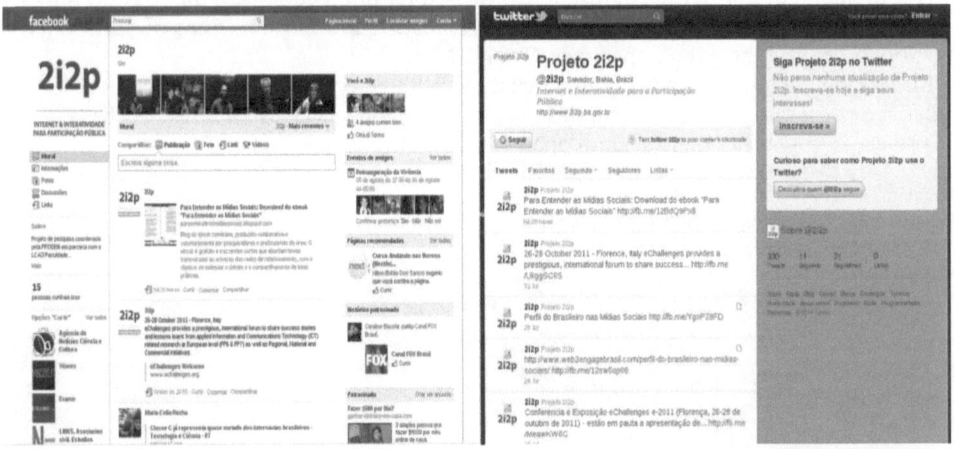

Figure 3. Presence of 2i2p project in social media.

The 2i2p project also used various social media to manage different purposes, as a way of experimentation.

To document the project and share ideas, it used a more authorial blog (www.2i2p.ba.gov.br). To communicate briefly and immediately these considerations, we created a profile on Twitter (http://twitter.com/2i2p).

To establish contact with the social network interested in the theme of the project and quickly disseminate experiences, events, opinions related to the subject, created a Facebook page (http://www.facebook.com/pages/2i2p/126873257356960). Images of this presence are shown in Figure 3.

4 SOME RECOMMENDATIONS FOR INTERACTION IN SITES

Exploring the Internet as a channel for interaction with citizens, with the now popular use of geographic information (GI), although in a challenging environment conducive to public participation and cooperation, has necessarily to explore understanding of key aspects for successful communication, among which are: requirements for communication between state and citizen, organizing information to facilitate search and issues for a good Web page design. Taking communication efficacy into account the 2i2p project addressed some aspects.

4.1 *Aspects of State-citizen communication*

The source of this reflection came from the PHd thesis of a member from the project team, briefly presented here. According to Silva (2009) the State communication must meet the following historically democratic requirements: a) Publicity: The State must become more visible and transparent to citizens, b) Responsiveness: The State must establish a more dialogic communication with citizens, c) Porosity: The State must become more susceptible to public opinion.

These requirements may be mediated through relationships or communicative forms ranging from a simple, more informative, up to a complex relationship, more argumentative. The author's work was used to frame the government's communication with use of the IG, according to the three cited democratic requirements, shown in Table 1 below.

The research project used this scheme to appropriate the mechanisms of Web 2.0: Twitter to communicate; blog (Wordpress) to share the technical documentation of the project and allow the exchange of views; posts in RSS format, to share and enable sharing blog content; Facebook to connect with a social network interested in these issues. It also used the following free services: RadarUol to give access to information about the blog use and UserVoice to get feedback and rank user reviews.

About interaction with the GI, resources are used for interaction with the context, interaction with the data itself and interact with the representation of data, some of them offered by Google Maps, others developed over Google Maps API.

Enhancing the bottom-up perspective, case of volunteered geographic information, the so-called crowdsourcing just discussed, this possibility has only been exemplified by the creation of project 2i2p++ using the Ushahidi open source platform (see Figure 2).

Table 1. GI versus requirements for democratic citizen-State interaction.

Democratic Requirement/ Communication level	Publicity	Responsiveness	Porosity
Utilitarian/ Informative	Location equipaments e services		
Instructive/ Argumentative	Presentation of plans/projects	Presentation of alternatives	Election of alternatives
Imperative			Elaboration of alternatives
Involved technology	Static maps Interative maps	Static maps Interative maps 3D models	Interative maps 3D models simulation

Source: http://www.2i2p.ba.gov.br/wp-content/uploads/2009/09/Hamburg_seminar-apresenta%C3%A7%C3%A3o-resumida.pdf

4.2 *Aspects of information architecture*

Promoting public participation via Internet implies giving citizens the right to open their own ways to access information and take the initiative in dialogue not only with the central government but with the community that has an interest in taking decisions on the future of the territory. Thus the user must allowed to record their own reading paths and classify the knowledge expressed, or even created on the site.

Luca Rosati (2007) states that the classification of content gives rise to innovation since it helps the user to establish new relationships. For him, every information architecture has two main axes: a vertical (or paradigmatic) which represents the hierarchical relationship between elements, and a horizontal (or syntagmatic) which represents semantic relationships of contiguity between elements system (Carcillo & Rosati, 2007).

According to Rosati (2007), this second axis is important because it enables or enhances a type of evolutionary research typical of the site, very useful for: a) identify other elements of similar type that can be of further interest (not originally anticipated/sought), b) correct the course of research, in which case the element was not properly obtained or in cases where new needs arise from the research itself. To allow the correlation of information by the user it is a way to contribute to more dialogic systems.

Considering a map as a metaphor and the user as a tourist, Dario Corno (2007) starts from the picture which he called "electronic urbanistic" and proposes the use of a series of discrete units of strong symbolic/syntactic value in assembly of digital sites, facilitating its movement, such as are the landmarks in the real cities. He argues that a light text, organized according to autonomous units, though related, gives rise to the satisfaction of a plurality of information needs and encourages interaction with site users, through the mobilization of tacit knowledge in the process of constructing their own textual paths (Corno, 2007).

Although he said the site design must face the fundamental problem of providing "full and quickly" the information needed by simulating a more general kind of question—Where to click? Where do I find?—it not have to predict the steps the user will give, but has to ensure him that the necessary information is in the site.

According to the perspective adopted by the author, to give greater clarity and lightness to the text presuppose open the site to the implicit knowledge "outside", allowing users to review their information and formulate new hypotheses when their beliefs do not prove to be true. Thus the design of the site should provide an increased quota of metarepresentations—representations that focus on other representations—because it is through representations of second order that the user will seek further explanations to interact with the textual environment of reference (Corno, 2007).

To assist the user in constructing meaningful reference around a specific theme, he recommends that the architecture of the site provides the linkage of information in a format not top-down nor bottom-up, but networked—from concept to concept.

Thus, the reference will be constructed by isolating a series of textual environments that can exchange signals with each other, which are semantically autonomous in certain respects. The user operates their own textual path by connecting different units, which together account for your research needs (Corno, 2007).

The research project, while not tested all the solutions to address the suggestions of the two authors cited above, recommends to cover those aspects. Such recommendations are outlined in the Guide 2i2p (http://www.2i2p.ufba.br/guia2i2p), in Portuguese, among which we highlight here some recommendations concerning access to information:

It is recommended that sites and portals of Public Administration adopt measures with respect to ease of access to data and information:

a. Make available steadily at the top of the site, an internal search engine for the immediate reach of the content related to the topic of interest;
b. Adopt a multidimensional semantic classification system (facet) to give greater efficiency searching for information;
c. Adopt classification systems that meet user requirements and not be just confined to the organization and functions of Public Administration;

d. Give visibility to search engines, considering, among others, the following parameters:
   - The amount of text content on the subject sought by the user.
   - The presence in the texts of keywords that presumably users use as search terms in the engines
   - It is desirable that public sites promote each other by entering, each time it is useful and meaningful, link to other sites of public interest in the subject treated or reference.

4.3 *Design aspects*

Design issues are addressed in a very tangential way in specific recommendations section of the Guide 2i2p. They are tips, examples and links to use free tools to implement web sites in accordance with the general recommendations of the Guide. However the first thing to emphasize is the importance of the user participation to begin at the design process and it goes until websites evaluation, not to mention his collaboration with content elaboration.

In case of the 2i2p project, users voluntarily arrived at the experimental site comments on the initiative, some praising, others demonstrating their frustration with the limited possibilities of interaction initially allowed (a beta release in January 2010 allowed only choice/activation of thematic maps for Salvador and its metropolitan region, control its transparency and on/off subtitles). After signing the User Voice service, this movement is no longer registered, a fact not discussed by the research group. Within the group, there was discussion about the design at various stages of its design and prototyping. A presentation of one of these exercises is found in 2i2p blog (http://www.2i2p.ba.gov.br/exercicios).

Regarding accessibility, although not having developed a prototype following the recommendations of international initiatives, the Guide 2i2p highlighted that it is essential to comply with accessibility principles defined in international forums. It is mandatory for Public Administration and it is recommended for sites of any group or entity to use standards as those adopted by Brazilian federal government (http://www.governoeletronico.gov.br/acoes-e-projetos/padroes-brasil-e-gov). It is also recommended to engage citizens with disabilities, whether in monitoring the accessibility of the site, or to identify best practices for accessibility (http://www.2i2p.ufba.br/guia2i2p/index.php?option=com_content&view=article&id=9&Itemid=4).

Another issue to have in mind relates to the interaction paradigm used that, according to Polillo (2010), evolves from "point and click", a mouse-based desktop metaphor, widely difused in the 90 s of last century, to the model "stand up and walk" established with the advent of iPhone in mid-2007, that significantly changed the design and use of mobile phones.

The iPhone integrates a variety of technologies—GPS, web browser, multimedia player. Its multi-touch screen has good resolution, that lets you control functions with a variety of gestures. Its sensing systems allow automatically collection of information about itself, about the user and the environment, providing contextual information (Polillo, 2010).

The more devices are endowed with intelligence and access to data and information available on the Net, the more physically integrated to the environment, more they disappear from our view, more apparent it becomes the user interface (Polillo, 2010). In this context, opens up new possibilities for exploration of data sets, becoming possible to leverage the visualization—dynamic, interactive, user-controlled—and cognition of space.

The model "stand up and walk around" comes closest to exploring the environment, as wanted Parsons (1994) at the end of last century, speaking about the various forms of geographic data. He pointed out that the advantage to using images, video and sound to represent the spatial data is that they have "high information content and not based on abstract symbolism often present in conventional cartographic mapping" (Parsons, 1994).

Thus it is possible to envisage that, for a better view of geographical space, we must have applications that provide representations of alternative abstraction levels: a) one of these provides information of a more qualitative, descriptive attributes about the area, facilitating the location associated with experience of the individual and b) another that supports the knowledge of the entire space and the relations between objects and spatial phenomena.

Such levels of representation should be well integrated so that knowledge of space establishes a permanent support for the situated knowledge that, without the first, could lead to disorientation. In short, we must contextualize the spatial locational reference, taking into account the need to maintain spatial references socially relevant at both levels of representation. The insertion of this type of reference can guarantee the reduction of the degree of abstraction of the first level mentioned. On the other hand, it promotes the integration between the two levels, when we added a spatial reference that is meaningful, it establishes an explicit link between them. It is needed to relate location and semantics, since the location must be imbued with meaning (Shum, 1990).

In 2i2p project the reference to context was conceived as being provided by diverse media such as photos (geotagged images), views, videos, or even three-dimensional models, and although this last aspect has not been implemented in the project, could be developed through the technology of "augmented reality" that permits to view through a device (camera, displays, smartphone) digital models superimposed on physical reality.

5 CONCLUDING REMARKS

The Internet, basic infrastructure and central factor of the new social space, is today a key arena for promoting a fruitful dialogue between public administration and citizens. Web 2.0 applications (blogs, wikis and social networks) provide environments in which the main idea is dialogue, exchange and shared rewriting. Over the Network is possible to increase or establish new arenas, helping to boost public participation, particularly with the use of these applications, since we rely on network speed.

The exploration of some of these environments has been the subject of reflections of the project "Internet and Interactivity for Public Participation", the 2i2p project, many of them presented here.

The information organization of government sites is a key not only to facilitate the acquisition of information sought, but also for the "creation" of information, by facilitating the construction of references through which the user add significance to the information object on the routes they choose. That is, the government website should be able to promote the construction of pathways significant by citizens, often marked by others who have left their mark there, encouraging—why not?—the fortuitous meeting of a newness that can encourage them to connect to (other) communities of interest.

Giving voice and time to (often) amateur, folk taxonomies are used to label shared resources on the Web, including texts, images, addresses, profiles. In terms of geographical space, geotagging is used to provide location information, allowing its recovery through maps (Goodchild, 2007). But at this moment the 2i2p project experimental site doesn't addressed this issue.

Regardless of pre-ordered schemes, keywords are used in classifications (tags) spontaneous and collaborative without wishing to be complete. Different tags placed by different users at the same features create an alive content network and multiply mental pathways, establishing new connections, putting in contact ideas, ideals, people, communities, in a way not predictable (Maistrello, 2007).

Moreover, the facility that user has now to upload geographic information is an example of how technology helps to spread the use of data what, before Google, it was almost exclusively of experts, in general, employees of government agencies engaged collecting spatial data, that not always are reported via web, often through maps previously designed, providing little or no ability to interact with spatial data (if only for representation of data).

The use of interactive maps gives the user the power to explore phenomena related to geographical space, according to its own timing and need for reflection, nowadays the cartographer is not anymore the exclusive responsible for access and management of geographic databases, despite the eventual production of maps of poor quality (Kraak, 2001). One might even say that the map now no longer be supply driven to be demand-driven, amplified by access to remote databases afforded by the web.

Aspects of interface design and usability should receive careful treatment. When it comes to maps, the design analysis must consider the characteristics of geospatial data, defines the content and the levels of perception, scale and accuracy required, one should check for standard symbols, and these factors can influence each other. But nowadays, the use of free mapping tools available on Internet gave rise to the birth of the so-called neogeographers communities, because the knowledge of cartographic techniques to produce legible maps are embedded in the software. The great contribution becomes the knowledge of the territory that the citizen has and can now provide.

In short, if usability is one aspect that should be assessed carefully in web applications, particularly when it comes to maps, whose design should adequately convey content to the site objectives, research efforts must now go beyond the use of maps as the only way to present geographic information to explore the use of other media and technologies such as interactive models, augmented reality and social networks. Comercial applications are beginning to provide this kind of content not only for desktop browsers, but for portable devices such as smartphones or tablets.

The 2i2p project presented here tested some free Web services and thereby highlighting the exploitation of geographic information as a means of promoting participatory processes. It aims to familiarize the data processing company of state with this innovative approach to help involve citizens actively in defining issues that govern the life of their community in planning strategies, in implementing policies and managing their own territory (Porrelo, 1983, apud Debernadi & Rosso, 2007).

Local issues are of greatest impact on the lives of citizens. Citizens can bring to the Administration knowledge (on parks, neighborhoods and city) that is distributed among social groups that interact in urban space. This knowledge could be easily accessible through the Network, which in itself already constitutes a driving factor in the development of social and urban space.

In institutional terms, considering the urban space in Brazil, the concern with the participation process is explained in the City Statute, approved by Brazilian Congress in 2001. The statute sets the guidelines for the country's urban policy at the federal, state and municipal levels and provides in its Article 2, section II:

> democratic management through people's participation and associations representing various segments of the community in the formulation, implementation and monitoring of plans, programs and projects for urban development (BRASIL, 2001).

Chapter IV of the Act, Section 43 shows how the democratic management of the city will be guaranteed. It is included among his tools, urban policy councils, debates, public hearings and consultations, conferences on topics of interest at law projects and urban plans, programs and projects of popular initiative. In Article 45 of this chapter, the Act says:

> The management bodies of regions and urban agglomerations include mandatory and meaningful participation of people and associations representing various segments of the community in order to guarantee the direct control of its activities and the full exercise of citizenship (BRASIL, 2001).

The constitutional principle of people participation was established. The means proposed to do it happen—public hearings, publication of the documentation and access to the same population—are insufficient (often the studies underlying the technical and political decisions are not even released). They do not guarantee the effective participation of the population, particularly in large cities or metropolitan areas where they live today over a third of the population. The 2010 census data from IBGE released confirm that the concentration of population in large metropolitan areas of the country continues to growth.

Thus we come to the end of the first decade of this century, in a consolidation scenario of a new culture of ICT use in an urban world and in a country—Brazil—that concentrates its population in large cities, with virtually only a case of effective public participation for cite: some experiences of Participatory Public Budgeting in three or four states.

In November 2004, Ministry of Cities in its publications, realeased in the National Policy Seminar on Urban Development in Brasilia, reiterated the need to intertwine the processes of popular participation in urban regional planning from the "participatory reading" (MINISTERIO DAS CIDADES, 2004a). In another document emphasizes the need to incorporate approaches and practices for the "construction of the institutional conditions that permit the expansion of popular participation in the definition of urban policy" (MINISTERIO DAS CIDADES, 2004b).

The State must appropriate the benefits of new information technology to enhance citizenship, addressing the transparency, participation and collaboration as stated in the Open Government Directive.

To become effective the access to information and computer technologies we must give special attention to familiarize the citizen with both, the public task as with the geographical area under discussion, to be able to capture and even help build the collective memory of urban space, allowing speculation about possible futures.

Sites with focus on public participation in local affairs (planning or management of public space) can now benefit from bottom-up perspective afforded by new technologies, bringing its contribution to the transition that the company is doing in his vision of the role of public information. As stated by Ernesto Belisario (2010), the centrality of citizen participation, universal access to public data and web usage (especially mobile devices) are distinctive features of an open government.

To consider the city a social network with each node connected on the Internet means designing a map of local realities. City Networked would be a classic example of application based on social networks, a social network whose scope would not only find friends, post photos and videos to look for work, but becoming a protagonist of the territory and represent the complexity (Mastrello, 2007).

REFERENCES

Belisario, E. Open gov, la vera innovazione è il cittadino. APOGEOnline. 20 set. 2010. Available in http://www.apogeonline.com/webzine/2010/09/20/open-gov-la-vera-innovazione-e-il-cittadino. Accessed September 23, 2010.

BRASIL, 2001. LEI n. 10257, July 10, 2001. Regulamenta os art. 182 e 183 da Constituição Federal, estabelece diretrizes gerais da política urbana e dá outras providências.

Carcillo, F. & Rosati, L. Dalla classificazione per i Cittadini la classificazione dei Cittadini. Il caso TaggaTO. In: VENIER, Federica (cura). *Rete Pubblica: Il dialogo tra Pubblica Amministrazione e cittadino: linguaggi e architettura dell'informazione*. Perugia, Guerra Edizioni, 2007. pp. 113–123.

Corno, Dario. I cittadini in rete. Uma prospettiva "ingenua" sul dialogo tra amministratori e utenti. In: Venier, Federica (cura). *Rete Pubblica: Il dialogo tra Pubblica Amministrazione e cittadino: linguaggi e architettura dell'informazione*. Perugia, Guerra Edizioni, 2007. pp. 39–49.

Crampton, J.W. Interactivity Types in Geographic Visualization. In: *Cartography and Geographic Information Science*, v. 29 pp. 85–98, 2002.

Debernardi, L. & Rosso, E. *Governance e Sistemi Urbani*. Rome, Carocci Editore, 2007.

Goodchild, M.F. Citizens as sensors: the world of volunteered geography. *GeoJournal*, Netherlands, v. 69, p. 4, pp. 211–221, Aug. 2007.

Goodchild, M.F. Commentary: whither VGI? *GeoJournal*, Netherlands, v. 72, n. 3–4, pp. 239–244, Aug. 2008.

Hudson-Smith, A. 2009. *Digitally Distributed Urban Environments: The Prospects for Online Planning*. 2009. Thesis (Doctor of Philosophy)—Center for Advanced Spatial Analysis, University College London.

Kraak, Menno-Jan. Settings and needs for web cartography. In: Kraak, Menno-Jan; Brown, Allan (eds). *Web Cartography—developments and prospects*. Taylor & Francis, USA. 2001.

Laurini, R. Groupware and Public Participation for Urban Planning. *Proceedings of the International Workshop on Groupware for Urban Planning*, Lyon, France, 4–6, 2001.

Maistrello, S. *La parte abitata della Rete* . Milano, Tecniche Nuove, 2007.

MINISTÉRIO DAS CIDADES, 2004a. MINISTÉRIO DAS CIDADES. Planejamento territorial urbano e política fundiária. Brasília: Ministério das Cidades, 2004a.

MINISTÉRIO DAS CIDADES, 2004b. MINISTÉRIO DAS CIDADES. Política nacional de desenvolvimento urbano. Brasília: Ministério das Cidades, 2004b.

Monaci, S. *La conoscenza on line: logichi e strumenti*. Le Bussole, 322. Rome, Carocci Editore, 2008.

Nembrini, A. & Joerin, F. Proposition for a Participative GIS leading to a concerted diagnostic, application at the neighborhood scale. In *UDMS'02, 23, 2002, Praga. Proceedings* ... Praga: Urban Data Management Society, 2002.

Parsons, Ed. "GIS visualization tool for qualitative spatial information". In: Hearnshaw, Hilary M. & Unwin, David J. *Visualization in Geographical Information Systems*. Chichester: John Wiley e Sons, 1994. pp. 201–210.

Pereira, G.C., Davis, C. & Rocha, M.C.F. Establishing a sub-national SDI in Bahia state (Brazil) - its limits and possibilities In: *Urban and Regional Data Management*. 1 ed. London: Taylor & Francis Group, 2009, v. 1, pp. 151–158.

Polillo, R. *Facile da Usare—Una moderna introduzione alla ingegneria dell'usabilità*. Edizioni Apogeo, 2010. Available at: http://www.rpolillo.it/index.php/libri/facile-da-usare/. Access in Sep 2010.

Rosati, L. Classificare per innovare. La classificazione come investimento e scoperta. In: Venier, Federica (cura). *Rete Pubblica: Il dialogo tra Pubblica Amministrazione e cittadino: linguaggi e architettura dell'informazione*. Perugia, Guerra Edizioni, 2007. pp. 91–99.

Shum, Simon. Real and Virtual Spaces: Mapping from Spatial Cognition to Hypertext. *Hypermedia*, v. 2. London: Taylor Graham Publishing, 1990, pp. 133–158.

Silva, S.P. da. Papéis Democráticos e Dimensões Analíticas da Comunicação Política do Estado: Um estudo de caso sobre o Portal da Presidência da República. *XVIII Encontro Anual da Associação Nacional dos Programas de Pós-Graduação em Comunicação (Compós)*. Belo Horizonte: PUC-MG, jun. 2009. Available at <http://www.compos.org.br/data/trabalhos_arquivo_coBx71ptTiwD6.pdf>. Accessed July 6, 2009.

Urban and Regional Data Management – Zlatanova, Ledoux, Fendel & Rumor (eds)
© 2012 Taylor & Francis Group, London, ISBN 978-0-415-67491-1

Towards connected governance

L. Schroeder, L. Hvingel, H.S. Hansen & B.H. Jensen
Aalborg University, Aalborg, Denmark

ABSTRACT: The role of information and communication technologies (ICT) in e-Government seems to be changing into a new e-Government-as-a-whole concept which focuses on the provision of services at the front-end supported by integration, consolidation and innovation in the back-end processes and systems to achieve cost savings and improved service delivery. In Europe the adoption of the INSPIRE Directive emphasises geographic information (GI) and a well functioning spatial data infrastructure (SDI) as fundamental elements in e-Government. Due to the UN especially the Nordic approach to establishing digital infrastructures is an important conceptual way to efficient e-Government. In Denmark the development within the GI-community is characterised by an informal, bottom-up approach with little attention regarding organisational aspects. This paper will refer to a number of cases based on examples from the municipal of Rudersdal illustrating how the collaboration and innovation among different players within the GI-community actually is a determining factor for a successful development of spatial data infrastructures with the potential of enabling connected governance concepts.

1 INTRODUCTION

Considering the role of information and communication technologies (ICT) in e-Government a change seems to be on its way leading to a new *e-Government-as-a-whole* concept which focuses on the provision of services at the front-end supported by integration, consolidation and innovation in the back-end processes and systems to achieve cost savings and improved service delivery (United Nations 2008).

The UN e-Government Survey from 2008 refers to the vision of *whole-of-government* as a holistic approach to ICT-enabled public sector governance and it is stated how the process of developing e-Government can be framed by three phases, where the first phase is focused on creating information infrastructures gradually moving towards a second phase characterised by integration of information, services and governance models followed by a phase of transformation (United Nations 2008). Denmark and the other Nordic countries are topping the list in the UN ranking of e-Government readiness and the Nordic approach to establishing digital infrastructures is emphasised as an important conceptual way to e-Government (United Nations 2008)(World Economic Forum 2009).

Geographical information (GI) and spatial data infrastructure (SDI) has gained increasing awareness among researchers as well as among executive level managers and politicians. Thus a well functioning spatial data infrastructure is seen as an important element in e-Government and online self-service systems. In Europe especially the adoption of the INSPIRE Directive has put spatial data infrastructure on the agendas of the parliaments combined with national e-Government strategies (Hansen et al., 2010).

Rajabifard et al. (2002) points out how the complexity of SDI with interrelated and interconnected technical and institutional elements and the multiplicity of stakeholders involved imply that governance is an important aspect of the institutional framework necessary to support decision making about all aspects of an SDI. Masser et al. (2007) notes that hierarchical governance structures are required to enable the participation of national and local governments and the private sector in decision making. Hierarchical structures typically operating top-down, with authority flowing from higher to lower levels is also the usual Danish way of implementing reforms and innovation in the public sector—from the national state level to regions and municipalities. Nevertheless, a combination of consensus-based decisions and mutually binding public-public and public-private partnerships has traditionally been the driver for the development of SDI in Denmark (Hansen et al., 2010).

Referring to the UN the holistic approach is essential and it is emphasised how this new kind of *connected governance* demands a re-engineering of technology and processes as well as skills and the mindsets of public sector employees to be able increase the efforts of governments towards organising government agency functions and services along with improving citizen e-Services (United Nations 2008). Dealing with the important challenge of how to establish an organisational culture based on more collaborative and participative mindsets to facilitate the necessary transformation the Danish GI-community is an interesting example. In Denmark the further development of "connected governance" is supported by an innovative GI-community, though the lack of awareness regarding the spatial dimension as an integrative infrastructural element is still a challenge when developing e-Government in traditional parts of the public administration.

Referring to specific cases from Denmark the aim of the research presented in this paper is to analyse the innovation potentials within the GI-community based on knowledge sharing, cooperation and consensus creation between the various partners within a GI-based holistic framework as means of enabling the on going transformation. Following the introduction is a description of the theoretical framework focussing the different aspects relating to front-end verses back-end processes referring to the conceptualisations of "government-as-a-whole" and "connected governance". After a brief introduction to the SDI-related projects we introduce a model elaborating the conceptualisation of back-end verses front-end processes. To analyse how the Danish municipalities and the national government gradually are connecting within a holistic digital concept the municipality of Rudersdal and aspects of digitalising municipal planning procedures will be described within this framework referring to the different levels of data management and digital procedures. We summarise our findings in a discussion based on a SWOT-analysis.

2 CONNECTED GOVERNANCE

2.1 *ICT enabled governance*

The UN e-Government Survey 2008 points out how the concept of *connected governance* is based on "a systematic approach to collection, reuse and sharing of data and information" and characterised by the combined efforts of improving cooperation between government agencies as well as deepening consultation and engagement with citizens (United Nations 2008). A well functioning spatial data infrastructure is increasingly seen as an important element in e-Government and online self-service systems, which is expressed by the vision of *spatially enabled governments*, which concerns establishing "an enabling infrastructure that will facilitate the provision of the place or location to all human activities as well as government actions, decisions and policies" (Masser et al., 2007). Masser et al., emphasises the great challenge due to the lack of awareness regarding spatial concepts besides a small elite of spatially aware professionals and points out how more inclusive models of governance to enable the very large number of stakeholders from all levels of government are needed (Masser et al., 2007).

2.2 *Infrastructure—integration—transformation*

Due to the UN survey the process of developing e-Government can be framed by three phases, where the first phase is focused on creating information infrastructures gradually moving towards a second phase characterised by integration of information, services and governance models followed by a phase of transformation (Fig. 1).

Due to the UN "the concept of connected government is derived from the whole-of-government approach, which is increasingly looking towards technology as a strategic tool and as an enabler for public service innovation and productivity growth" (United Nations 2008). Due to this understanding Figure 2 illustrates how government based on traditional modes of service delivery via e-government changes into connected government with an emphasis on an integrated approach focusing on enhancing the value of services to the citizen.

As pointed out by the UN it is necessary to integrate and redesign government organisation and processes in order to achieve greater value in service delivery as well as reducing costs. So it is emphasised how "genuine cost savings and quality improvements will occur only if there is a re-engineering of the internal structures and processes of the administration towards a connected form of governance" which demands re-engineering at various levels as illustrated in Figure 3 (United Nations 2008).

1. Infrastructure: Creating an information infrastructure both within the public sector and across society at large based upon reliable and affordable Internet connectivity for citizens, business and all stakeholders in a given jurisdiction.

2. Integration: Leveraging this new infrastructure within the public sector and across society in order to share information and bundle, integrate, and deliver services through more efficient and citizen-centric governance models encompassing multiple delivery channels.

3. Transformation: Pursuing service innovation and e-Government across a broader prism of community and democratic development through more networked governance patterns within government, across various government levels and amongst all sectors in a particular jurisdiction.

Figure 1. Three phases framing the process of developing e-Government (United Nations 2008).

Traditional government	→	**e-Government**	→	**Connected government**
↓		↓		↓
Traditional modes of service delivery	→	**e-Services**	→	**Value of services**

Figure 2. Evolving approach to public service delivery (United Nations 2008).

1. ***Intra-Government Process Re-engineering → efficient, responsive and tailored government to reflect citizen needs***
2. ***Inter-Government Process Re-engineering → efficient, joined-up borderless government:***
    a. ***Vertical cooperation/integration between levels***
    b. ***Horizontal cooperation/integration between agencies at the same level***
    c. ***Multi-stakeholder cooperation (with private and third sectors)***
3. ***Re-engineer legacy technology, processes, skills and mindsets***

Figure 3. Levels of re-engineering internal structures and processes (United Nations 2008).

2.3 *Modelling key elements of the re-engineering process*

The OECD points out how the member countries increasingly are focussing their efforts to broaden the focus on e-Government programs as means of enhancing efficiency and effectiveness as well as improving on going service delivery (OECD 2010).

The UN emphasises that "an increase in the value of services is not possible without consolidating the way the back-end systems and processes work to bring about the front-end service delivery" (United Nations 2008). So it is pointed out that to be able to fulfil the visions of an effective connected government a "bigger and better" front-end with a "smaller and smarter" back-end is a necessity (United Nations 2008).

Furthermore it is pointed out how moving towards connected governance demands a "re-engineering of technology and processes as well as skills and mind sets of public officials in the government within a holistic framework" (United Nations 2008), which can be understood as a change from a technical view to optics known from theories concerning organisational learning, knowledge management, and human factor based innovation (Wenger et al., 2002).

Figure 4 illustrate some of the central aspects related to a simple value chain considering the distinction of back-end verses front-end processes within a holistic whole-of-government concept referring to a GI-environment.

The model is based in the understanding presented by the UN, so the front-end processes are referred to as "services" characterised by the element of contact as opposed to the back-office processes, which are more production-oriented and referring to "the internal operations of an organisation that support core processes that are not accessible or visible to the general public" (United Nations 2008).

The model is intended to be able to take into consideration the actual actors in the innovation process as well as the kind of use centred innovation and will be further elaborated and exemplified in the following.

2.4 *The Danish GI-community—a community of practice*

The whole-of-government approach is characterised by "government agencies and organizations sharing objectives across organisational boundaries, as opposed to working solely within an organisation" (United Nations 2008). This approach is one of the characteristics of the Danish GI community (Schroeder et al., 2010), which can be compared to the picture of communities of practice as formed by people who engage in a process of collective

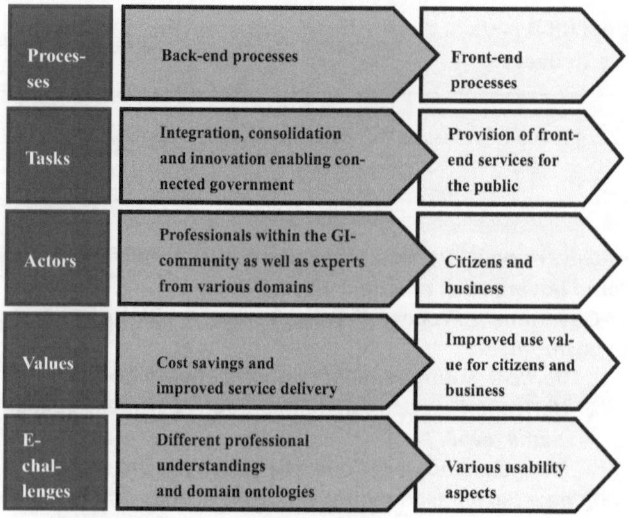

Figure 4. A value chain considering the distinction of back-end verses front-end processes within a holistic whole-of-government view.

learning in "a shared domain of human endeavour" (Wenger et al., 2002). Though, among the great challenges in the on going development are the various professional understandings and differing domain ontologies across organisational borders as well as a general lack of awareness regarding the GI-infrastructural dimension in central traditional parts of Danish government especially at the national level. The reorganisation of central GI-based back-office processes interacting with the national effort regarding the digitisation of the procedures regarding municipal plans will be further illustrated by means of the elaborated model.

3 GOVERNMENT-AS-A-WHOLE IN DENMARK

3.1 *SDI initiatives*

Denmark has been using ICT in government for several decades—firstly focussing mainly on process automation to achieve efficiencies in public administration shifting towards more elaborated e-Government, which can be seen as corresponding to the UN phase one and two. On the European scene the need of accessing environmental data has since 1998 resulted in legislation that focuses on public access to information about and the possibility of participation in the decision making in local, national and international environmental cases (UNECE 1998).

The focus on the interrelationship between the public sector and the citizens lead to the implementation of the Public Sector Information directive in 2005 (Commission of the European Communities 2003) and the necessity of seamless interoperability of European data lead to the INSPIRE directive in 2007 (Commission of the European Communities 2007). INSPIRE does not only focus on the standardisation of geographical data, but also focus on a Spatial Data Infrastructure.

During the previous ten years the Danish society has moved steadily towards an SDI by collecting data, making institutional arrangements and adopting standards—but without any overall strategy. However, the INSPIRE process has facilitated the development of a formal national SDI. December 2008 the Danish Parliament adopted a new law concerning the setup of an Infrastructure for Geographic Information, and geographic information is now considered as a key component in e-Governance (Danish government 2008).

3.2 *The Danish e-Government approach*

A very informal, bottom-up approach is in Denmark a distinctive feature of the development of e-Government. The organisational aspects in itself has not been an issue of much attention at the management level, though the actual situation shows that the way the different players actually collaborate is a determining factor for a successful development in the field of e-Government as illustrated by Figure 5.

During summer 2009 the University of Aalborg and Geoforum Denmark—the Danish national GI organisation—carried out a web based survey amongst Danish local, regional and national institutions to study the level of readiness for INSPIRE implementation in Denmark. This survey shows that the Danish model of organising the actors result in a close connection between the technicians and the decision makers, and also results in a general overall knowledge of new initiatives, for instance the INSPIRE Directive (Schroeder et al., 2010) (Hansen et al., 2010). This way of developing e-Government corresponds to the notion of connected governance frameworks, where intergovernmental processes can be integrated vertically between various government agencies and/or horizontally between agencies at the same level and/or with the inclusion of private sector and other stakeholders (United Nations 2008).

In the Danish strategies for e-Government the main focus is on a better public service, a higher degree of efficiency and on stronger collaboration (Danish Government et al., 2007). In regard of the focus on better public service, the understanding is that a high service-level is also a reflection of a high degree of efficiency as expressed by Figure 6.

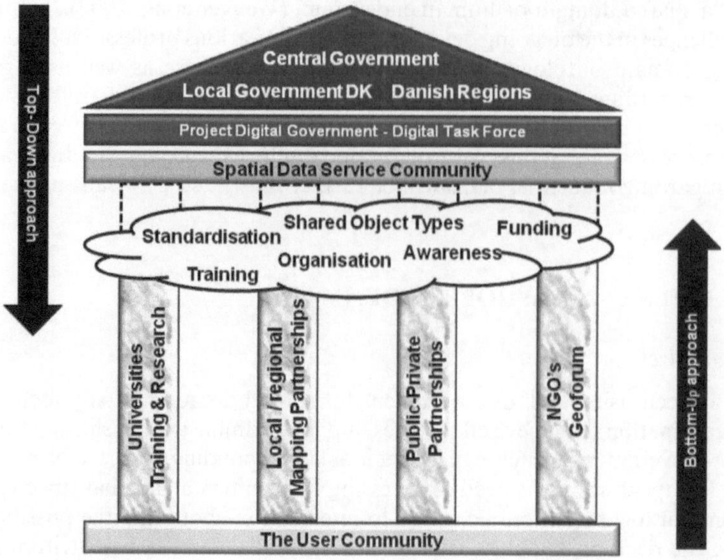

Figure 5. Building the Danish Spatial Data Infrastructure (Hansen et al., 2010).

This understanding can be elaborated by referring to the UN understanding of back-office verses front-office processes. The needs of the different types of users at different levels have to be addressed in different manners and more distinctive GI-aspects of the back-end processes especially regarding the understanding of the value of suitable data as part of the Danish spatial data infrastructure. According to this understanding some geographical data plays a more profound role by as illustrated in Figure 7, which shows how basic data are understood as metadata, reference data as well as multi-sector data, which provide the foundation for use of data within the various specific sectors (Servicefællesskabet for geodata 2004).

3.3 *Elaborating the model*

To be able to frame the complexity of the value created at different levels of a GI-based e-Government environment the concept of front-end verses back-end processes presented in Figure 4 is elaborated due to the need for distinguishing:

- *Back-back-end processes* referring to the basic infrastructural elements handled by professionals insuring SDI-aspects as basic data sets, interoperability, quality.
- *Back-end processes* referring to the internal operations of an organisation that support core processes, which are not accessible or visible to the general public.
- *Back-end/front-end processes* referring to front-office services mostly used by professionals and linked to back-office functions within local governmental organisations.
- *Front-end processes* referring to front-office services designed for citizens and business.

In Figure 8 the content of these processes are illustrated as a value chain based on the enhanced use of spatial data and information and the actors involved at the different levels, which will be exemplified by:

- *Common Object Types (FOT)* as a back-back-end process handled by municipal and private GI-environments ensuring the provision of basic data sets—93 of the 98 municipalities in Denmark have now joined the network.
- *Plansystem.dk* as a back-end process organised at the ministerial level and a keystone in Danish e-Government functioning as portal to all municipal plans in Denmark.
- *Digital municipal plans* as a front-office service for citizens and business functioning as a central part of local e-democracy services—and an essential back-office function embedded in a GI-based management system and working environment in the individual municipals.

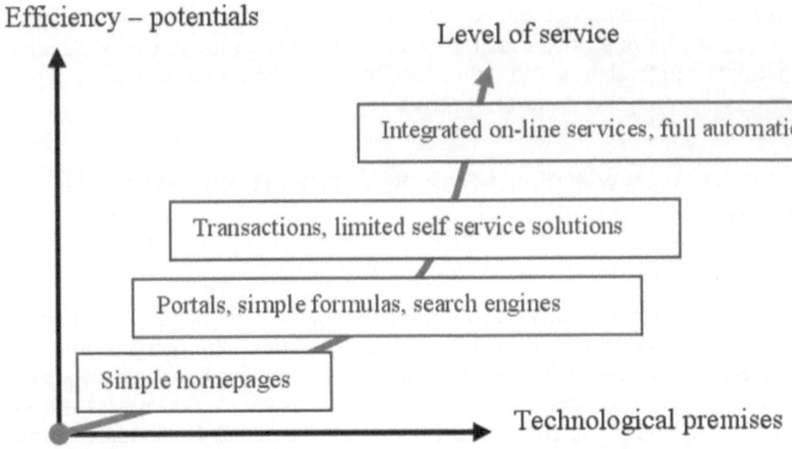

Figure 6. Development of e-Government (after Danish Government, 2003).

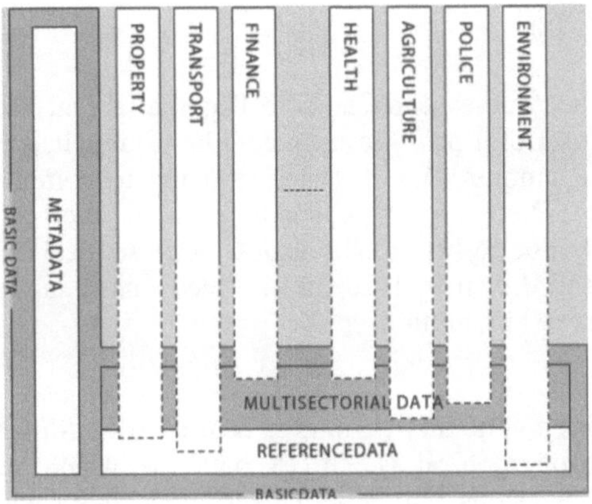

Figure 7. The basic concepts of data interrelationship (Servicefællesskabet for geodata 2004).

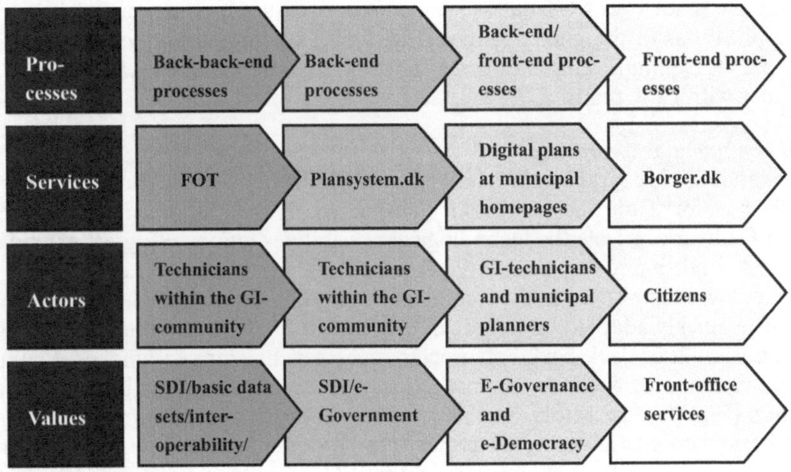

Figure 8. Aspects of the value chain referring to the processes from back-end basic infrastructures to front-end services.

- *Borger.dk* as a front-office service intended to be functioning as the main e-Government entrance for citizens among others including a specific service for house owners linking to central national geodatabases—an online service where the house owners can apply for building permits is among the newest initiatives.

4 GOVERNMENT-AS-A-WHOLE REGARDING THE MUNICIPALITY OF RUDERSDAL

4.1 *E-Government initiatives in the municipality of Rudersdal*

As a result of the Danish municipality reform in 2007, where 271 municipalities were reduced to 98, the former and Birkeroed and Soelleroed municipalities were merged into the new municipality of Rudersdal. The municipality is located in the northern capitol area has a strong focus on digitisation of back-end processes as well as front-end services supporting the vision of creating good living conditions for its 55.000 citizens. Rudersdal is chosen to exemplify the model as it is actually the only Danish municipal, which has been part of the innovation processes at all four levels.

4.2 *Back-back-end processes*

4.2.1 *FOT*

Together with the National Survey and Cadastre The Spatial Data Service Community and The National Organisation of Local Governments have launched FOTDenmark. FOT is a Danish acronym for Common Object Types referring to common public geographical administration data (FOTDanmark 2010). The topographic themes and geographic names are complicated—mainly due to historical reasons. Because local mapping has historically been widely dispersed and driven by independent needs, national maps of Denmark have been difficult and expensive to maintain.

4.2.2 *The concept of FOT*

The purpose is to establish a national base map to be used at all administrative levels by combining the nationwide topographical TOP10DK database, which has a specified accuracy of 1 meter in all three dimensions, and large-scale technical maps with an accuracy requirement of 10–20 cm used by local government administrations (Dael et al., 2008). In 2008 77 municipalities were parts of the network and the local network HRKS/Midt including the municipalities Egedal, Frederikssund, Furesoe, Allerød, Rudersdal and Hoersholm who were among the first to get FOT data were also part of a development project regarding the browser applications in the municipalities (Fig. 9). At the moment 93 of 98 municipalities have joined FOTDenmark and about one third of the municipalities have already covered their territory with FOT maps.

4.2.3 *GI-communities of practice*

The understanding of geographically based data as an inherent part of e-Government is a result of a persistent effort from the Danish GI-community. The development of e-Government and SDI in Denmark reflects the technological possibilities and is to a great extend a result of engagement and shared responsibility among the players (Schroeder et al.,). This way of working implies a short way from the technicians to the decision makers and often ideas born on one level is quickly adapted on the other levels as the FOT-project illustrates.

FOT is an example of back-back-end processes where technicians within the GI-community (municipalities, mapping agencies, interest groups) are working together, based on a common set of values (Fig. 8). The actors on this level sees the spatial data infrastructure and the notion of basic data sets as an essential mean to data interoperability and quality and efficiency in e-Government. Though, it is crucial that FOT geodata is established nationwide in the coming years to ensure that FOT becomes an integral part of the national spatial data infrastructure and, further, the e-Government (Jarmbæk 2009).

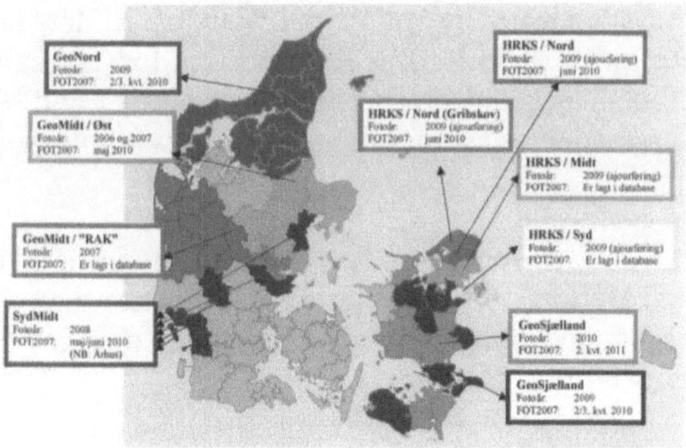

Figure 9. The map shows the Public-private partnerships across the country—HRKS/Midt including the municipalities Egedal, Frederikssund, Furesoe, Allerød, Rudersdal and Hoersholm were the first to get data into work.

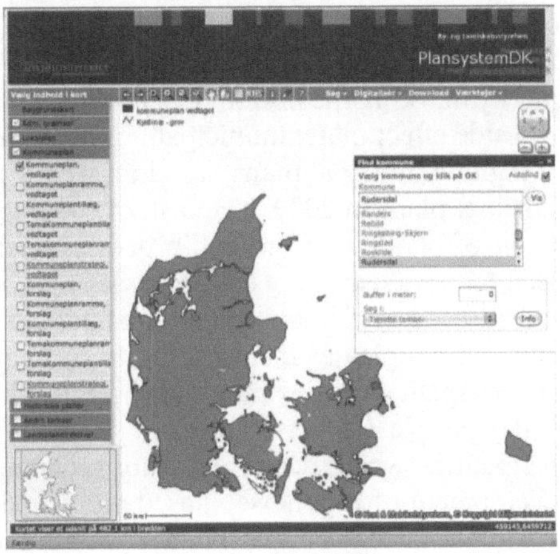

Figure 10. Mapbased access to the municipal plan of Rudersdal via PlansystemDK. (www.plansystemdk.dk)

4.3 *Back-end processes*

4.3.1 *PlansystemDK*

It is mandatory for the Danish municipalities to upload their municipal plans and local urban plans to the central register PlansystemDK (www.plansystemdk.dk). Accessibility to data, metadata, as well as user-friendly and efficient access and distribution systems represent fundamental building blocks in an SDI, which is emphasised in the INSPIRE principles. PlansystemDK is a fine example of a back-end process that applies to these demands.

PlansystemDK was recognised as an outstanding approach to local SDI in Europe and awarded by the eSDI-Net+ project in 2009. The reason for the reward was firstly the data infrastructure, hence the way that data is uploaded to and can be accessed in the system. This infrastructure is based on the recommendations of the Danish Spatial Data Service Community, who in many years have recommended that data are kept and maintained in one and only one place. Furthermore PlansystemDK is based on open source.

4.3.2 *Functionality of PlansystemDK*

PlansystemDK contains plans produced in accordance with the Danish Planning Act. The geoportal ensures that planning data are standardised and universally accessible, provides easy access to municipal and state spatial plans, and is a simple way for the municipalities to report their plan proposals and adopted plans to the national public authorities. Fig. 10

illustrates the mapbased access to the municipal plan of Rudersdal via PlansystemDK. The register feed other portals with this information for instance the public portals Public Information Online—targeted at citizens (www.ois.dk) and The Danish Nature & Environment Portal—targeted at the professional users (www.miljøportalen.dk).

4.3.3 *Spatially enabled governance*

PlansystemDK is a result of cooperation between the ministries, the municipalities and the Local Government in Denmark. PlansystemDK illustrates the potentials of innovation within a GI-based working environment based on visible common values due to a shared understanding of the daily administration.

4.4 *Back-end/front-end processes*

4.4.1 *Digital municipal plans*

Since 2006 all Danish municipalities has been obligated to upload their municipal plans and their local urban plans to a central register called PlansystemDK (www.plansystemdk.dk). This demand is met by uploading a copy in Portable Document Format (PDF) to the site and for the local urban plans by marking the boundaries of the plan and adding metadata for this area in the register. As a side effect many municipalities including Rudersdal has chosen to develop more advanced digital municipal plans based on web-technologies as part of the general revision of the municipal plans in 2009. The task of making useful digital municipal plans is an interesting example of back-end/front-end processes referring to e-Governance in general as well as more specific issues regarding e-Democracy.

Ensuring public involvement is a key-element in making municipal plans. Accordingly the digital plan has an important role to play as means of addressing and involving the citizens in the future plans for the municipality. On one hand it is important to make the plans accessible and readable for everyday people. On the other hand the plans serve as the basis for the administration in the municipalities, which demands a thoroughness and fitness to meet the professional needs. The development of digital web based municipal plans is not mandatory. Hence there is no overall national strategy or specific demands in the field.

4.4.2 *The plan09 project regarding municipal plans*

As a follow up initiative after the municipal reform in 2007 a common forum discussing the innovation and development of municipal planning in a broader perspective, called Plan09, was established by the Ministry of the Environment (Plan09 2010). The municipality of Rudersdal was one of eight municipalities participating in a subproject encouraging the exchange of experience in developing digital municipal plans. Identifying the needs of front-end as well as back-end users is a specific topic in the projects (Fig. 11).

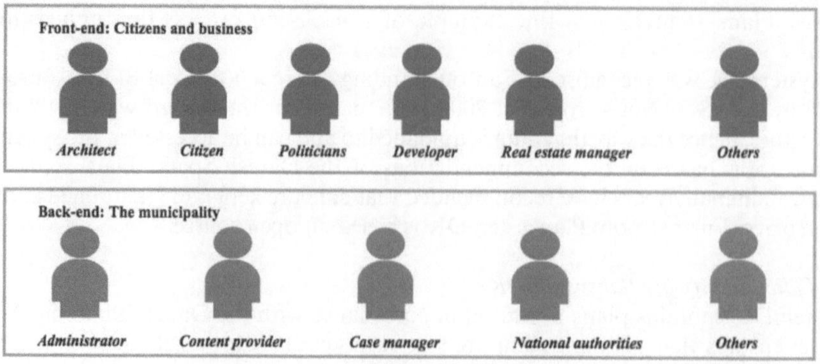

Figure 11. Front-end and back-end users of the digital municipal plans (after Plan09 2010).

4.4.3 *Prototyping processes*

A central conclusion in the project is that user specification and prototyping is essentials elements in the process though it is a difficult task to meet the needs of the very different end-users in one solution. Many of the municipalities are working with Content Management Systems (CMS) in order to integrate the workflows in the digital plans. The result is often a product with a low appeal to the citizens who find the solutions too incomprehensible.

The first generation of digital municipal plans leaves an impression of a narrow focus on digital representation of the municipal plans instead of rethinking and reorganising the way of working with municipal planning. There is also of lack of reusing data and services from other public registers. For instance there is no active link between the mandatory plans in PlansystemDK and the digital municipal plans.

4.5 *Front-end processes*

4.5.1 *Borger.DK*

One of the initiatives pointed out in the UN survey as well as the OECD report is the portal for citizens Borger.dk (www.borger.dk). This portal is together with the portal for enterprises Virk.dk (www.virk.dk) key issues in the Danish development of front-end services. The two portals have been tools to achieve fully digital communication between the citizens, the private sector and the public sector (OECD 2010).

Borger.dk has two entrances for information. First of all the portal is an entrance to general information about public service and as a portal the site links to other relevant websites. Secondly Borger.dk delivers targeted information to the individual citizen by the use of a personal sign-on (in Denmark there is a single sign-on signature for all public information as well as e-banking). After logging into the system the portal displays the information registered in public registers—tax, property information, health.

4.5.2 *Digital application for building permits*

Among the services is a specific service for house owners linking to central national geodatabases—six municipalities including Rudersdal is part of a new initiative experimenting with an online service, where the house owners can apply for building permits (www.boligejer.dk/byggetilladelse). The interface gives access to sketching tools as well as specific property information—the map shown in Figure 12 is based on FOT-data.

4.5.3 *Creating value of services*

The main portal Borger.dk is a typical example of front-end services with a strong focus on the citizens as end-users, and citizen needs have been incorporated by means of personas

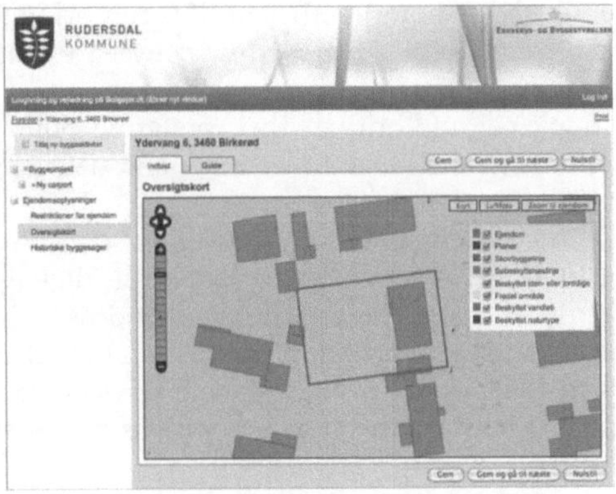

Figure 12. Applying for building permit at a specific address in the municipality of Rudersdal.

representing different types and life-stiles, while the approach to the project concerning digital building permits is more technical due to its economic potential.

5 BRINGING GI TO ITS FULLEST POTENTIAL

5.1 *Innovation potentials in GI-communities*

The understanding of geographically based data as an inherent part of e-Government is a result of a persistent effort from the Danish GI-communities. By forming powerful interest groups ad hoc the GI-communities have succeeded in introducing geodata to a series of public initiatives and hence, in the latest national strategy for e-Government, both geodata and the way of working in communities with a common focus area is an inherent part of the understanding of the way to a more efficient e-Government. Likewise the need of standardisation, the use of geographical information systems and IT-architecture is a result of the interrelationship between the GI-communities and the formal national stakeholders. In order to obtain a deeper insight in the innovation potential of GI-communities a SWOT analysis is carried out in order to identify the strengths, weaknesses, opportunities and threats.

5.2 *Strengths*

Referring to the UN the *collaboration between actors* is a matter of great importance in the process of confronting the digital and developmental divides regarding technological as well as organisational matters. The four Danish cases illustrate how GI-communities function as communities of practice where practitioners develop a shared practice based on a shared repertoire of resources. FOT is in this respect a very good example of different actors committing to a project based on *common values*.

Rudersdal is actually the only municipal, which could serve as an example in all four cases, though in general the different back-end related aspects illustrated by the cases is confirmed by a web questionnaire to Danish authorities which conveys a picture of a model with a strong emphasis on the sub-national levels where implementation is done by technicians in cooperation with leaders (Hansen et al., 2010). Even before the recent law on geographic information (Inspire) more than half of the municipalities had a strategy for the use of geographical information and the development of better administration systems and in the end better public service.

A specific strength is the GI-based working environment itself enabling *prototyping processes* due to visible and to some extend tangible artefacts.

5.3 *Weaknesses*

The citizen portal Borger.dk shows how an efficient SDI can bring forth new and targeted information. Though there seems to be *a missing link* between the back-end and the front-end innovation processes. OECD refers to the results of a recent survey where 50% of the respondents state their municipal authority as their preferred point of contact while only 20% mention the citizen portal as their favourite entry-point despite government intentions of Borger.dk as the main entry to all public services.

Additionally almost all respondents (98%) are delivering services through the website of their own organisation (OECD 2010). Similarly the case of the digital municipal plans shows how it is difficult to meet the needs of the very different end-users in the specific technical solutions even if user specification and prototyping are essentials elements in the process of development. Likewise there is a lack of reusing data and services from other public registers and portals as the missing link between the mandatory plans in PlansystemDK and the digital municipal plans illustrates.

Also problems caused by *lack of documentation* of the dynamical processes of informal innovation environments could be a matter for further concern, which also relates to the

OECD pointing out how a comprehensive picture of the potentials of initiatives across the whole public sector is missing (OECD).

5.4 *Opportunities*

Due to the UN as well as the OECD there is a *need for a more holistic approach* as means of navigating in the transformational mode of developing connected governance. The OECD expresses further more the *need for common visions and pictures*. This could lead to a broader focus on SDI potentials as means of enhancing capacities within the public sector.

On one side the integrative and communicational potentials of spatial information due to the visual, physical and geographical components relating to common understandings as means of conceptualisation and prototyping processes regarding daily life as well as professional working environments (FOT 2010).

On the other side core SDI aspects as the need for improving data standardisation and the use and flow of public sector information within and across levels of government to meet the needs of citizens, business or government and to improve cross-governmental collaboration by clarifying the primary holder of core data and the conditions for access or reuse by multiple parties among the central conclusions in the OECD review (OECD 2010).

Especially combined with the suggested development of core capacities and skills in the public sector to meet the growing demand for project and programme management and design—particularly in the case of large ICT projects—and to support successful e-government implementation, full exploitation and leveraging of e-government projects and advances in the modernisation agenda.

5.5 *Threats*

Considering the increasing number of common components in large scale IT-project the OECD points out how a greater co-ordination in developing and maintaining the focus on clearly defined objectives as well as envisaging the development of new ones will allow motivated and efficient decision making and avoiding any waste of resources. Though a fixed and well-organised structure leaves *limited room for experiments* and perhaps put a damper on innovation regarding technological as well as organisational matters. In the longer term this may reduce competiveness in a globalised world (Hansen et al., 2010).

6 CONCLUSION

The aim of this research was to analyse the potentials within the GI-community as means of facilitating the actual move from government dispensing services via traditional modes to an emphasis on an integrated approach focusing on enhancing the value of services to the citizen. Though a crucial point is that an increase in the value of services is not possible without consolidating the way the back-end systems and processes work to bring about the front-end service delivery.

The four Danish cases relating to different levels of e-Government re-engineering processes and GI-based working environments indicates the potential when taking advantage of employee experiences as means of user centered innovation creating value in front-end services as well as back-end procedures. Among 98 Danish municipalities only Rudersdal could serve as a specific example represented in all four cases, which indicates the importance of an organizational culture facilitating innovation as well as strategic efforts.

The GI-working environments facilitate collaboration and prototyping processes due to the integrative nature of tasks as well as technology possessing the potential of making common values visible and iterative processes tangible. Considering re-engineering processes in a larger scale for instance SDI projects at the national or inter-municipal level function as prototypes making visions visible and future identities and working conditions sharable among practitioners.

REFERENCES

Commission of the European Communities (2003). The reuse of public sector information, Directive 2003/98/EC of the European Parliament and of the Council, *Official Journal of the European Union*, L345, pp. 90–96.

Commission of the European Communities (2007). Directive 2007/2/EC of the European Parliament and of the Council of 14 March 2007 establishing an Infrastructure for Spatial Information in the European Community (INSPIRE), *Official Journal of the European Union*.

Dael, S., Frederiksen, P. & Jørgensen, L.T. (2008). The Danish way to a national spatial data infrastructure, *The International Archives of the Photogrammetry, Remote Sensing and Spatial Information Sciences*, Vol. XXXVII. Part B4, pp. 1721–1728, Beijing 2008.

Danish Enterprise and Construction Authority (2007). *Programme for user-driven innovation*, Ministry for Economics and Business Affairs, http://www.deaca.dk/userdriveninnovation

Danish Enterprise and Construction Agency (2008). *Medarbejderdreven innovation* (employee driven innovation), Ministry for Economics and Business Affairs, http://www.ebst.dk/file/15165/medarbejderdreven_innovation_aftale.pdf

Danish Government (2003). *Whitepaper on IT architecture*, Danish Ministry of Science, Technology and Development, Copenhagen (in Danish).

Danish Government (2008). *Lov om infrastruktur for geografisk information* (implementation of Directive 2007/2/EC of the European Parliament and of the Council of 14 March 2007), Ministry of the Environment, Copenhagen.

Danish Government, Local Government Denmark and Danish Regions (2007) Towards Better Digital Service, Increased Efficiency and Stronger Collaboration—The Danish e-Government Strategy 2007–2010, Copenhagen.

FOTDanmark (2010). *Specification FOT 4.0.*, FOTDanmark, Copenhagen. http://www.fotdanmark.dk/Materiale/Files/Specifikation+4.0+p%c3%a5+engelsk

Hansen, H.S, Schroeder, L., Hvingel, L. & Christiansen, J.S. (2010). Capacity building for SDI Implementation—A Danish Case Study, *International Journal of Spatial Data Infrastructure Research*.

Jarmbæk, J. (2009). The Danish Way: Development of the Danish Spatial Data Infrastructure through binding collaboration, Paper presented at *GSDI 11*, Rotterdam, http://www.gsdi.org/gsdiconf/gsdi11/papers/pdf/179.pdf

Masser, I., Rajabifard, A. & Williamson, I. (2007). Spatially Enabling Governments Through SDI implementation, *International Journal of Geographical Information Science*, 22(1), 5–20.

OECD (2010). OECD 2010, Efficient e-government for smarter public service delivery Denmark 2010, *OECD e-Government Studies*.

Plan09 (2010). *Digital Kommuneplan: Kravsspecifikation gennem brugerinvolvering*, (Digital municipal plan: User centred requirement specification). http://digitalkp.plan09.dk/download/fase1/Digital%20 Kommuneplan%20Oplaeg%20v%20 Klaus%20Bundvig.pdf

Rajabifard, A., Mary-Ellen, F., Feeney, M.-E. & Williamson, I. (2002). Future Directions for SDI Development, *International Journal of Applied Earth Observation and Geoinformation*, vol. 4, pp. 11–22.

Schroeder, L., Hvingel, L. & Hansen, H.S. (2010). E-Government and Geographical Information Based Collaboration Patterns. *Lecture Notes in Computer Science*, vol. 6267, pp. 204–218.

Servicefællesskabet for geodata (2004) *Basisdata* (basic datasets), Copenhagen.

UNECE (1998). *Aarhus Convention*—UNECE convention on Access to Information, Public Participation in Decision-making and Access to justice in Environmental Matters.

United Nations Department of Economic and Social Affairs (2008). *UN E-Government Survey 2008—From E-Government to Connected Governance*, United Nations Publication, New York.

Wenger, E., McDermott, R. & Snyder, W. (2002). *Cultivating Communities of Practice: A guide to managing knowledge*, Harvard Business School Press.

World Economic Forum, INSEAD (2009). *The Global Information Technology Report* 2008–2009, Geneva.

Urban and Regional Data Management – Zlatanova, Ledoux, Fendel & Rumor (eds)
© 2012 Taylor & Francis Group, London, ISBN 978-0-415-67491-1

Engaging with local communities: A review of three years of community mapping

C. Ellul
Department of Civil, Environmental and Geomatic Engineering, University College London, UK

L. Francis
Mapping for Change, London, UK

M. Haklay
Department of Civil, Environmental and Geomatic Engineering, University College London, UK

ABSTRACT: Web mapping provides a powerful way to engage with local communities and Mapping for Change (MfC) has set up a series of local Community Maps around the UK, with each map highlighting local issues, planning concerns, development sites and environmental issues. The precise content of each of the maps differs from site to site and has been determined in full consultation with each Community. Communities add information to the map themselves and act as sensors in their local environment, highlighting both issues and positive factors.

This paper describes the evaluation of fifteen Community Maps with regard to the number of active users, people viewing the data, the types of data being added and the relationship between activity on the map and activity within the groups. The results obtained are of interest to local government community engagement processes—MfC represents one of the longest lasting Community Mapping activities to date.

1 INTRODUCTION

1.1 *The importance of web mapping*

A UK Government White Paper entitled "*Communities in Control—Real people, real power*" (GOV 2008) published in July 2008 reflects ongoing efforts by central and local government to increase community participation and generate a more vibrant local democracy. Access to information forms an important part of this process. More recently, the newly elected Prime Minister, David Cameron, introduced the concept of the *Big Society*, which encourages local communities to take more responsibility for issues in their neighborhood, allowing charities and social enterprises to participate to a greater extent and devolving power from government to citizens. People should "*feel both free and powerful enough to help themselves and their own communities*" (Cameron 2010).

Internet-based Geographical Information Systems (Web GIS) have long been seen as part of this information dissemination and process, as well as being a tool to encourage participation in local decision making, falling under the general umbrella of Public Participation GIS (PPGIS, which focuses on the use of GIS by non-experts and occasional users, Haklay and Tobon 2003). They have importance both in the top-down 'information' dissemination processes and the bottom-up 'participation' processes forming part of local activities (Talen 2000). In particular, community-sourced mapping is of importance—information regarded as relevant by community groups is not necessarily the identical to that provided by local government (perhaps due to issues of scale, currency or trust). Community data provides a useful insight into the issues and concerns of local groups, givingan on-the-ground live monitoring system for the urban environment. It also overcomes the fact that Government-issued information may tend to ignore local knowledge (Elwood, 2007).

This paper evaluatesan ongoing 3.5 year Community Mapping project where a flexible web based mapping system (*Community Maps*) was developed to provide a customizable resource for a growing number of diverse community groups in the UK. Issues including frequency of use of the Community Maps, the activity levels of various members and activity levels in each group are examined for a total of fifteen Community based projects, using data gathered between April and September 2010.

1.2 *Mapping for change and community maps*

Mapping for Change (MfC 2010) is a social enterprise that exists to support the development of sustainable communities. The organization was set up as a partnership between University College London and the London 21 sustainability network, following a very successful partnership that commenced with the establishment of the online London Green Map in 1997 (LGM 2010). MfC specializes in providing participatory mapping services to communities, voluntary sector organizations, local authorities and developers using a suite of innovative tools for communication. The organization's focus is on engaging and empowering communities to make positive transformations to their localities and MfC works in partnership with local government, third sector groups and other organizations.

Currently, MfC is working with seventeen community groups in the UK (fifteen of which have been evaluated here). Each group has been provided with a customized *Community Map*, known as a Minisite (a full list of these can be found at http://communitymaps.org.uk—clicking on a link takes the users to the selected Minisite). As will be seen in Section 3, the groups are diverse both in their interests (which appear as *themes* or layers on the maps) and consequent data requirements for the map and in their geographical location. It should also be noted that online mapping forms one of the many engagement activities used by MfC and partners with the groups.

2 THE COMMUNITY MAPS WEBSITE

A number of free community mapping options can be identified—sites such as Google's My Maps option (Google 2009), Platial (2011) and Microsoft's Bing Maps (Microsoft 2010) allow the addition of points, lines or polygons on the map. Flickr (2011) allow users to geo-tag photos and share them with others. The Environmental Systems Research Institute (ESRI) allow users to contribute to a centrally stored community map (ESRI 2010). However, initial discussions with community groups highlighted a number of limitations of existing approaches, and a user-centred design approach was taken to capture requirements for the Community Maps Web GIS (see Norman and Draper 1986 and Haklay and Tobon 2003 for details of user-centred design). *Users* in this case refers both to members of the Mapping for Change team and to the end-users of the website—i.e., the community groups and individuals who want to share knowledge about their locality. Initial functional requirements were gathered through a series of discussions and via workshops with the various groups, who were encouraged to express views as to what types of information should be included on the map. Following first deployment, meetings with the groups resulted in feedback on the site and suggestions for improvements and additional functionality. Issues addressed by the custom development are summarised below (and described in greater detail in Ellul *et al.*, 2009). A series of administration tools have also been developed to allow non technical members of MfC staff to automatically add new group maps and new themes to these or existing maps.

2.1 *Moderation*

A number of the issues raised by community members—for example noise, planning applications or pollution—may be considered controversial either within the community or by external bodies with whom the community is working to change the situation. A process of moderation is required to ensure that no inappropriate data or commentary appears on the maps. Additionally, users must register before adding data.

2.2 *Themes*

Many existing community map sites allow users to enter data in an open format, usually by allowing free text entry into one field. However, this approach does not provide any mechanisms to group the data into a more logical grouping (a grouping called a *layer* or a *theme*) in GIS terms. Without a *theme* approach, users cannot view only data of interest. A free text approach (such as that taken by Google's My Maps, where the user simply types information into a single text box) also limits the standardisation of data capture (and may complicate subsequent analysis of the data). MfC therefore uses a more structured approach to allow the community to collect and present information in a coherent format and to ensure that all required (mandatory) information is gathered as part of the capture process. Importantly, each Community Group is free to identify which themes they require on their map—this is done when the map is set up, through a process of consultation with MfC staff. The types of content they wish to be shown and are interested in are thus the content shown on their map. Thus, each individual Minisite will have a different list of themes from all the others, depending on Community requirements. Themes are displayed in alphabetical order. Figure 2c shows the legend listing the layers on one Minisite.

2.3 *Search tools*

Using search functionality on sites such as Google Maps (Google 2009) provides a range of results not necessarily related to the content, geographical area or theme of the current map and does not search database content. For MfC, searches are limited to the community site in question.

2.4 *Community identity*

A feeling of community can be more readily created through the use of a site specific to each community, with the name of the group clearly displayed. Equally importantly, the concept of a 'boundary' for the area of interest of each community can help to reinforce the sense of geographic community that characterizes MfC activities. Communities can make use of

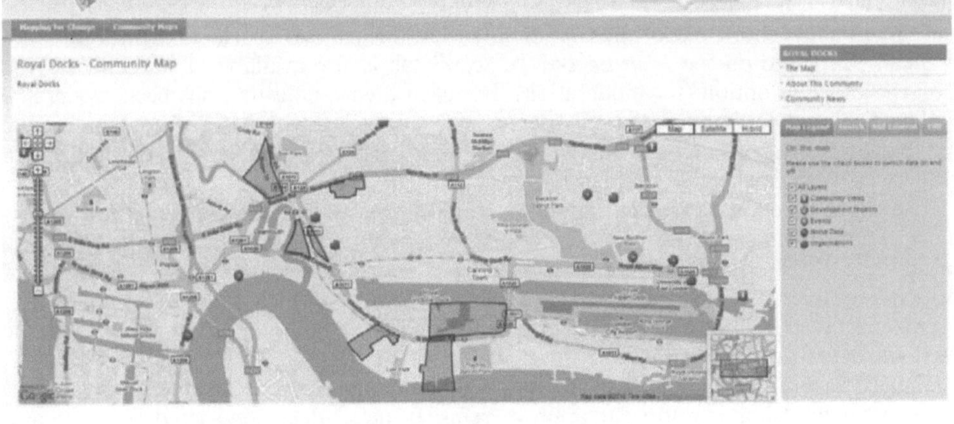

Figure 1. An overview of the royal docks Minisite.

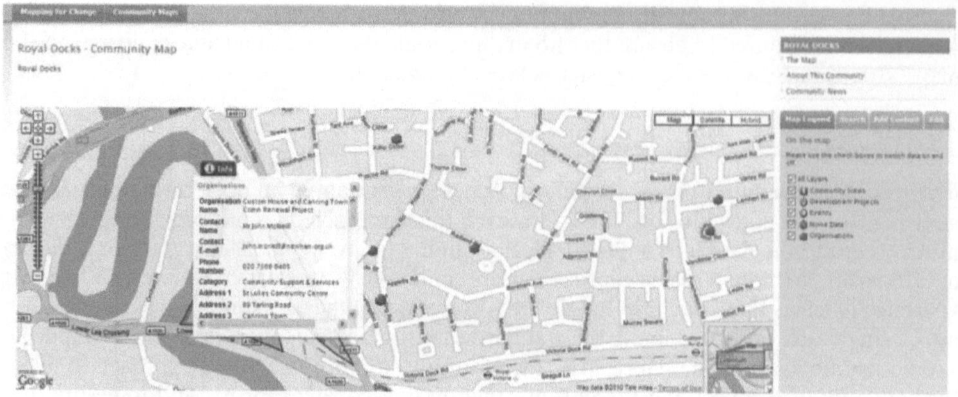

Figure 2a. The map showing an Information pop-up.

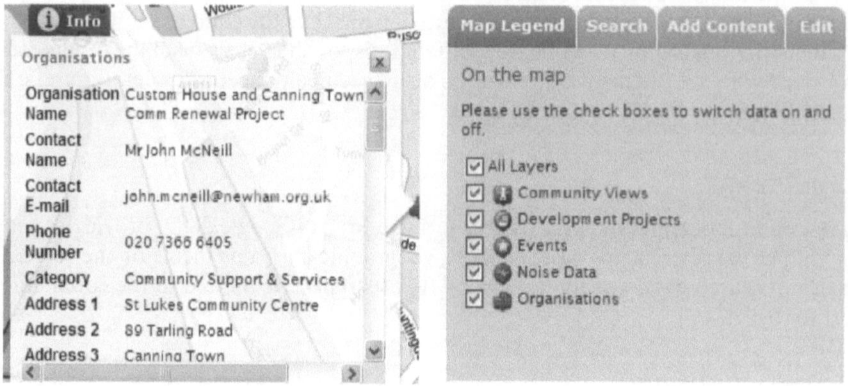

Figure 2b, c. Detailed view of the information pop-up and legend.

their own branding and colour schemes, facilitating integration of the map with other online information.

The Google Maps Application Programmers Interface (API) provided an appropriate tool kit for MfC, in particular due to its dominance (Google reports a total of 50,000 Google Maps-based mash-ups in 2007, Tran 2007) and Web 2.0 high performance map display, usability, ease of implementation, integration of data from multiple sources (Haklay *et al.*, 2008). The website makes use of PHP and a MySQL database.

Figure 1 above shows one of the Community Maps Minisites. Associated themes are automatically displayed on the *Map Legend*. A *Search* tab is also available, along with an *Add Content* and *Edit* option (for moderation). For each theme, data that has been moderated will be displayed on the map and any user, whether registered with MfC or not, can browse the data on the map. Clicking on an item on the map will display information about this item (Figure 2a, with details in Figure 2b and c).

The search function can be configured to search all themes or a specific theme and returns a list of matching items (performing a free text search on the data in the database). Clicking on an item displays it on the map with associated information. On clicking the *Add Data* option, registered users are presented with a list of themes to which data can be added. The user is prompted to draw a point, line or polygon on screen and then fills in a form with the attribute data for the new entry.

Finally, the *Edit* tab on each map provides functionality for local or MfC administrators to moderate data. Administrators can approve, refuse or delete data. Registered users can also make changes to their own data.

3 EVALUATING THE COMMUNITY MAPS

3.1 *Evaluation method*

As part of the development process, the *Community Maps* website incorporates Google Analytics tools. These record visitors to the site, including number, length each visitor remains on the site and geographical location of the visitor. In conjunction with this, the bespoke *Community Maps* code captures the details of any user who adds data to the site, and when this data was added. Both these sources of information were evaluated in conjunction with known activity for each group (provided by the MfC team). For the Google Analytics processes, data was collected for the period between April and September 2010. Data additions were evaluated over the entire period of the life of each Minisite and a review was carried out in Summer 2010 to summarize the types of content requested by users on the maps.

3.2 *General growth of community mapping*

MfC's Community Mapping tools were initially developed as part of a research project targeting four groups in the London (Archway, Royal Docks, Hackney Wick and Thames Ward). Each group was selected due to the rapid changes taking place in their respective local environment—for example, the Hackney Wick group is located adjacent to the London 2012 Olympic Site—and the research project ran from 2007 to 2009. Since that period, however, both the number and the scope of community groups involved in Mapping for Change have grown. The number of Minisites now totals 17 (with more imminent) and focus has moved beyond the built environment to encompass noise issues, environmental issues, sustainability, schools, education, local authorities and university engagement with local communities. In recent months, MfC has also been approached by a housing association, has been involved in a challenge programme for innovators and is currently participating in health-related, traffic-safety related and pollution-related funding applications. Alongside growing scope, the geographic focus has also moved beyond London to cover diverse areas in England and geographic extents range from very small urban communities to more rural groups extending over large proportions of Counties. A list of the evaluated Minisites and their location and specific interest is given in Table 1.

3.3 *Number of data points*

The increase in interest in Community Mapping is also reflected in the growth in the amount of data added to the maps—a total of 282 data items were added in 2008, 253 in 2009 and 592 have been added to date this year (September 2010). The significant increase in 2010 could at first be thought to be the result of the number of Minisites added during this period—a total of five have been added between January and September, compared to a total of approximately six in 2008/2009.

Interestingly, however, it is the first pilot Minisite, 'Royal Docks', established in September 2007, that is currently showing most activity. The Royal Docks area is adjacent to one of London's major international airports (London City Airport) and the high number of points relates to an intensive noise mapping activity that was carried out in the area during the eruption of the Icelandic Eyjafjallajökull volcano and subsequently. The activity provided useful data in relation to the specific impact of airplane noise on the neighborhood, as all planes were grounded for some days. The continued involvement of the Mapping for Change team with the Community Group permitted the community to continue taking noise readings across a wider area over a prolonged period following on from the initial project in 2007. As a result, they were in a position to take advantage of the overnight change in the noise situation caused by the volcano, and deploy a team of community members to undertake noise mapping. A total of 200 points have been added to the Royal Docks site this year. Figure 3 shows the general trend in number of points per theme –this is a relatively flat graph with the majority of the themes having had under 20 points added. Additional details can be seen in Table 2—as can be seen, it is the noise dataset with around 225 points that contains

Table 1. MfC Minisites forming part of this evaluation.

Name	Location	Est.	Established by	Target user group
Archway	London	2007	Better archway forum	Better archway forum
Arnhem wharf school	London	Feb 2010	Humanities education centre (tower hamlets local authority)	Arnhem wharf school
Brixton	London	Dec 2008	Local GP/environment group	GP patients and local residents (current feedback—not used because not enough data on the map currently and no time to add the data myself (GP)
Columbia school	London	Feb 2010	Humanities education centre (tower hamlets local authority)	Columbia school
Fortismere school	London	Feb 2008	Externally funded by city bridge—part of environmental justice project	Fortismere school
Hackney wick	London	Nov 2007	MCSC—funded programme	Local residents
John scurr school	London	Feb 2010	Humanities education centre (tower hamlets local authority)	John Scurr school
London climate resource map	London	Oct 2009	Mapping for change/low carbon communities network	London based individuals and organizations
London faith map	London	April 2009	Conservation foundation	London faith based organizations
Lycee Djignabo	Senegal	Sep 2008	Part of environmental justice project by MfC	Lycee Djignabo school
Marks gate	London	May 2008	Externally funded by city bridge—part of environmental justice project	Local residents
Newcastle climate action	Newcastle (NE England, Rural)	Jan 2010	MfC and Newcastle CVS (funded externally)	Local residents/groups
North dorset climate action	Dorset (SW England, Rural)	Dec 2009	MfC and dorset agenda 21 (funded externally)	Local residents/groups
Royal docks	London	Sep 2007	MCSC—funded programme, currently heavily utilized by FTF	Local residents
Thames ward	London	May 2008	MCSC—funded programme	Local residents

the most data (the second most popular theme is a general catch-all theme called '*Community Views*' with a total of 125 points).

Examining Figure 3and Table 2 in further detail reveals an interesting outcome of the flexibility provided by the MfC administrator toolkit (which allows the administrator to create a new themes for each minisite without having any programming or database skills)—namely, the presence of a large number of themes containing very little data. There are 61 themes in total and this large number of themes reflects not only the diverse interests of the Minisites, but the consequent requirement for each community group to capture slightly differently-focused information. For example, the Hackney Wick Minisite includes a theme called 'historical data'. A similar theme, 'History and Culture' has been created for the Brixton site, with a different data structure. Although the MfC administration tools make the creation of these themes a very simple tasks, differences make it more difficult for sites to access each-other's content, limiting the scope for interaction between groups who perhaps have similar interests.

3.4 *Visitor locations, numbers and length of visit*

Over the period between April and September 2010, a total of 1569 new visitors were measured, with an additional 935 returning visitors—these are identified by the *Google Analytics* tools via a cookie placed on the machine (Google, 2011). The Firefox browser was

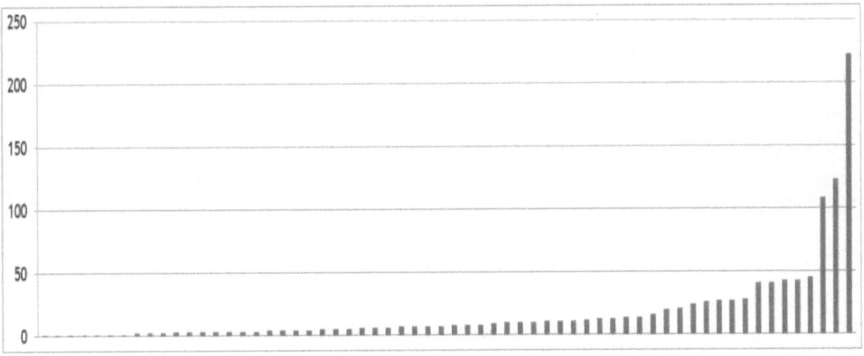

Figure 3. General trends in number of points per theme (see Table 2 for detail).

Table 2. Top 20 themes by number of points added.

Rank	Theme	# Points	Rank	Theme	# Points
1	Noise	222	11	Food related issues (Climate map)	27
2	General user comments	123	12	Things we would like to change	24
3	Things we love	108	13	Local projects (Hackney)	24
4	Environment groups	45	14	Buildings and energy (Climate map)	21
5	Community facilities	43	15	Interesting places	20
6	Community organisations (climate map)	43	16	Points of historical interest	16
7	Schools—memories	41	17	Green areas	14
8	Things we do not like	41	18	Schools	14
9	Questions	28	19	Recycling (Climate map)	13
10	Transport	27	20	Food related issues (Brixton map)	13

used by 38% of visitors to the site, closely followed by Internet Explorer at 37%. The slight dominance of Firefox reflects a trend towards the statistics collected by W3Schools (2010), who indicate that 30.7% of internet users use Internet Explorer and 45.8% use Firefox.

The range of countries from which visitors originate is also of interest, given the UK focus of the current Minisites. This is reflected in the visits, but a minority of visitors also originate elsewhere (Table 3)—anecdotally, many of these visits can be traced to specific events or blog-posts relating to the site or the activities of Mapping for Change.

Figure 4 shows the average number of seconds that users spent on each Minisite for the period of April to September 2010, as reported by the Google Analytics tools linked to the Minisites—the tools permit a breakdown to this level of detail as they monitor each URL, including parameters, submitted by users. As can be seen, out of the 15 Minisites active during this period, atotal of 8 show an average visit duration of over two-and-a-half minutes. Although this average cannot be differentiated by task (were the visitors browsing the data, adding data or site administrators?) it is significantly higher than a casual visit would require (Weinreich *et al.*, 2008 report a median stay of 12.4 seconds for first visits to a web page, having examined over 60,000 first page visits).

Table 3. Origin of visitors to the community maps sites.

Country/Territory	Visits
United Kingdom	1970
United States	111
Ireland	67
Germany	48
Canada	48
Spain	23
Italy	20
Australia	19
France	19
Philippines	16

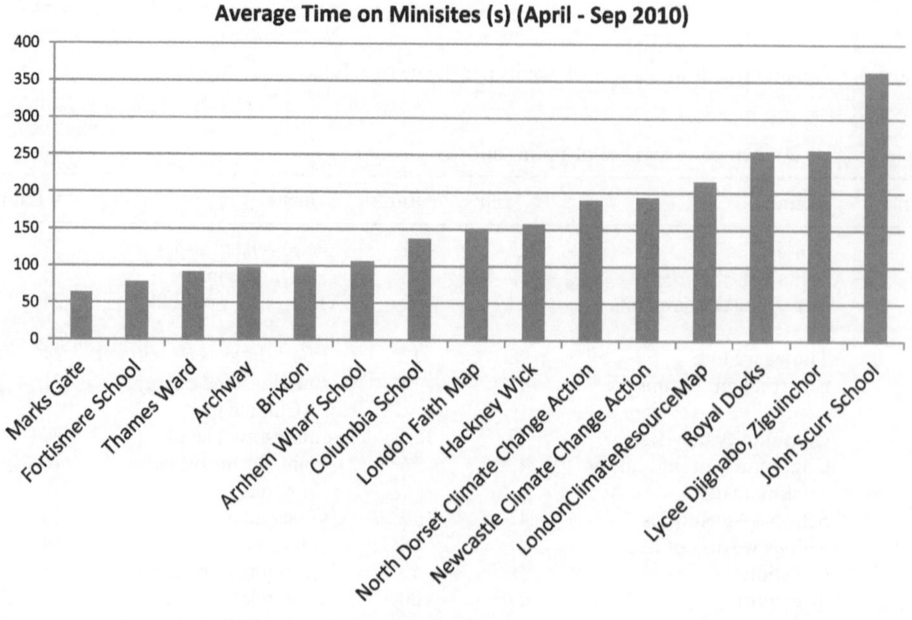

Figure 4. Average time on Minisites in seconds.

The ordered list of Minisites in Figure 6 can be compared, however, with the Unique Page Views statistic shown in Figure 5. This gives a measure of the number of different people who have visited each Minisite, and shows that Lycee Djignabo, which had an average visit duration of 250 seconds, was only viewed by 27 persons over the last six months, whereas the North Dorset and Newcastle Climate change sites had a total of over 300 unique visitors each.

The Page View statistics allow some initial inferences as to the activity of the local groups. For example, the four Minisites at the lower end of the Unique Page Views scale are those where there is no current active involvement from Mapping for Change and no commitment from the local group to maintain the site (indeed, the Lycee Djignabo site was set up during a visit from one of the team to a school in Africa, and has had no further interaction with MfC since that time). The lack of unique page views on the other school sites may be a result of summer holidays, which occur between July and September.

Activity levels could also be measured by examining the number of data points added to each site by individual users—the trend is shown in Figure 6, which plots number of points added by each active user. This reveals a challenge currently being faced by Mapping for Change (and reflected in the field of Volunteered Geographic Information)—i.e., the fact

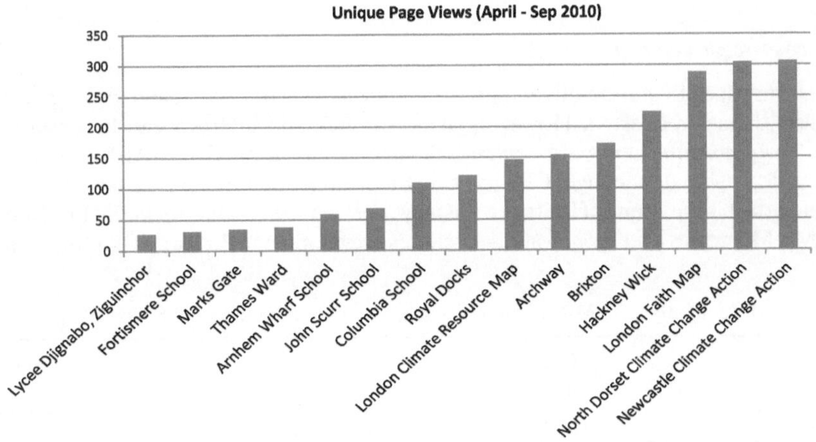

Figure 5. Unique page views.

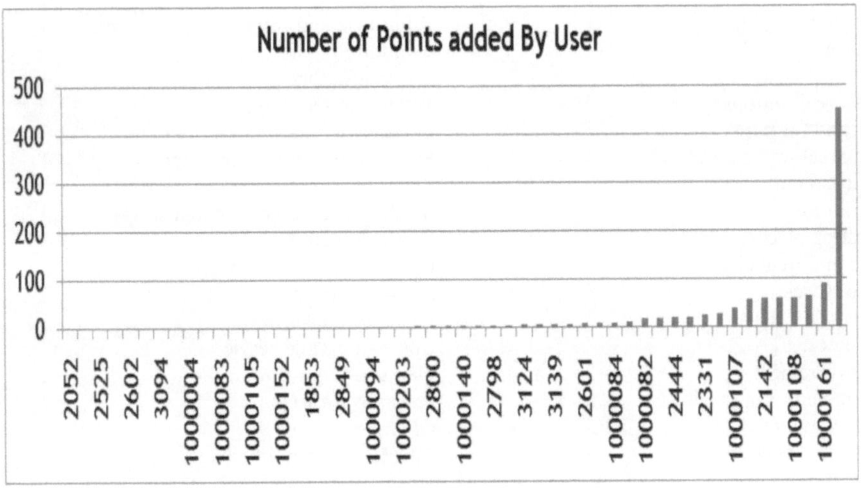

Figure 6. Trend in number of data points added per user (by userID).

that a few users are very active whereas most contribute little or no data. UserIDs 1000106 and 1000161 (second and fourth from right in the above graph) relate to two users from the Royal Docks group, adding a large number of noise readings. In addition to this, UserID 2139 (rightmost on the graph above) is in fact the MfC administrator. The high number of entries for the administrator highlights the two models for Minisites that have emerged when working with community groups, each of which reflects the differing skills and technical confidence levels of group members. In the first model, the community is actively involved in adding and moderating data. In the second model, the community members are more comfortable with less technically-focussed activities such as adding points to paper maps to capture data. This data is then added to the site by the MfC team (hence the statistics for UserID 2139).

Although the relatively complex process required to add data could be seen as a factor leading to the two models above, discussion with the users has highlighted that his is not in fact the case—in fact general feedback on the workflow has been good, in particular when the user guide is used as a training tool. The experience of the MfC team has highlighted that one of the main issues is that maps having less data capture activity have generally been commissioned by external forces and rarely by the communities themselves. This may lead to the result that the communities have no sense of ownership of the map, and no vested interest in keeping it up to date or in adding information.

3.5 *Themes requested by the groups*

Given the very specific nature of each group's requests, it is not possible to aggregate directly the general themes requested. However, general trends and interests can be observed—for example *environmental issues* are covered by themes including Green Areas, Noise, Waste and Recycling, Climate Change, Climate and Environmental Groups, *historical and cultural issues* include Culture and Heritage, History and Culture, Historical Data, Interesting Places, My Memories and *urban change* includes Development Projects, Olympics, Hackney

Table 4. Most frequently used MfC community maps themes (adapted from Sharp 2010).

Theme name	Occurs on # maps	Theme name	Occurs on # maps
Events	9	Conservation areas	2
Things we love	7	We wish	2
Community views	6	Response/answer	2
Transport	6	Noise data	2
Community facilities	5	Education and training	2
Questions	5	Community and voluntary organisations	2
Things we do not like	5	Waste and recycling	2
Food (and farming)	5	Classes	1
Green areas and natural environment	4	Home improvement and repairs	1
Organisations	5	Made here and independent shops	1
Interesting places	4	Youth activities	1
Health and wellbeing	4	Children and young people	1
Resuse/recycle	3	Health services	1
Answer	3	History and culture	1
My memories	3	Local shops and services	1
Our environment	3	Gainsborough landmarks	1
Development projects	3	Hackney projects	1
Historical data	3	Olympics	1
Climate and environmental groups	3	Local economic/time-bank	1
Buildings and energy	3	Renewable energy and advice	1

Projects, Global Development and Area Regeneration. The large majority of the maps also have themes for local events and listings of local organizations. Table 4 lists the count of the themes—as can be seen, many themes relate to only one map, indicating that the information is only of interest to one specific group.

4 KEY OUTCOMES AND FURTHER WORK

4.1 *The communities must drive the maps and content—flexibility is key*

Both the growth in reach (the diverse types of groups) and in geography (from London to other parts of the country, and from Urban to Rural) of the MfC Minisites serve to confirm the potential of Community Mapping. The above measures of activity highlight the fact that it is the community-driven maps (i.e., those created at the request of the community rather than external groups), along with maps having current activity in terms of community engagement (in particular with the MfC team), that are more succesful. For example, the Schools maps were integrated into curricular activities relating to local environment and geography, and once these had been covered in class, became relatively inactive. Maps such as the Dorset and Newcastle Climate Action maps, created at the request of existing groups, show a more consistent level of activity—the maps were set up by the groups to support their projects, and they also recruited participants. Anecdotally, ongoing engagement from the MfC team is also important to the success of a map—whether this takes the form of formal events or simple follow-on phone calls.

Work with the various groups has highlighted the importance of Community driven information as there may be a certain element of distrust in information provided by local authorities, in particular in relation to contentious issues such as new developments. The range of themes requested by the users, and the diverse nature of these themes from group to group, highlight the importance of a customised approach to community engagement. Group members are, as a whole, prepared to act as sensors of their local environment provided that they have an underlying interest in the issues at hand. Indeed, the diverse nature of the data provides a very valuable resource for the corresponding local authorities, who are able to gauge issues in their area without the confrontation that may arise through a more direct approach.

4.2 *The map is not enough*

MfC's experience shows that the map should be seen as only one in a range of community engagement tools, which may range from flyers to workshops or on-street activity to gather community perceptions and include regular dialog between MfC and each group. This need for a suite of engagement approaches is further necessitated by the digital divide. Indeed, current estimates state that approximately 66% of the population of the United Kingdom has access to the internet (Changing Media 2007) (although this is increasing year on year). However, this number decreases to 59% for the age group between 55 and 64, which is conversely more likely to participate in community activities (Dutton and Helsper 2007, Kitchen *et al.,* 2006), resulting in a fairly significant digital divide. Of those with internet access, a number may not have the confidence to perform tasks beyond searching and following hyper links (RAC News 2006). These basic browsing skills can be contrasted with those required to make use of a community map—zooming, panning, clicking on the map for information (this contrast is sometimes known as the secondary digital divide, Dutton and Helsper 2006, Hargatti 2006).

To overcome this issue, the MfC toolkit offers the ability to send text messages notifying users of local events and of planning applications in their neighbourhood. The use of paper-based mapping activities followed by data capture by the MfC team leads to additional activity on the Minisites in many cases, as users are then keen to see their data on the digital maps.

4.3 *Further work*

A number of additional avenues for further work can also be identified—the *Community Maps* concept has now been incorporated into two additional research projects—one on auto-ethnography of suburbs, and a second making use of *citizen science* (where members of the general public collects, analyses and discusses scientific data relevant to their life) to allow cyclists to measure the levels of pollution to which they are exposed during their commute. As part of this, a *Community Map* application for mobile phones will be developed.

Storing the data captured by users in a MySQL database opens up the possibility of further analysis of the data. Questions relating to the quality of user-generated spatial data can be examined, as well as spatial analysis of the data captured to and text analysis of any attribute information.

The user-centred approach taken towards website development is an ongoing process. Requests from users vary from the ability to set their own passwords to the ability to create entire Minisites themselves. Upload of video data, the creation of a discussion board and blog, the addition of data from newly released official data sources (for example planning applications, pollution information—these and many other datasets have been released as part of an ongoing government Open Data initiative in the United Kingdom, http://data.gov.uk/) and the ability to make use of both Google and Open Street Map mapping have also been identified as users have realised that data missing on one map source may be present on another (Joyer 2010). Such requests further highlight the level of engagement of users with the site, and with Community Mapping.

Finally, usability analysis will be performed on the existing web site to ensure that the low activity in terms of data addition is the result of natural community group dynamics (where one or two users are very active and others perhaps less involved) rather than the website design.

REFERENCES

Cameron, D. 2010. *Big Society Speech,* [online] Available from: http://www.number10.gov.uk/news/speeches-and-transcripts/2010/07/big-society-speech-53572, [Accessed 30 December 2010].

Changing Media (2007) *UK Internet Usage*, report prepared for the BBC Trust [online] Available from: http://www.bbc.co.uk/bbctrust/assets/files/pdf/regulatory_framework/service_licences/service_reviews/uk_internet_mkt_research.pdf, [Accessed: 25th May 2011].

Dutton, W.H. & Helsper, E. 2007. *Oxford Internet Survey 2007 Report: The Internet in Britain*, Oxford Internet Institute, [online], Available from: http://www.oii.ox.ac.uk/microsites/oxis/publications.cfm, Accessed: 25th May 2011.

Ellul, C., Haklay, M., Francis, L. & Rahemtulla, H. 2009. A Mechanism to Create Community Maps for Non-Technical Users. *The International Conference on Advanced Geographic Information Systems and Web Services—GEOWS*. Cancun, Mexico.

Elwood, S. 2007. Grassroots groups as stakeholders in spatial data infrastructures: challenges and opportunities for local data development and sharing *International Journal of Geographical Information Science* 22:1, 71–90.

ESRI 2010, *ArcGIS Online Community Maps Programme*, [online] Available from: http://www.esri.com/software/arcgis/arcgisonline/community-maps.html [Accessed 30 December 2010].

Flickr 2011, *Flickr—Share your Photos, Watch the World* [online] Available from: *http://www.flickr.com/* Accessed 25th May 2011.

Google 2011—Cookies & Google Analytics, Available from: http://code.google.com/apis/analytics/docs/concepts/gaConceptsCookies.html, [Accessed 25th May 2011].

GOV 2008, *Government White Paper—Communities in Control, Real people, real power*, [online] Available from: http://www.communities.gov.uk/publications/communities/communitiesincontrol, Accessed: 25th May 2011.

Haklay, M., Singleton, A. & Parker, C. (2008). Web mapping 2.0: the Neogeography of the Geoweb, Geography Compass, forth coming Haklay, M.E., Tobón, C. (2003). Usability Evaluation and PPGIS: Towards a user-centered approach. International Journal of Geographical Information Science 17(6), 577–592.

Hargittai, E 2002. *Second Level Digital Divide: Differences in People's Online Skills*, First Monday, Vol 7, No. 4 [online], Available from: http://chnm.gmu.edu/digitalhistory/links/pdf/introduction/0.26c.pdf, Accessed: 25th May 2011.

Joyer, J. 2010. OpenStreetMap for Community Maps, MSc. Disseration, Department of Civil, Environmental and Geomatic Engineering, UCL (Unpublished).

Kitchen, S., Michaelson, J., Wood, N. & John, P. (2006). *Department for Communities and Local Government—2005 citizenship survey—Active Communities Report* [online], Available from: http://www.communities.gov.uk/publications/communities/2005citizenshipsurveyactive, Accessed: 25th May 2011.

LGM 2010, The London Green Map [online], Available from: http://greenmap.london21.org/ [Accessed 30 December 2010].

MfC 2010, Mapping for Change [online] Available from: http://www.mappingforchange.org.uk/ [Accessed 30 December 2010].

Microsoft 2010. *Microsoft Bing Maps* [online] Available from: http://www.bing.com/maps/ [Accessed 30 December 2010].

Norman, D.A. & Draper. S.W. 1986. User Centered System Design; New Perspectives on Human-Computer Interaction Lawrence Erlbaum Associates, Inc. Mahwah, NJ, USA.

Platial 2011, *Platial—Who and What's Nearby*—[online] Available from: *http://platial.com* Accessed 25th May 2011.

RAC News 2006. *Map Skills lost on younger generation*, [online] Available from: http://www.aviva.co.uk/media-centre/story/2836/map-skills-lost-on-younger-generation/, Accessed: 25th May 2011.

Sharp, R. 2010. *The Effective Use of Public Sector Information for Community Engagement,* Unpublished MSc. disseration, Dept. of Civil, Environmental and Geomatic Engineering, UCL.

Talen, E. 2000. Bottom-Up GIS—A New Tool for Individual and Group Expression in Participatory Planning, *APA Journal*, Vol. 66, No. 3.

Tran, T. 2007. Google Maps Mashups 2.0, Google Lat-Long Blog, posted 11/7/2007, http://google-latlong.blogspot.com/2007/07/google-maps-mashups–20.htmlAccessed 25th May 2011.

Weinreich, H., Obendorf, H., Herder, E. & Mayer, M. 2008 Not quite the average. *ACM Transactions on the Web*. Volume 2 (1):1–31 [Accessed May 6, 2011].

W3Schools 2010. *Browser Statistics—Web Statistics and Trends* [online] Available from: http://www.w3schools.com/browsers/browsers_stats.asp [Accessed 30th December 2010].

Part III: Environmental monitoring and assessment

Urban and Regional Data Management – Zlatanova, Ledoux, Fendel & Rumor (eds)
 ISBN 978-0-415-67491-1

Specifying spatial attributes and relations in Urban Heat Islands: A generalizable model applied to Trento, Italy

D. Vettorato
University of Trento, Italy

D.C. Prosperi
Florida Atlantic University, Fort Lauderdale, USA

ABSTRACT: The Urban Heat Islands (UHI) affect the quality of life and energy consumption in urban settlements. Urban morphology has an important role in shaping of UHI but the knowledge of the specific relationships and impacts of morphology elements on this phenomenon is still unclear. The paper present a method based on technical aspects of data assembly, measurement and estimation procedures to make specific these relationships. In particular GIS and multivariate statistical techniques are used to find correlations between surface temperature and urban morphology aspects. The methodology is calibrated for Trento, Italy, but it is perfectly general. The methodology's test shows interesting empirical results that suggest some spatial strategies for the urban design and planning in order to contribute to the reduction of the Surface UHI effect. The spatial perspective of the analytical framework that allow direct consideration of the spatial distribution of the UHI effect is finally discussed in the perspective of an intelligent city planning that can neutralize the intensified UHI effects brought about by unlimited growth of cities, raising densities, and climate change. Outlines of ongoing modeling issues conclude the paper.

1 INTRODUCTION

The Urban Heat Island phenomenon (UHI) is defined as the difference in temperature between urban areas and the surrounding rural areas (Oke, 1987). The overheating of urban areas adversely affects the quality of life in urban settlements by, among other things, increases the need for additional energy use, for example, to cool buildings. Considering that more than 75% of the world's population lives in urban and/or suburban areas (UN, 2002) and that global warming is becoming more and more evident, urban heat islands are an actual and emergent problem to study and ameliorate.

Despite the global importance of the UHI phenomenon and its implications in terms of sustainability and quality of life of cities, very few operational instruments are available to design local climate mitigations strategies and plans at the urban scale (CABE, 2002). Methodologies based on simple remote sensing observations are simply descriptions and analytical studies are plagued with place-specific results. The purpose of this paper is to contribute a potential methodology that is both more general and more spatial, to more fully detail the spatial attributes of the problem.

The paper is organized as follows. In the next section, we briefly describe the Urban Heat Island (UHI) concept, including both its manifestation and its relationship to the built environment, including a statement of ongoing research needs. After describing the specific research problem of calibrating a methodology focused, but not specific to Trento, Italy, the methodology section contains elements related to model structure, calibration, and analysis. The fifth section presents the results of the analysis. Finally, the paper concludes with a discussion of ongoing modeling issues, which are briefly outlined.

2 THE UHI PROBLEM AND ITS SPECIFICATION

2.1 *Simple descriptions*

For some time, the only way to spatially visualize the UHI phenomena was interpolating different temperature surveys recorded by terrestrial sensors. The low spatial density of these sensors, even if their precision were very high, never permits to obtain maps with a high spatial resolution. According to Gartland (2008), the magnitude and the spatial importance of a heat island were not fully appreciated until they were first visualized from the air in the 20th century thought the urban surface temperatures maps. Satellites and specially equipped aircraft are able to map temperatures on the Earth's surface and have found very distinguishable hot spots in and around urban areas all over the world. "Remote sensing can be used to find temperatures and other characteristics of *surfaces*, for example, roofs, pavements, vegetation and bare ground, by measuring the energy reflected and emitted from them. Specialized equipment on aeroplanes or satellites is used to take pictures of the visible and invisible energy radiating from cities and their surroundings" (Gartland 2008).

Various studies have been done to determine how surface temperatures affect air temperatures in urban areas (Fukunagam 1990, Imamura, 1989). Relationships have been found between the remotely sensed surface temperatures and the air temperatures for different cities. These relationships have been found to be highly dependent on spatial patterns, geography and weather conditions, so in cloudy, windy weather, the effects of surface temperatures on air temperatures are diminished. However, even if windy conditions could mitigate the overheating phenomena of urban areas, the structural determinants of the UHI effect can be spatially identified and mapped.

Unfortunately the relationships between surface temperature and urban characters generally apply to a particular urban area only, so the correlation for Paris or London (Watkins et al., 2002) cannot be extended to a city with different climate, geography or development patterns (Gartland, 2008). Heat islands in urban areas tend to display typical characteristics, but the intensity and timing of heat islands vary at each location. The heat island in each community ultimately finds its own unique balance between temperatures and energy flows, based on the area's terrain, construction and weather conditions.

2.2 *Urban morphology literature*

Urban morphology has an important role in shaping of urban heat island phenomena. Intelligent city planning can neutralize, for instance, the intensified heat island effects brought about by unlimited growth of cities, raising densities, and climate change effects. Intelligent use of cladding materials and surfaces can decrease temperatures in zones where it is required, e.g., on roof tops in summer and above pavements (Rovers, 2007). For this reason the urban design and planning disciplines should be involved in the research and/or the solutions to this problem.

Table 1. Theoretical impacts (modified from Gartland, 2008).

Morphological attribute	Energy impact
Lack of vegetation	Reduces evaporation
Widespread use of impermeable surfaces	Reduces evaporation
Increased thermal diffusivity of urban materials	Increases heat storage
Low solar reflectance of urban materials	Increases net radiation
Urban geometries that trap heat	Increases net radiation
Urban geometries that slow wind speeds	Reduces convection
Increased energy use	Increases anthropogenic heat
Low land altitude	Increases average temperatures

Several authors (Gartland, 2008; Oke, 1981, 1982, 1987, 1988; Rohinton, 2009; Rovers, 2007) reported in fact the relationship between structural and morphological aspects of urban settlements. These relationships include:

- The air temperatures are driven by the heating of urban surfaces, since many man-made surfaces absorb more of the sun's heat than natural vegetation does;
- The urban building materials are impermeable and watertight, so moisture is not readily available to dissipate the sun's heat;
- The areas with the least vegetation and greatest development tend to be hottest;
- Temperature of dark, dry surfaces in direct sun can reach very high temperature during the day;
- Vegetated surfaces with most soil under the same conditions reach only lower temperature;
- Dark materials in concert with canyon like configurations of buildings and pavement collect and trap more of the sun's energy;
- Heat islands tend to become more intense as cities grow larger.

From an urban morphology point of view, these characteristics can be sorted into the four main causes of heat island formation: (1) reduce evaporation; (2) increased heat storage; (3) increase net radiation; and (4) reduced convention. These can, according to Gartland (2008) be shown as in Table 1.

2.3 *Open research questions*

Even if some causes-effects of the UHI phenomena were coded by the literature and related to urban morphology aspects, what is still not clear is the proportions the urban morphology elements spatially affect the overheating of urban areas. Open questions are:

- How to geo locate those city's areas that are more prone to the overheating phenomena?
- How to understand and describe what happens, from the urban morphology point of view, in those urban areas that corresponds to UHIs?
- How to drive or suggest solutions to the UHI phenomena from an urban design and spatial point of view?

The understanding of these issues could improve the efficacy of the urban design and planning solutions to mitigate the UHI phenomena. It could address more efficient interventions in specific sectors and areas.

3 RESEARCH PROBLEM

The aim of the research is to contribute to an easy and replicable methodology to analyze the surface temperature of a city and its correlation with various urban morphology elements. The analysis is tested on a case study located in Italy: an area of the city of Trento. This city is settled in a valley floor and is classifiable as a medium density center.

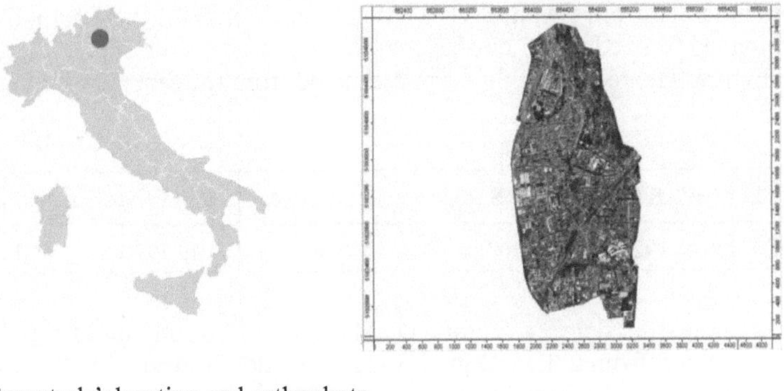

Figure 1. Case study's location and orthophoto.

4 METHODOLOGY

4.1 *Overall methodological approach (The proposed methodology)*

Development of a methodology to specify the relationships between temperature surface variations and the aspects of urban morphology that influence variations in surface temperature proceeds through four steps.

The first is to clearly identify variables of interest (derived from the theoretical literature). Based partially on the discussion above, the major urban morphology elements include: (1) reflective capacity of surfaces; (2) visibility of the sky; (3) vegetation presence and density; (4) solar irradiance; (5) terrain elevation; and (6) building compactness.

The second is to construct measurements for these variables in a geo database using available remote sensing and GIS tools. For each of these, one or more indicators is eventually needed. According to the analytical framework published in Vettorato (2010), different aspects of UHI are related to different spatial scales. To preserve the multi-scalar dimension of the UHI the indicators are calculated, where possible, at different spatial resolutions. Measurement of these variables is the major technical challenge.

The third part is to examine relationships among these variables. Throughout, the major variable of interest—surface temperature—is "embedded" within an overall analytical framework. Multivariate Statistical Analysis applied to raster images (MSARI) is selected. The MSARI technique allows exploration of relationships between many different data layers or types of attributes (Jolliffe, 2002). The Multivariate statistics is a form of statistics encompassing the simultaneous observation and analysis of more than one statistical variable. In particular the Principal Component Analysis (PCA) involves a mathematical procedure that transforms a number of possibly correlated variables into a smaller number of uncorrelated variables called principal components. It is mathematically defined as an orthogonal linear transformation that transforms the data to a new coordinate system such that the greatest variance by any projection of the data comes to lie on the first coordinate (called the first principal component), the second greatest variance on the second coordinate, and so on (Jolliffe, 2002).

Finally, relationships between variations in surface temperature and each of the urban morphology attributes are explored using a four class categorization of all variables.

4.2 *Variables*

The six urban morphology elements (UME) and the surface temperature (ST) are measured by several variables, as shown in Table 2.

4.3 *GeoDataset construction and measurements*

The overall geo database is constructed.

Measurements for ALB (reflective capacity), NDVI (vegetation) and the ST (surface temperature) were obtained from LANDSAT ETM+ scene. The used LANDSAT ETM+ (reference 8) (downloaded from edcsns17.cr.usgs.gov) was captured on the 26th of August 2001, as shown in the Figure. The recording time was estimated from the Sun elevation and Azimuth on 10:45 am.

Data acquired	WRS Path	WRS Row	Cloud Cover	Sun Elevation	Sun Azimuth
2001/08/26	192	028	5.34	49.7916695	145.7333527

Measurements for SFV (sky view factor), DEM (digital elevation), and BC (building compatness) were obtained from a 1 m Digital Surface Model derived from a LiDAR survey provided by the Province of Trento (2008).

Table 2. Theoretical concepts and operational variables.

Morphological aspects	Morphological attributes	Spatial indicators	Variable names	Common spatial resolution
Reflectance Capacity of the Surface	Increased thermal diffusivity of urban materials; Low solar reflectance of urban materials.	ALBEDO	**ALB**	30 m/px
Visibility of the Sky	Urban geometries that trap heat; Urban geometries that slow wind speeds.	Sky View Factor	**SVF**	1 m/px
		Sky View Factor Density	**SVFD**	30 m/px
Vegetation Presence and Density	Lack of vegetation; Widespread use of impermeable surfaces.	Normalized Difference of Vegetation Index	**NDVI**	30 m/px
Solar Irradiance	Increased thermal diffusivity of urban materials; Increased energy use; Low solar reflectance of urban materials.	Beam Irradiance in Watt/m^2	**IRR**	1 m/px
Terrain Elevation	Low land altitude.	Digital Elevation Model	**DEM**	10 m/px
Building density	Urban geometries that trap heat; Urban geometries that slow wind speeds.	Surface/Volume Ratio	**BD**	20 m/px
Temperature		Surface Temperature	**ST**	60 m/px

4.3.1 *ALBEDO (Reflective capacity of surface)*

The Albedo was calculated using the procedure published by Liang and Chen (2002) following the formula for the calculation of the albedo for the total shortwave:

$$\alpha_{short} = 0.356\alpha_1 + 0.13\alpha_3 + 0.373\alpha_4 + 0.085\alpha_5 + 0.072\alpha_7$$

where:
α_i represents the land surface ETM+ spectral reflectance.

4.3.2 *SVF + SVFD*

The sky view factor (SVF) was calculated according to the formula provided by Brown, Grimmon and Ratti (2001), reelaborated by Boehner and Antonic (2009), and implemented in the open source SAGA GIS. The sky view factor density (SVFD) was calculated starting from the sky view factor analyzing the variation of the cells values, in the analysis window, form the center cell value.

4.3.3 *NDVI (Normalized difference of vegetation index)*

The NDVI was calculated using the ETM+ bands 3 (Red, 30 m/px) and 4 (Near InfraRed, 30 m/px) and applying the formula published in literature by several authors (Tucker, 1979):

$$NDVI = (NIR - VIS)/(NIR + VIS)$$

4.3.4 *IRR (Solar irradiation on surfaces)*

Measurement of IRR (irradiation) was obtained from an irradiation model implemented in GRASS GIS, which was itself based on the Digital Surface Model (DSM) above (for the complete procedure see Vettorato & Geneletti, 2009). In synthesis a 1 meter resolution DSM has been used as an input for the r.sun (JRC, 2009) irradiation model embedded in GRASS GIS 6.4, obtaining a map in W/m^2/year per pixel.

4.3.5 *BD (Building density factor)*
The building density was derived from the Surface to Volume ratio of the buildings analyzing the variation of the cells values, in the analysis window, from the center cell value using a using the Center Versus Neighborhood function (Olaya, 2004).

4.3.6 *ST (Surface temperature)*
The Surface temperature was calculated using the ETM+ band 6 (2-lowgain) and applying the formula and the coefficients provided by Landsat (Landsat 7, 2005) that converts the Radiance (given in Digital Numbers) to Kelvin (is the inverse of the Planck function), and that converting the Kelvin to Celsius. The formula is:

$$T = \frac{K_2}{\ln\left(\frac{K_1}{CV_R} + 1\right)}$$

where:
T is degrees in Kelvin
CVR is the cell value as radiance
K_1 is a given conversion coefficient that for ETM+ is 666.09
K_2 is a given conversion coefficient that for ETM+ is 1282.71

The temperature value where compared with the air temperature recorded at two meteorological stations (Borgo valsugana and Trento sud) on the same date and time. The results shown surprisingly a very good matching between the Surface Temperature values reordered by the satellite and the air temperature values recorded by the meteorological stations with an average difference of + 0.2°C. For this reason it was possible to hypothesize that for the case study area and for the climate and meteorological conditions of analyzed moment the surface temperature given by the remote sensor was a good proxy for the estimation of the air temperature.

Table 3. Database construction.

Source	1st Transformation	2nd Transformation	3rd Dtransformation	4rd Transformation
LiDAR	DSM 1 m/px	SKY VIEW FACTOR 1 m/px	SVF DENSITY 1 m/px	**SVF DENSITY 30 m/px**
			SKY VIEW FACTOR 30 m/px	
		BEAM IRRADIATION 1 m/px	**BEAM IRRADIATION 30 m/px**	
		BUILDINGS VOLUME 30 m/px	SURFACE/VOLUME RATIO 30 m/px	**BULDING'S DENSTY 30 m/px**
	DEM 1 m/px	**DEM 30 m/px**		
LANDSAT ETM+	BANDS 3 & 4; 30 m/px	**NDVI 30 m/px**		
	BANDS 1, 3, 4, 5, 7; 30 m/px	**ALBEDO 30 m/px**		
	BAND 6; 60 m/px	SURFACE TEMPERATURE 60 m/px	**SURFACE TEMPERATURE 30 m/px**	

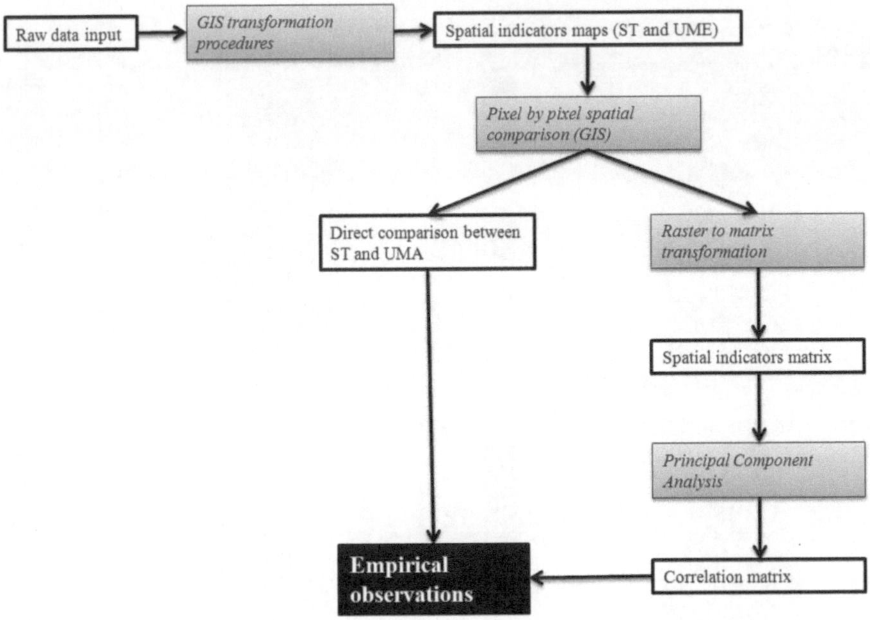

Figure 2. Flowchart of the procedure.

Fuller technical specifications, reporting the transformation sequence between the data source and the final elaborated products (in bold), are shown in Table 3 while the flowchart of the overall procedure is presented in Figure 2.

4.4 *Simple mapping of the variables*

Simple mapping of each of the morphological aspects demonstrates a key point of this analysis. These variables have non-uniform spatial distributions. Parts of Trento are "hotter" than other parts. As would be expected, sky view, vegetation, and building compactness vary by land use pattern.

These simple maps, shown in Figure 3 provide a useful first impression of some of the dynamics involved in understanding the UHI phenomenon.

4.5 *Multivariate statistical analysis*

The different patterns evident from the maps in Figure 2 clearly call for some sort of multivariate statistical analysis to determine correlations between the "dependent variable" ST and the urban morphological elements as well as discussion about the interactions or redundancies of the morphological aspects themselves.

The obtained raster maps were normalized and transformed in a matrix where each pixel was corresponding to a record and each indicator was corresponding to a variable. The first analysis is a simple correlation analysis for the entire variable set, shown in Table 4. As expected the results show that any strong correlation exists between one single morphology indicator in particular and the surface temperature. The DEM (negative) and the Building Density (positive) shows a medium correlation with the surface temperature. The Sky View Factors shows a high positive correlation with the Solar Irradiation.

The second multivariate analysis is a principal components procedure to determine redundancies among the morphological variables. The screen plot shown in Figure 4 shows that four factors could describe the 78,4% of the variation among the cases. But perhaps more interesting is the right hand side of Figure 4 which shows how the dependent variable, ST, is represented in the principal components factor structure.

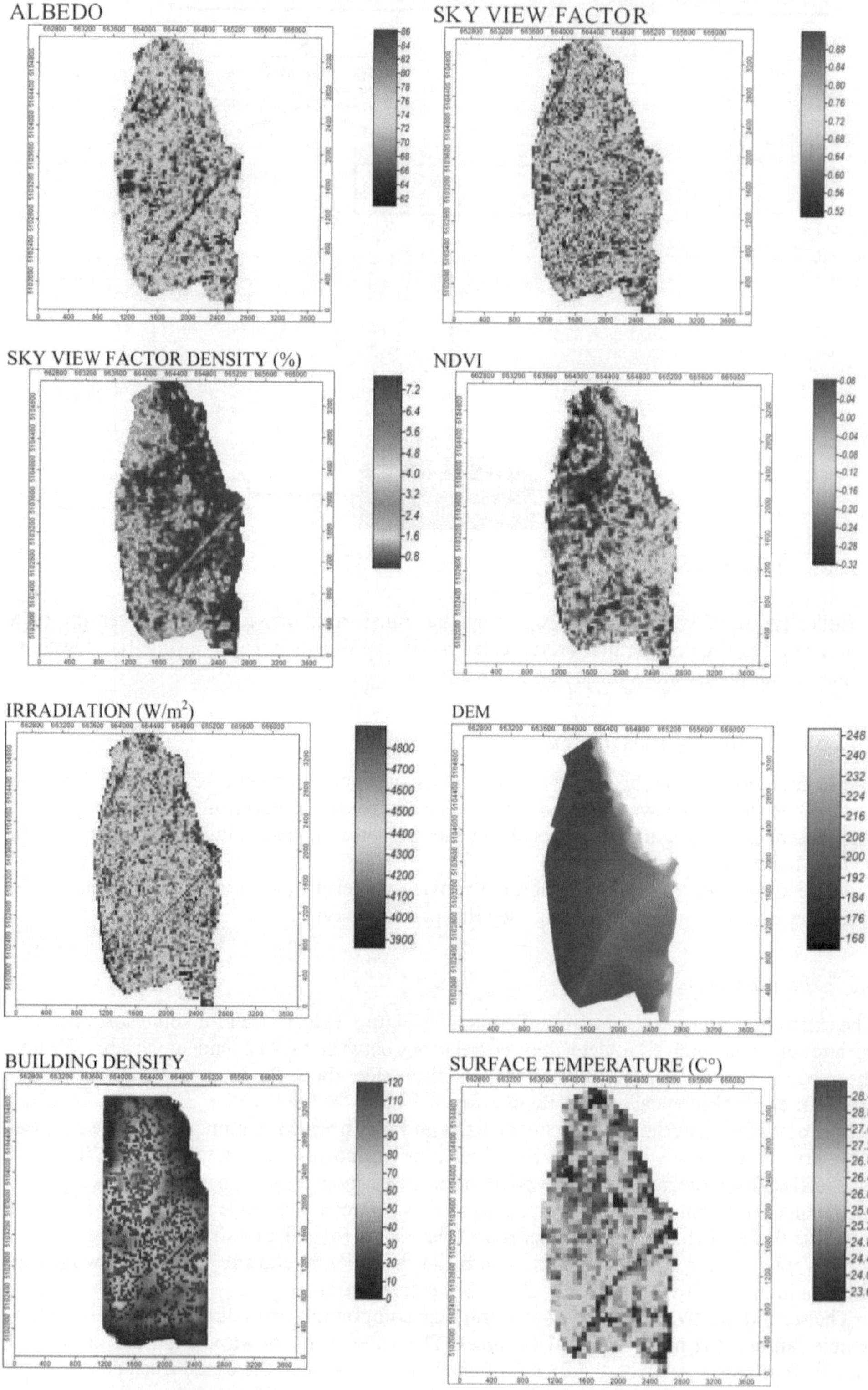

Figure 3. Maps of the variables.

Table 4. Correlation matrix.

Variable	ALBEDO	DEM	IRR	SVFD	SVF	BC	NDVI	*TEMP
ALBEDO	1,000000	-0,139692	0,126354	0,240010	0,283885	0,023959	0,002845	0,075851
DEM	-0,139692	1,000000	0,143669	-0,327342	0,016590	-0,225340	-0,012850	-0,269476
IRR	0,126354	0,143669	1,000000	0,327096	0,713103	-0,193000	0,033430	-0,020337
SVFD	0,240010	-0,327342	0,327096	1,000000	0,430688	-0,099331	0,064378	0,090130
SVF	0,283885	0,016590	0,713103	0,430688	1,000000	-0,280751	0,024608	-0,067209
BC	0,023959	-0,225340	-0,193000	-0,099331	-0,280751	1,000000	-0,009210	0,224044
NDVI	0,002845	-0,012850	0,033430	0,064378	0,024608	-0,009210	1,000000	-0,029107
*TEMP	0,075851	-0,269476	-0,020337	0,090130	-0,067209	0,224044	-0,029107	1,000000

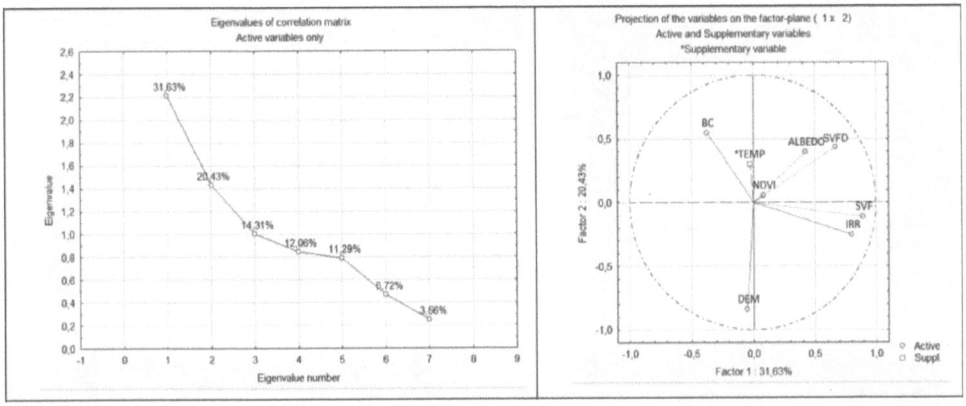

Figure 4. Screenplot of the correlation matrix and projection of the variables in the factor plane 1–2.

5 RESULTS

The analyses above are then used to examine differences in ST within the study area. For ease of presentation, the ST map is transformed into a four class variable and mapped, shown below in Figure 5. It is reasonably clear that the T4 (hottest) zone contains the historic core of the city, while the T1 zone (coolest) are found on the periphery.

As expected different surface temperature show differences in the related urban morphology composition. These relationships are shown graphically in Figure 6.

These relationships can be summarized as follows:

- the T1—low temperature area corresponds to: high albedo, low building density, medium presence of vegetation, high solar irradiation factor, medium sky view factor density, high sky view factor, medium elevation.
- the T2—medium-low temperature corresponds to: medium albedo, low building density, high presence of vegetation, high solar irradiation factor, high sky view factor density, medium sky view factor, medium elevation.
- the T3—medium-high temperature corresponds to: very low albedo, low building density, very low presence of vegetation, low solar irradiation factor, very low sky view factor density and very low sky view factor.
- the T4—high temperature corresponds to: high albedo, high building density,very low presence of vegetation, low solar irradiation factor, high sky view factor density, low elevation.

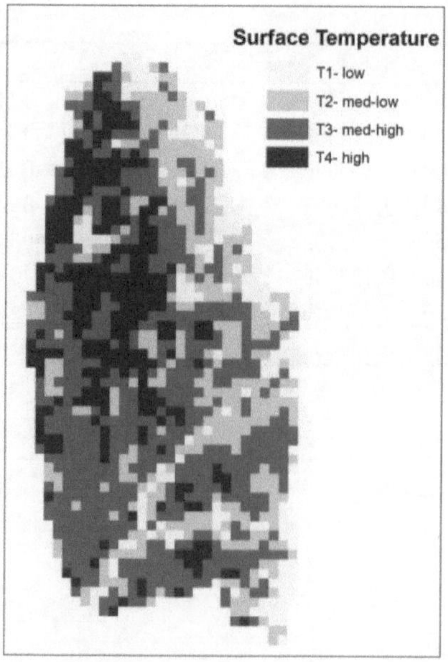

Figure 5. Surface temperature classes.

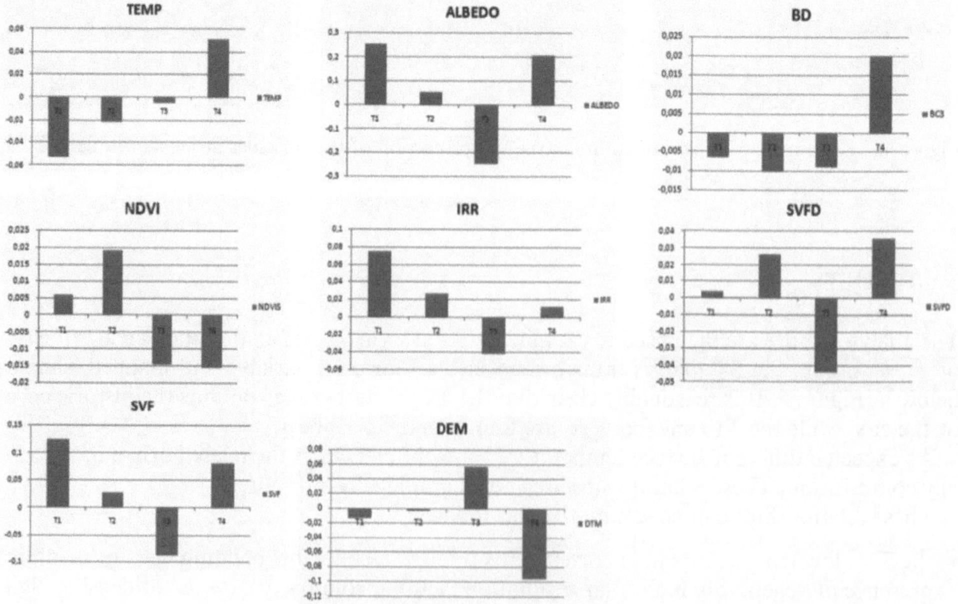

Figure 6. Average of the normalized values of each indicator for the 4 temperature classes (T1, T2, T3, T4).

At a more general level, there appears to be seven empirical results that, while specific to the Trento case study, shed some light on the UHI phenomenon. These are:

- the elevation is negatively related to the surface temperature, but it is not a determinant factor to reduce the surface temperature;
- the high albedo can reduce the surface temperature but it is not sufficient if the building density is high;

- the building density with low vegetation and low albedo is positive correlated to high surface temperatures;
- the presence of vegetation is related to low temperatures. In the area T1 we have low temperature and medium presence of vegetation but the area includes a river;
- the irradiance factor is negatively related to the surface temperature;
- the high sky view factors can reduce the surfaces temperature but only in presence of a vegetation cover or a very high albedo;
- it is possible to make the hypothesis that the more the surfaces are pleated and the building density is high and the vegetation is low, the more the surface temperature is high.

With these methodology applied to the case study it is possible to suggest some spatial strategies for the urban design and planning in order to contribute to the reduction of the Surface UHI effect in Trento. These are:

- The areas T3 and T4 show to be the most affected by the Surface Heat Island Effect. In the area T3 probably the albedo should be improved as well as the presence of vegetation and the sky visibility. The relative high elevation of the T3 area didn't show many benefits.
- In the T4 area even if the albedo is high the building density affects the temperatures. The thermal comfort should be probably improved enhancing the vegetation cover. In those areas characterized by a high sky view factor (big roads, big squares, the railway station area, etc.) the vegetation cover is determinant to reduce the surface temperature.
- Learning from the T1 and T2 areas it is possible to state that the low building density reduce the surface temperature only if the surfaces are covered by vegetation or water.

6 DISCUSSION

The proposed methodology can contribute to the debate on the effect of the urban morphology on the UHI effect. In particular the spatial perspective of the analytical framework to allow direct consideration of the spatial distribution of the UHI effect.

On the technical end, the multivariate statistical analysis applied to raster images shows to work properly and to be a powerful analytical instrument. The construction of the geo database and the dataset availability was a determinant factor for the quality of the analysis. Fortunately many dataset (i.e., the LandSat images) are available on-line at a very low cost. The availability of some open-source GIS software that embedded several spatial analysis tools reduced the elaboration time. Further improvements of the procedure will include several thermal images covering a wider time-scale, during the same day and in different seasons. Even the selection of the spatial indicators will be probably subject of a critical review: some other spatial indicators could be introduced in the analysis (i.e., water presence, leaf area factor, relative humidity, urban canyon temperatures) to obtain a more precise description of urban morphologies and microclimate.

The introduction of urban morphology spatial indicators in the analysis of UHI shows the potential results obtainable in terms of urban design and planning strategies elaboration. The methodology is replicable and customizable and allows better understanding of spatial relations among the urban morphology variables and the UHI phenomena in other cities and geographical contexts. In particular this methodology can contribute to answer to some policy makers and urban planners questions like:

- How to easily geo locate those city's areas that are more subjected to the overheating phenomena?
- How to understand and describe what happens, from the urban morphology point of view, in those urban areas that corresponds to UHIs?
- How to drive or suggest solutions to the UHI phenomena from an urban design and spatial point of view?

Starting from the results obtained with the proposed method the urban morphology and vegetation design could be successfully used to promote a climate sensitive and sustainable urban design for the critical zones. In particular the possibilities are: to optimize the morphology to enhance the shading of buildings and public spaces during thermally critical times of the day (by manipulating urban geometry); to use the urban vegetation for sheltering and shading buildings and public spaces; and/or to consider in the design the role of vegetation also for CO_2 absorption, ecosystem connections and water retention.

REFERENCES

Boehner, J. & Antonic, O. (2009). "Land-suface parameters specific to topo-climatology". In: *Hengl, T. & Reuter, H. (Eds.): 'Geomorphometry—Concepts, Software, Applications'*, Vol. 33. Elsevier.

Brown, M.J., Grimmond, C.S.B. and Ratti, C. 2001. "Comparison of methodologies for computing sky view factor in urban environment". *International Society of Environmental Hydraulics Conference, Tempe, AZ.* Internal Report Los Alamos National Laboratory, Los Alamos, NM. LA-UR-01-4107.

CABE, 2002. *The value of good design: how buildings and spaces create economic and social value*, London: Commission for Architecture and the Built Environment.

Gartland, L. 2008. *Heat islands: understanding and mitigating heat in urban areas.* London: Earthscan.

Fukunaga, K. 1990. *Introduction to Statistical Pattern Recognition.* London: Elsevier.

Imamura, R. 1989. "Air–Surface Temperature Correlations", *Controlling Summer Heat Islands.* Berkeley,CA, Lawrence Berkeley Laboratory.

Jolliffe, I.T. 2002. *Principal Component Analysis*, Series: New York: Springer Series in Statistics, 2nd ed., Springer.

JRC, Joint Research Center, 2009. UE. *Solar Model R.SUN*, http://re.jrc.ec.europa.eu/pvgis/solres/solmod3.htm browsed on June 2010.

Landsat 7. 2005. *Science Data Users Handbook*, prepared by Landsat Project Science Office, http://ltpwww.gsfc.nasa.gov/IAS/handbook/handbook_toc.html, last updated 10 May 2005.

Landsat 7. 2010. www.landsat.org/ browsed on June 2010.

Liang and Chen. 2002. "Atmospheric Correction of Landsat ETM+ Land Surface Imagery: II. Validation and Applications" *IEEE Transactions on Geoscience and Remote Sensing*, 2002.

Olaya, V. 2004. A gentle introduction to SAGA GIS. Victor Olaya.

Oke, T.R. 1981. "Canyon geometry and the nocturnal urban heat island: comparison of scale model and field observations". *Journal of Climatology* No. 1.

Oke, T.R. 1982. "The energetic basis of the urban heat island". *Quart. J. Roy. Meteorol. Soc.* No. 108.

Oke, T.R. 1987. *Boundary layer climates*, London and New York: Routledge.

Oke, T.R. 1988. "Street design and urban canopy layer climate". *Energy and Buildings* no. 11.

Rohinton, E. 2009. "Sustainable urbanity and urban climate change: amelioration of uhi's as a quality-of-life agenda for tropical mega-cities" *The seventh International Conference on Urban Climate, 29 June – 3 July 2009, Yokohama, Japan.*

Rovers, R. 2007. "Exergy in Urban areas". *Inventory of ecological & planning concepts, models and energy potentials.* SenterNovem: RUG (Rijksuniversiteit Groningen).

Souza, L.C.L., Rodrigues, D.S. and Mendes, J.F.G. 2003. "The 3D SkyView extension: an urban geometry acces tool in a geographical information system. In *Klysik, K., Oke, T.R., Fortuniak, K, Grimmond, C.S.B. and Wibig, J. (eds.): Proceed Fifth Int Conf on Urban* Climate. Vol. 2. Lodz, Poland: University of Lodz.

Tucker, C.J. 1979. "Red and Photographic Infrared Linear Combinations for Monitoring Vegetation", *Remote Sensing of Environment*, no. 8.

United Nations. 2002. *World Urbanization Prospects, The 2002 Revision.* New York: Department of Economic and Social Affairs, Population Division, United Nations.

Vettorato, D. 2010. "Pianificazione ed energia: un approccio basato sull'analisi delle interazioni tra morfologia insediativa e comportamento energetico degli insediamenti". *Materiali della XIII Conferenza SIU (transcript non rivisti dai relatori), Roma, 25–26–27 febbraio 2010*, Società Italiana degli Urbanisti, Città e crisi globale: clima, sviluppo e convivenza, *Planum—The European Journal of Planning on-line*.

Vettorato, D. & Geneletti, D. 2009. "Estimation of potential solar energy at urban scale: an approach based on LiDAR images analysis", *Representation of Geographical Information for Planning*, Bologna: Esculapio.

Watkins, R., Palmer, J, Kolokotroni, M. & Littlefair, P. 2002. 'The London heat island—surface and air temperature measurements in a park and street gorges'. *ASHRAE Transactions* no. 108.

Urban and Regional Data Management – Zlatanova, Ledoux, Fendel & Rumor (eds)
 ISBN 978-0-415-67491-1

Sustainable urban growth for Athens

L. Mantelas & P. Prastacos
Foundation for Research & Technology—Hellas, Institute of Applied & Computational Mathematics, Regional Analysis Division, Greece

U. Mollay
OIR—Austrian Institute for Regional Studies and Spatial Planning, Austria

ABSTRACT: In this article an attempt is made to analyze and present alternative long term scenarios for urban growth in the Athens metropolitan region. The scenarios are partially based on the concept of sustainable metabolism and focus on the form the city can take in year 2050. The methodology used is relatively coarse because of the lack of data. Four different scenarios are estimated, one assuming that existing trends will continue, another that foresees increased densities to minimize energy consumed in transportation and finally another two that account for the lack of green space in many areas.

1 INTRODUCTION

1.1 *The concepts of urban metabolism and sustainable urban metabolism*

The concept of metabolism permits to understand and analyze the way societies use resources, energy and land and other elements of the environmental system, for maintaining and reproducing themselves. In urban areas the form of the cities, that is the way they have been being built and the way they are expanding or contracting is influencing the quantities and qualities of resources used in the urban environment. Urban development includes processes of growth in new areas, decay and abandonment and also restructuring and rehabilitation in parallel. While the dynamics of urban development have been studied and debated for a long time (Urban Audit, 2007), the relationship between urban development and urban metabolism with the environment is not completely understood (McDonald & Patterson 2007).

Although demographic growth in European cities is rather moderate and mainly due to migration, the most significant factors effecting urban spatial growth are the growing number of smaller households and the increasing space consumption by households, business units and of course the transportation infrastructure for connecting the various activities. In many cities, these dynamics extend far beyond urban centers and are heavily affecting rural areas (ESPON 2007). The continuous sprawl of the urban fabric into the periphery, even in cities where population is not increasing, is a substantial concern because of the environmental consequences and the waste of resources (European Environmental Agency, 2002).

Present patterns of resource and energy consumption are associated to three main factors, namely: the physical and technological qualities of buildings, the urban (spatial) structures—together forming the built environment—and the consumer lifestyles, greatly depending on the level of economic development. If these flows and consumption patterns can be transformed to quantitative levels and qualities less damaging to the ecological system, the long-term sustainability of urban systems can be improved.

1.2 *Scope and context of the article*

The work reported in this article is part of the project "Sustainable Urban Metabolism for Europe" (SUME—http://www.sume.at/). This FP7 Collaborative Research Project analyzes the spatial characteristics of the built urban systems and the impacts these "urban forms" can have on the quality and quantity of resources needed to maintain them. The analysis is conducted in five European cities (Athens, Munich, Oporto, Stockholm and Vienna) with different development scenarios assumed in order to assess what types of urban forms and built structures can lead to a reduction of resources and energy consumption. This article presents the Athens case. The Athens scenarios although following the same methodology used in the other four cities it is different in some respects because of the lack of data and concrete information on the various development projects being anticipated in the near future.

The basis for the analysis undertaken rests in the existing building stock and urban form of Athens and the population distribution of two different development scenarios. It must be pointed out that the scenarios in this analysis are not actual forecasts nor projections in the sense that they depict the expected or most probable configuration of the urban area in the future. Instead, the objective is to demonstrate the range of alternative development paths between a trend scenario (baseline scenario or BASE scenario) and two more ambitious ones; the aim of the first one is to decrease the energy consumed for transport (SUME1 scenario) while for the second one the objective is to improve the quality of life by providing more open space (SUME2 & SUME3 scenarios). In doing this, the four scenarios should not be perceived as extreme stand points, but rather as the range of realistic options from two different perspectives; one assuming that existing development trends are continuing in the future and another one that envisions a more ambitious development policy considering metabolic interdependencies.

Drawing up different scenarios, attempts to answer the question: "*Is it possible to considerably affect the metabolic performance of urban regions by urban development policies and planning, and how much*?"

2 SCENARIOS

2.1 *Data*

The Athens Urban Metropolitan Area (UMA)—221 km^2 and approximately 3.5 million population—was segmented in 238 cells, each one having on the average about 150,000 inhabitants. For each of the zones several datasets were assembled allowing the analysis of densities, the mix of uses (residential vs. economic), the building typology and age structure and the accessibility to the main transport lines. Supplementary information was collected with respect the limitations for further development, such as protected zones, historic areas, parks etc. Most data were obtained from the 2001 Census of Population and Housing and for simplicity reasons it was assumed that they correspond to year 2000 as well. More specifically, the datasets used include information on the following:

- population and number of households (Census 2001)
- number of jobs (Census 2001)
- number of dwellings tabulated by the year they were constructed (Census 2001)
- number of non-residential buildings (Census 2001)
- area of continuous and discontinuous urban fabric (km^2, source: CORINE 2000).

Year 2050 was selected as the end year for the scenarios. It is true that 2050 is 40 years away from the present and therefore changes in technology and socio-economic conditions could alter the way the city develops. However, the objective of the study being to estimate changes in the urban area arising from planning policies related to densification it was felt that having a longer period of forecasts will demonstrate the impact of these changes more evident.

Estimation of population and number of jobs for year 2050 was made in two steps. For year 2020 initial estimates for population and jobs for each cell were obtained from Attiko Metro S.A. that performed a study on the future of the Athens area in 1996. These figures were modified since they were prepared before the results of the 2001 Census and were rather overoptimistic with respect the growth of the urban agglomeration. For year 2050 the change in the regional population between 2020–2050 was allocated to the various cells using the methodology described below. Total population for the UMA was obtained from the Hellenic Statistical Authority.

2.2 *Methodological approach*

Allocation of population and jobs within the UMA is performed in 2 steps. In each of the steps, a portion of the estimated population and jobs change is allocated. It was assumed, that this allocation takes place within the UMA; only the remaining surplus of future population (and jobs) which cannot be accommodated within the UMA according to the density rules is to be allocated outside of the UMA.

More specifically, Step 1 calculates the portion of the changes in population and employment for 2000–2020 (as estimated by the Attiko Metro) that can be hosted in existing dwellings on a cell basis when we assume that there is a 10% increase of floor space per person. This assumption is made for both the BASE and SUME1 scenarios. In the following Step 2, for the allocation of residual population and jobs within the total UMA (that is the remaining population and jobs from Step 1 plus the estimated change for 2020–2050), the existing housing characteristics are considered as important while no further change in floor space per person is assumed.

For the allocation of population in Step 2, the "on-site-density" (OSD—defined as (population + jobs/2)/km^2—urban fabric as shown in CORINE) was defined as an upper limit for future densification. In this classification, 7 classes are defined: single family house, Mixed area (low density), Mixed area, Dense mixed area, Low multi storey, Multi-storey and Dense multi-storey.

The rationale for the algorithm used in Step 2 is that prevailing conditions in an area with respect housing types usually remain stable over long periods and change only through major development impulses (e.g., accessibility increases due to new transport infrastructure, changing (higher) land values, specific development projects etc.). As a general assumption, cells with a distinct housing type (i.e., mainly single-family houses) will not change fundamentally over time, without a specific reason. Thus, without a major impact (e.g., accessibility provided by new high-capacity transport lines) each cell is expected to host as much population as possible without moving to a higher density class or to exceed its maximum density threshold.

On the other hand, under certain assumptions, housing density in some cells may exceed the current values (thresholds) by allowing the cell to upgrade its class thus permitting the accommodation of more population and jobs. It was assumed that this could happen in areas where there is:

- Existence/anticipation of large urban development projects,
- Proximity to existing transport axes and/or future high-capacity public transport infrastructure.

2.3 *BASE and SUME1 Scenarios*

The objective of the scenarios estimations is to demonstrate the differences between the allocation of jobs and housing in the base year (2000) and the future allocations for year 2050, which are the result of different development strategies. The BASE scenario is viewed as a continuation of current planning policies supporting past spatial development trends (densities and configurations of the urban fabric). On the other hand, the SUME1 scenario is based on principles leading to a more sustainable spatial development in terms of energy consumed in the transport sector. This takes place by changing current planning policies towards a clear guidance of urban development on major transport routes, providing compact and integrated new developments and the aim is to increase the densities in areas of good accessibility.

The BASE scenario was set up, using the building typology and densities throughout the UMA as a reference. The alternative SUME1 scenario is based on the concept that policies will be adopted that would:

- increase densities in existing built-up areas, if there is or will be good accessibility of public transport
- increase the rate of renovation and replacement of existing older buildings, thus reducing the energy needed for heating and cooling them.

BASE and SUME1 scenarios are differentiated—as the result of the development 1 applied in Step 2—in terms of the:

- size of population spreading beyond the UMA border,
- densities applied (for the newly urbanized areas),
- spatial distribution/proximity to the lines of public mass transport.

The BASE scenario provides this estimate under a trend-like development strategy, while the SUME1 scenario shows the potential magnitude of policy impacts to reduce development outside the UMA until 2050 by focusing development close to public transport.

2.4 *Evaluation of the scenarios—the UDP Indicator*

The scenarios generate the distribution of population and jobs among the cells within the UMZ and estimate the population and jobs to be located outside the UMZ in year 2050. These spatial patterns are evaluated with respect their potential impact on transportation and the energy consumption needed for transport. The approach is a qualitative rating of each cell's location with respect to its proximity to the center of the metropolitan area where there is a heavy concentration of jobs and access to high-capacity public transportation. To quantify these, the Urban Diversity Pattern (UDP) indicator was defined which consists of the following three sub-indicators:

- Indicator 1: Accessibility to high capacity public transport infrastructure
- Indicator 2: Centrality (proximity to center functions)
- Indicator 3: Diversity (mix of economic and residential functions).

The accessibility to high-level public transport infrastructure (Indicator 1) is analyzed by implementing different circular catchment areas (<5, <10 and >10 km), which have their midpoint in the centre of the city. Each cell in the UMZ is evaluated according to which catchment area belongs to and according to its proximity to public transport. Cells touching directly a metro or urban railway line are evaluated with more points than those without direct connection since it can be assumed that there will be higher share of car use (Newman & Kenworthy 1989).

Indicator 2 is also evaluated on different catchment areas (<2.5, <5 and <10 km) which have the center of the city as midpoint. The closer a cell is situated to the historical centre, the more functions are located in this cell. Therefore, the population will have to cover less distances to reach attractive destinations in a city. As a result, the closer a cell to the historical center is, the higher its rank. It must be recognized that the definition of this indicator is based on the assumption that the city is monocentric, an assumption that might not be completely valid.

The OSD impacts the urban diversity pattern as higher densities raise the number of potential users (demand) of activities. A high job and population density is evidence of an enlarged number of functions available for users in short distance (ECOTEC, 1993) and therefore an indicator for a larger variety of the offers and attractiveness of a cell. In this respect, the density is linked with urban functions and therefore Indicator 3 uses four density categories which allow to rate the diversity pattern of a city. In general the cells with the highest density get the best rating.

Each cell in the UMA is evaluated for each indicator through a rating procedure. The highest rating a cell can reach for each of the sub-indicators are 4 points and therefore the maximum value the UDP indicator can take is 12, the sum of the three sub-indicators. To obtain the overall UDP indicator for each scenario the individual cell UDPs are summed up each one weighted by the population of the cell.

2.5 *Results*

The 221 km² of urban fabric within Athens UMA are very densely inhabited; 65% of the population (2001 data) lives in areas with primarily multi-story buildings with only 2% of the population living in areas with single family houses. The remaining population lives in areas where there is a mix of multi-story and single family housing. During the period 2000–2050 a total of 300,000 people (9% growth) are allocated in the BASE and SUME1 scenarios. Due to the high average densities and land availability, all population is allocated within the UMA for both, BASE and SUME1 scenarios.

Thus the maps referring to density patterns in BASE and SUME1 scenario (Figure 1) do not show major differences. Nevertheless, noticeable differences appear when the percentage densification is compared for the two scenarios (Figure 2). As a result of these small differences, the SUME1 scenario scores a UDP indicator that is higher than both the BASE scenario and the one of the 2000 Athens (Table 1).

The BASE scenario scores a lower UDP compared to 2000 something that could be explained by the fact that the new population is allocated in areas with lower proximity to

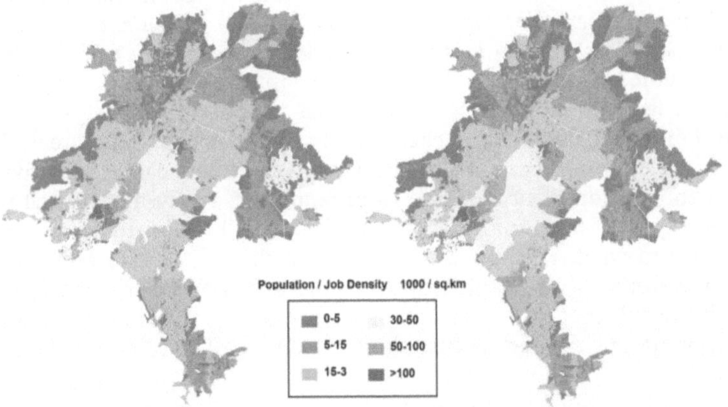

Figure 1. Population & Job densities for the BASE (left) and the SUME1 scenarios (right).

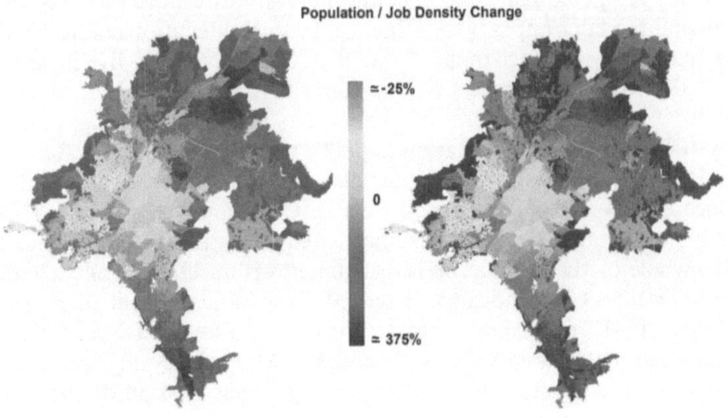

Figure 2. Population & Job densities change for the BASE (left) and the SUME1 scenario (right).

Table 1. The values of the UDP indicators.

	Indicator 1	Indicator 2	Indicator 3	UDP
2000	3.29	2.95	3.60	9.85
BASE	3.23	2.75	3.42	9.26
SUME1	3.46	2.91	3.72	9.87

transport lines. Nonetheless, the differences between the three UDP's (actual 2000, BASE and SUME1 scenarios) are small thus indicating that the high existing densities and the land availability in Athens allow the area to accommodate the population growth until 2050 in a rather sustainable way regardless of the development path followed.

3 ALTERNATIVE SCENARIOS

3.1 *Changes in methodology*

It seems that the dense and compact existing form of the Athens UMZ guarantees a sustainable future development if sustainability is defined in terms of proximity to public transportation. However, there are also issues of quality of life and access to green areas. Many areas lack any open space. Hence, a question that arises is whether it is *possible to redistribute population within the Athens UMZ in such a way so that lower densities and more urban green areas can be achieved thus leading to a more sustainable development path.* In an attempt to answer this question, two additional scenarios were populated.

The common feature in these scenarios is that it is assumed that 10% of the built areas (urban fabric in CORINE) of each cell is used as urban green areas. There is no doubt that this assumption has some limitations; it considers each cell individually as a microcosm without addressing the issue to proximity to a neighboring cell where there is already open space. Furthermore, it also applies the 10% additional open space to all cells irrespective to whether there is open space in the cell or not. However, since the objective of the scenario was to provide a macroscopic view of a scenario leading to a greener city it was felt that the 10% assumption of open space for each cell was justified. In terms of population this means that about 340,000 of the 2000 population will have to be relocated.

3.2 *Results*

Two alternative scenarios were calculated (Figure 3) with different assumptions concerning the floor space per person and the maximum density threshold parameters. The results obtained show that the later appears to influence the final allocations more heavily. This can be partially explained by the fact that 30% of the 2001 population live in areas with OSD greater than 30.000 and the average OSD for all cells is 80.000. In both the scenarios it is assumed that 10% of the urban fabric will be used as urban green areas which means that approximately 10% of the 2001 population (340.000 people) have to relocate.

Scenario SUME2, that assumes a 10% decrease in floor space per person 30.000 density threshold, allocates only 230.000 people outside the UMZ and scores a high UDP indicator. On the other hand, scenario SUME3 assumes a 10% increase on floor space per person,. As a result, the whole of the population is distributed within UMA while a high—yet lower than scenario SUME2—UDP indicator is scored. This indicates that theoretically population within Athens UMZ can be redistributed resulting to less dense cells with more green urban areas and high sustainability levels. Of course, the necessary adjustments in real world will require more or less demolishing and rebuilding a vast portion of the city and would require massive relocation of the population.

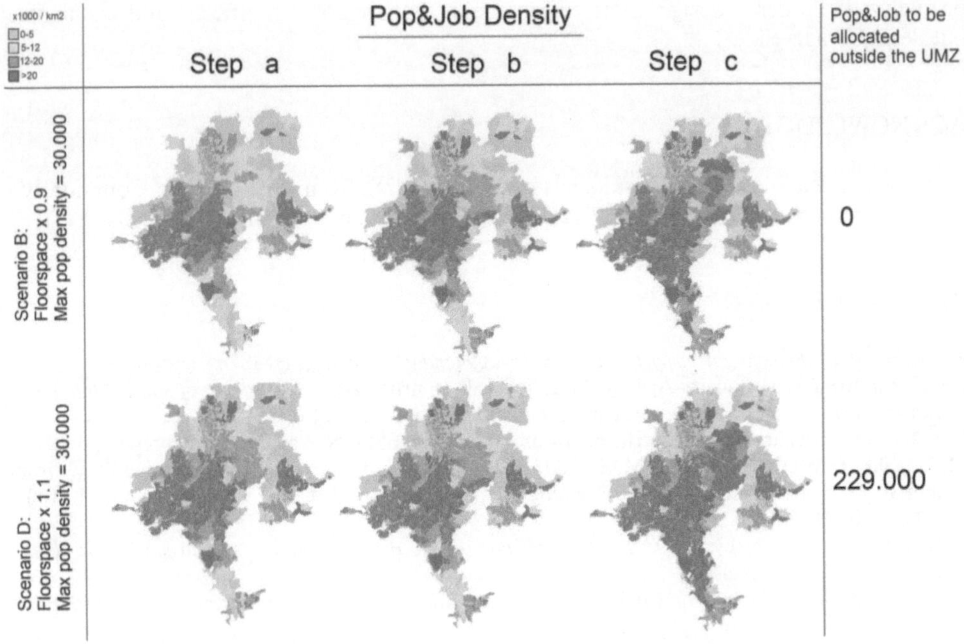

Figure 3. Population & Job densities for the alternative scenarios populated.

Table 2. The values of the UDP indicators for the alternative scenarios.

	Indicator 1	Indicator 2	Indicator 3	UDP
SUME2	3.47	2.95	3.73	10.15
SUME3	3.47	2.95	3.80	10.23

4 CONCLUSIONS

It is a well acknowledged fact that high population densities and compact urban forms are considered to be sustainable in terms of energy consumed in the transportation sector. For the case of Athens as discussed in this paper for year 2050, the existing high densities already support a relatively efficient transportation energy locational pattern. Changing the densities of some parts of the city and using available land results in locational patterns which based on the indicators used, describe and improved situation in terms of sustainability.

However, the issue in Athens is that existing densities are high and most important there is lack of open space and green areas in most parts of the city. The city can be considered sustainable in terms of energy consumption in transportation but there is a need for a better environment for its citizens. As shown by the alternative development scenarios that were populated, by 2050, 10% of the built area of each cell in the UMZ can be used as urban green areas and despite the population growth, population densities can get lower than 2001 without allocating population outside the UMZ. At the same time this can be achieved while improving the performance of Athens' urban configuration as shown by the change in UDP indicators.

On a higher level of analysis, the concept of sustainable urban development appears to be a key instrument in shaping sustainable cities. Nevertheless, its concept should be expanded in order to provide a more holistic view of the urban development process. To this end, various aspects concerning the quality of life, socioeconomic and cultural conditions should be analyzed in order to design urban environments that are not only efficient and sustainable but desirable as well.

ACKNOWLEDGEMENTS

The research leading to these results has received funding from the European Community's Seventh Framework Programme FP7/2007–2013 under grant agreement n° 212034.

REFERENCES

ECOTEC, 1993. *Reducing transport emissions through land use planning*, HMSO, London.

EEA—European Environment Agency. 2002. Towards an urban atlas. *Assessment of spatial data on 25 European cities and urban areas*. Environmental issue report N°30. Copenhagen.

ESPON, 2007. Scenarios on the territorial future of Europe. ESPON Project 3.2 Final report, p. 60.

McDonald, G.W. & Patterson, M.G. 2007. Bridging the divide in urban sustainability: from human exemptionalism to the new ecological paradigm. In: *Urban Ecosystems*. DOI: 10.1007/s11252-006-0017-0.

Newman, P. & Kenworthy, J. 1989. *Cities and automobile dependence: An international sourcebook*. Gower, Aldershot, England.

Urban Audit. *State of European Cities Report, Adding value to the European Urban Audit*, May 2007.

Urban and Regional Data Management – Zlatanova, Ledoux, Fendel & Rumor (eds)
 ISBN 978-0-415-67491-1

Urban forestry strategic fire protection via a susceptibility model

S. Canale, A. De Santis, D. Iacoviello, F. Pirri & S. Sagratella
Sapienza Università di Roma, Dipartimento di Informatica e Sistemistica, Roma, Italy

P. Di Giacomo
Protezione Civile, Comune di Roma, Roma, Italy

ABSTRACT: We present an adaptive system for automatic assessment of both physical and anthropic fire impact factors on peri-urban forestries. The aim is to provide an integrated methodology exploiting a complex data structure. This structure is built upon a multi resolution grid gathering historical land exploitation and meteorological data, records of human habits along with suitably segmented and interpreted high resolution X-SAR images, and several other information sources. The contribution of the model and its novelty relies manly on the definition of a learning schema lifting different factors and aspects of fire causes, including physical, social and behavioral ones, to the design of a fire susceptibility map of a specific urban forestry. The outcome is an integrated geospatial database providing an infrastructure that merges cartography, heterogeneous data and complex analysis. It establishes a digital environment where users and tools are interactively connected in an efficient and flexible way.

1 INTRODUCTION

In this paper we present the main guidelines of a new approach towards an integrated model for analysis and assessment of vulnerability to fire. We apply our approach to the protected forest of Castel Fusano, near Rome, see Figure 1. The green frame vertices have the following coordinates in WGS84 decimal degrees: Lat. 41.762054 Long. 12.287140 NW, Lat. 41.762054 Long. 12.362341 NE, Lat. 41.701957 Long. 12.362341 SE, Lat. 41.701957 Long. 12.287140 SW.

Fire prevention in peri-urban forests, namely, wildlife ecosystems enclosed within inhabited regions or interfacing metropolitan areas, such as parks or protected woodlands, is crucial

Figure 1. The region of interest Castel Fusano, Rome, selected on Google Map.

for urban management. In fact, several reasons affect a balanced growth of these areas and fire risk is one of the major causes. Despite of the good intentions, fire risk prevention often alters biodiversity, likewise the utmost cost for civil protection and maintenance have a negative impact on the destination of urban areas to forestry.

Indeed, while urban forestry is a great economic and social asset in metropolitan areas for air pollution reduction, climate control, and recreational and social benefits (see (Konijnendijk, Nilsson, Randrup, and Schipperijn 2005)), a lack of accurate protection plans can cause a deterioration of these areas. Deterioration and lack of well modeled protection plans can cause dreadful consequences for these historical rich legacies and also great risks for the surrounding urban areas.

A crucial scientific contribution to urban management in European metropolis, where large urban forests and parks form a beautiful interface with the metropolitan settlement, is to provide a strategic view of vulnerability to fire.

From several studies on European fires (e.g., (Martinez, Chuvieco, and Martin 2004; Reineking, Weibel, Conedera, and Bugmann 2010)) anthropic factors have been proved to be determinant in the fire impact on land. Numerous fires can start by careless human actions such as sparks from equipment, arced power lines, campfires, burning debris, discarded smoking products, and also because of pyromaniacs. The main difficulty in these studies is to conceal different data sources providing a suitable integration methodology that can face such a complex data structure. In particular, in this project we consider several layers of data at different resolution that can integrate physical, natural and social factors. We consider the high resolution data obtained from X-SAR images, provided by COSMO Skymed products, urban and land data provided by the Italian Civil Protection, meteorological data from the aeronautics services and finally behavioral data collected by different forensic sources.

We thus present the principal components of our methodology towards the design of a layered classification to obtain a fire susceptibility map. Our analysis is concentrated on a small area. The region of interest (ROI) is the ancient pinehood of Castel Fusano, near Rome, Italy, a protected area that has undergone a devastating fire in July 2000, see Figure 1.

The methodolody for the susceptibility model is described in five sections organized as follows. In Section 2 we present the basic projection grid of historical and analytical data collected and present the early data structure. These data survey, along with a history of fires, the social specification of fire risk assessment based on urban and land exploitation. These are also documented by the Italian Civil Protection and the Italian Fire Department. In Section 3, based on the acquired data, we specify how segmentation of X-SAR images provides a suitable clustering of regions according to features of humidity indices and vegetation status. In Section 6 we present the method for classification of the data integrated as described in Section 2 in order to obtain a consistent and homogeneous resolution for training a Support Vector Machine (SVM) based classification model. In Section 5 we discuss the behavior based methodologies assessing the contribution of the anthropic/human factors to fire susceptibility. Finally we end the paper with a short overview of the state of the art on these different issues.

2 DATA ANALYSIS AND LAYERS AGGREGATION

The goal of land and urban data analysis is to define accurate classification layers for the ROI to aggregate different resolution levels. These steps are needed for training a classifier that returns a map of fire susceptibility values not yet integrated with anthropic factors. These are specified by game theoretic methods overlaying the classification (see Section 5). Once the ROI different areas are specified on the basic tessellation, they can be assigned various tags, according to ancillary data.

Our designated region is tessellated by 30 tiles (see Figure 3). Each tile covers an area of $2500\,\text{m} \times 2500\,\text{m}$, namely 625 hectares. As there is a tile superposition of 100 m, the total surface covered by the basic tessellation is 16×10^3 hectares. The basic tile is consistent with most of the land, soil, and other data available historically. Each layer contributes with

Figure 2. Knowledge levels of soil characteristics, vegetation, land use, fire hazard, water river basins, administrative boundaries, etc., overlaid on the basic tessellation made of 30 tiles. On top the fire susceptibility map based only on physical information.

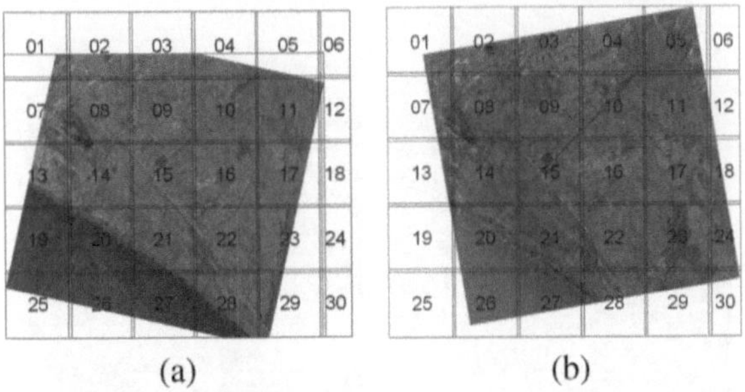

Figure 3. SAR images acquired, respectively the 14th of May 2010 at 4:56 p.m. and the 19th of May 2010 at 5:25 p.m. both overlaid on the basic grid map.

a knowledge level k_i, to the basic tessellation (see Figure 2). On the other hand the information gathered by SAR images is specified by a pixel covering $1\,m^2$ on ground.

The major problem with urban and land data, when these have rather different resolution, is to adopt suitable spatial statistics techniques to integrate missing variables and take care of unobservable errors in multivariate analysis. To integrate the different resolution levels we adopted a statistical method based on multilevel models to account for spatial heterogeneity, see for a review (Pez and Scott 2005). In this way, suitably interpreting a regression error, we can use the segmentation resolution given by SAR images interpretation.

3 LEVEL SET SEGMENTATION FOR SAR IMAGES

In this section SAR data acquisition, description and segmentatio are presented. The goal of segmentation is to extract specific signal features that describe the spatial distribution of the signal. The images are then partitioned into sub-domains homogeneous with respect to these characteristics. COSMO Skymed products are high quality and high spatial definition images (with 1 m of resolution in the SpotLight mode, level 1C-GEC) obtained in the X-band (wave length ranging from 2.4 to 3.8 cm). HH and VV polarization are available for SpotLight mode: two data samples acquired on May 2010 are shown in Figure 3. They provide a peculiar kind of information such as spatial, textural and geometric features of different segments (soil and vegetation). These informations are here considered for recognition and classification of targets meaningful to the fire susceptibility model (see Section 4). Moreover, monitoring the smooth changes in time, in the X-band response of these targets, well suits our study of the fire susceptibility dynamics. The X-band, indeed, is known to be sensitive to changes in the target moisture content as the response is lighter on increasing humidity.

The ROI is mainly composed of burned areas, different forest canopies, the undergrowth and the brush-wood. The full resolution signal has a very noise-like behavior. Indeed on the X-SAR data (Figure 4 (a)) the different patterns on the scene are hardly detectable if only the gray level intensity is considered. An improvement is obtained if a median filter is adopted that replaces the pixel value with the median of the values collected over a suitable neighborhood (Figure 4 (b)). The regions with different characteristics have a different texture and therefore are distinguishable by some properties of the gray level spatial distribution. To retrieve information from the pre-processed X-SAR images, a segmentation procedure is applied to a suitable transformation of the data. The most useful transformations are the uniformity and the local contrast. Uniformity is obtained as the sum of the square of the local image histogram bins value, while the local contrast is based on the local signal variance. For example, to identify the larger area burned in the fire event of year 2000, the uniformity transformation is well suited, allowing the discrimination between the burned area and the neighboring regions, Figure 5 (a).

A four levels discrete level set segmentation of the ROI applied on the transformed data provides a simplified representation of the original image preserving the information to be retrieved, see Fig. 5 (b).

(a) (b)

Figure 4. X-SAR data, on the left original image and on the right median filtered image.

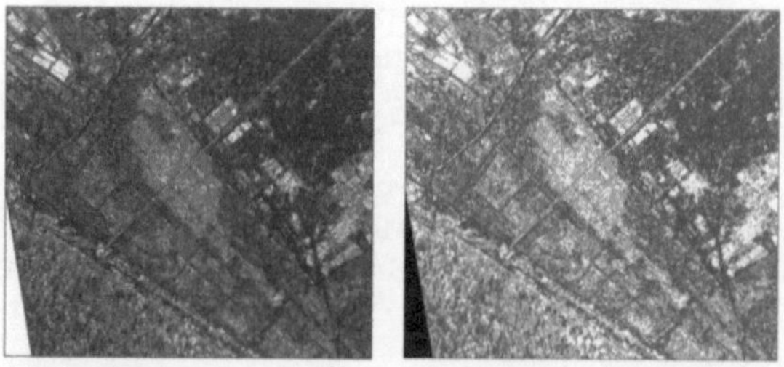

(a) Uniformity of image on Fig 4(b) (b) 4-levels segmentation of Fig 5(a)

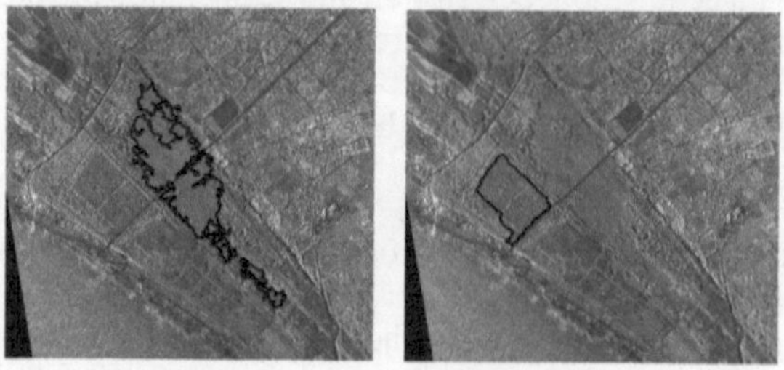

(c) Contour of the burned area (d) Contour of oak forest canopy

Figure 5. On the top SAR image processing; on the bottom some of the ROI identified areas.

A four levels segmentation is obtained by successive image binarizations: first the image I is partitioned into two distinct subregions (not necessarily simply connected) A_1, A_2, described by the *level set function* $\phi: D \to R$

$$A_1 = \{(i,j) \in D : \phi(i,j) \geq 0\},\\ A_2 = \{(i,j) \in D : \phi(i,j) < 0\} \tag{1}$$

Binarization is obtained by minimizing a fit error between the image and its binary representation, using a regularization term for the level set function ϕ. The problem has a unique global solution $(c_1, c_2, \{\phi(i, j)\})$, see (De Santis and Iacoviello 2007), where the $c_i, i = 1, 2$ are the signal mean values within the subregions A_1 and A_2. Once the first level optimal binarization is obtained, the regions A_1, A_2 are further binarized with the same procedure. Thus a four levels optimal segmentation is finally obtained.

The adopted segmentation method provides a good discrimination power of the different subregions, see the area burned in the fire event of 2000 in Fig. 5 (c). On the other hand, the measure of the local contrast allows, after segmentation, to identify other subregions like the typical canopy of the old oak forest, planted in rows, whose contour is shown on Fig. 5 (d). About ten different subregions can be identified in the above defined ROI. For each of them the high resolution X-SAR data allows for an accurate determination of the contour and extension, the fuel type, the gray level mean value that yields an estimation of average zone moisture content, one of the relevant physical parameter in the fire susceptibility model.

4 DATA CLASSIFICATION

In this section we present the data model and the machine learning based approach for data analysis. The data model we start with is the multi layer model specified in Section 2. Each layer mentions a quantitative attribute describing: soil and vegetation conditions, use and management; geographical information; past fire events; anthropic factors; risk/vulnerability degree. Further a dynamic and stratified ROI data representation is obtained, that is able to keep track of the evolution of changes at various layers being relevant to fire susceptibility assessment. In the past, in European countries like Greece and Spain, data regarding both soil and vegetation, and the meteorological conditions have been used for specifying an accurate and reliable fire risk index and a susceptibility map based on Data Mining and Machine Learning methods (see for example (Betanzos, Romero, Berdinas, Pereira, Andrade, Jimenez, Soto, and Carballas 2003), (Cortes and Morais 2007) and (Iliadis, Papastavrou, and Lefakis 2002)).

We define an input space $\mathbf{X}$ where each record of quantitative attributes describes an atomic area of the ROI. Then we assign to each record a class according to the segmentation level. The set of these records represents the training set. Our aim is to identify the classification rule defined over the training set and using this rule to assign the correct class to areas that have not classified yet. A novel and very efficient boosting technique combining kernel based Support Vector Machines has been used to learn the classification rule and to classify the different areas of the ROI.

The boosting techniques have been shown to be among the most promising and powerful methods for supervised machine learning. Usually, a boosting algorithm combines simple classifiers, the so called *weak learners*, in order to get a more precise classifier. The *dictionary* $D = \{h_1, \dots, h_k\}$ is the set of the weak learners we consider. Each weak learner is a Support Vector Machine separating data in an appropriate feature space. The choice of the kernel function for the feature map has been made considering the different kinds of available data. Our goal is that of finding the best coefficients $z_1, \dots, z_k$ of the linear combination providing the final classifier:

$$g(x) = sgn\left(\sum_{k \in D} z_k h_k(x)\right) \qquad x \in \mathbf{X} \tag{2}$$

Figure 6. ROI susceptibility map: different colors indicate different fire susceptibility levels, e.g., the red areas denotes oak trees with partial regular textures according to the foliage humidity levels.

As it is customary in machine learning models, the best combination is the one minimizing a loss function over the training set and a regularization function of the coefficients $z_1, \ldots, z_k$ in the combination (2). Alternatively, it is possible to bound the generalization capability of the weak learners in D, finding the best combination, by minimizing the loss function of the coefficients. As the family D is usually very huge, many heuristics have been proposed to identify the optimal combination (e.g., AdaBoost, (Freund and Schapire 1997)). We propose a linear optimization based approach to efficiently solve the problem of estimating the best classifier g in exact way. We have, thus, used enhanced data as specified in Section 2 to train a machine learning model based on Support Vector Machine (SVM). This model presents the fuel type information and the estimation of average zone moisture content extracted from the high resolution X-SAR images.

The final model trained on the available input data provides the susceptibility map. In this map different colors represent differently classified areas according to the features describing each pixel. By taking into account all the layers described in the present and previous sections, the susceptibility map yields an automatic classification of particular areas like oak tree forests, burned areas, Mediterranean maquis, and so on. An example of the final product of our model is shown in Figure 6.

5 MODELING ANTHROPIC FACTORS IN URBAN FORESTRY FIRE PROTECTION

In this section we present a brief overview of the methodology and main guidelines of anthropic factors modeling. The anthropic factors are particularly relevant in Mediterranean regions where the physical components have less impact with respect to regions having wide wild-lands areas. Several studies, such as for instance (Martinez, Chuvieco, and Martin 2004) and (Reineking, Weibel, Conedera, and Bugmann 2010) prove the relevance of human factors on fire ignition, for example simply considering the influence of population density on fire incidence. It is however quite hard to model the human impact and the interplay of different human behaviors, such as exploitation, random disruption, pyromaniac disorders, criminal offense, together with physical and meteorological influence, on the one side, andintervention and protection strategies on the other side, by fire operators and civil protection.

We have therefore devised a model based on Nash equilibrium, in which the anthropic factors are represented as game players. Here the game players are human behaviors using natural (e.g., wind or dryness) or non-natural (e.g., open fences, secluded paths) factors to build their profile strategies. Each game player has a payoff associated with her strategy profile set.

More specifically, let (S, f) be a strategic game with several game players, with $S = S_1 \times \ldots \times S_n$, S_i the strategy set of player i and $f = (f_1(s), \ldots, f_n(s))$, with $f_i : \times S_i \mapsto \mathbb{R}$, the payoff associated with the chosen strategy, and let $s_{-i} = (s)_{j \in N \setminus i}$ be the strategy profile of all the players except player i, a Nash equilibrium is a set of strategies profiles $(s_1^\star, \ldots, s_n^\star)$, $s_i^\star \in S_i$ one for each player, such that $\forall s_i \in S_i, f_i(s_i^\star, s^\star_{-i}) \geq f_i(s_i, s^\star_{-i})$, that is, no player has incentive to unilaterally change her action; namely, players are in equilibrium if a change in strategies by any one of them would lead that player to earn less than if she remained with her current strategy. Since the work of Kearns and colleagues (Kearns, Littman, and Singh 2001) the connection of Nash equilibria with graphic based statistic inference has become evident. In our case the strong connection (see also (Daskalakis and Papadimitriou 2006)) is with Hidden Markov Random Fields (Kuensch, Geman, and Kehagias 1995), stochastic processes taking values in a Euclidean space, and defined over a parameter space, as game player strategies and the neighborhoodsystem is spatially based. The advantage of a graphical abstraction of the interplay of several game player, in our case, is based on the fact that the payoff to each player is a function only of the strategy profiles of the player and her neighbors, rather than the strategies (actions) of the entire population. In particular, the payoff, namely the quantity each game player wishes to maximize, is defined according to the following basic criteria, see also (Vega-Garcia 1993), (note that the criteria implicitly mention contrasting canons):

- ease of ignition, in terms of probability to reach the desired location to start a fire in a cell.
- Ease of extinguishing, in terms of time to extinguish a fire before it has caused damages.
- Damages amount, in terms of burned area, fire power generated, with possible spread, importance of nearest buildings.
- Accessibility and escape from the affected area, in terms of probability to complete the action.
- Cell protection, in terms of probability to intervene in time.
- Cell prevention, in terms of probability to avoid ignition under different human, meteorological and fuel causes.

We are considering mixed strategy in which players randomize their choice, which have at least one Nash equilibrium. In the Markov random field model representation of the strategic game the graph is a lattice that has nodes labeled by the strategy profile set of each game players. The neighborhood system is defined by cliques, which is a set of points, all neighbors of each other, representing local game matrices. Namely, for any strategy $s_i \in S_i$ the payoff of game player i is specified with respect to the clique of her neighbors. That is, the probability of a choice is given as:

$$P(X = s) = \frac{1}{Z} \exp\left(\sum_i \sum_{j=1}^{m} w_j f_i(s_{ij}) \right) \tag{3}$$

Here Z is the partition function and w is the weight associated with the payoff criteria. Given the classification model, as described in the previous section, physical, meteorological, urban and forensic factors have been assessed at the same granularity of the segmented X-SAR images. This detail is clearly not adequate for the anthropic model, in which values to be maximized are defined according to behaviors, namely the space of a human action. In fact, several space data have been considered to fit the model: roads, including forest paths, but also small buildings, wood structures, walls, fences, railways, electric lines and fire-fighters facilities. All data has been geo-referenced. Considering the relatively small ROI (about $15\,\text{km}^2$), this has been divided into cells of $16\,\text{m}^2$. All the information available from data, related to the portion of area represented, is projected on each single cell of the lattice. Cell size is settled to maximize both the intra-cell homogeneity and the inter-cell heterogeneity referring to the related portion of the data model, according to the fitted model. From a computational point of view there is no difference between two adjacent cells with same characteristics so, in order to avoid *isomorphic phenomena*, adjacent cells are merged within the range $10-200\,m^2$. The inference problem amounts to finding the configuration that is most likely under the distribution $p(x)$. This configuration, overlayed on the classification, accounts for the anthropic factor contribution to the susceptibility to fire map.

6 CONCLUSIONS AND STATE OF THE ART

We have described an integrated framework that, from a basic structure of urban and land data, is lifted to anthropic data and finally includes fine clustering induced by X-SAR images segmentation. Here the X-SAR images have been provided by COSMO Skymed. These urban, land, and human behavioral data are used to build a susceptibility map for a pilot urban forest, namely the protected area of Castel Fusano near Rome, Italy. Our approach combining different methodologies is completely new in the literature. For the anthropic analysis, although several researchers have mentioned the relevance of this factor, and proposed evaluation and learning methods (see also (Li, Song, Ma, and K. 2009)), to our knowledge there is no contribution, so far, to assess a human behaviors model, of impact to fire, combined with other factors. On the other hand, several machine learning based approaches have been proposed in literature for fire risk analysis and prediction. In (Lapucci, Lombardo, Petri, and Santucci 2005) the fire risk areas were located in the Livorno district in Tuscany region by means of a synergetic use of data mining techniques to estimate the fire likelihood, given previous fire events, and a spatial multicriteria *analytical hierarchy process analysis*, for damage evaluation. An analogous experience using spatial data and data mining algorithms,to identify fire risk analysis in Portugal is described in (de Vasconcelos, Silva, Tome, Alvim, and Pereira 2001). The data on fire events, fire ignition locations and respective causes, have been collected by several Portuguese Forest Service field teams and two different machine learning models have been used and compared: logistic regression and artificial neural networks. The most interesting results showed that spatial fire ignition patterns, identified by the proposed approach, could be used for fire risk prediction and that these patterns could differ according to different fire ignition causes. Further studies and experiences in machine learning models for fire risk analysis can be found in (Betanzos, Romero, Berdinas, Pereira, Andrade, Jimenez, Soto, and Carballas 2003) and in (Iliadis, Papastavrou, and Lefakis 2002).

ACKNOWLEDGEMENTS

The work reported in this paper has been supported by ASI within the Cosmo-SkyMed Announcement of Opportunity (DC-OST-2009-116), SARFIRE Project 2288.

REFERENCES

Betanzos, A.A., Romero, O. Berdinas, B. Pereira, E. Andrade, M.I.P. Jimenez, E. Soto, J.L.L. & Carballas, T. (2003). An intelligent system for forest fire risk prediction and fire fighting management in galicia. *Expert Systems with Applications 4*(25), 545–554.

Cortes, P. & Morais, A. (2007). A data mining approach to predict forest fires using meteorological data. In *Proc. of EPIA 2007*.

Daskalakis, C. & Papadimitriou, C.H. (2006). Computing pure nash equilibria in graphical games via markov random fields. In *Proc. of the 7th ACM conf, on Electronic commerce*, EC '06, pp. 91–99. ACM.

De Santis, A. & Iacoviello, D. (2007). A discrete level set approach to image segmentation. *Signal, Image and Video Processing 1*(1), 303–320.

de Vasconcelos, M., Silva, S. Tome, M. Alvim, M. & Pereira, J. (2001). Spatial prediction of fire ignition probabilities: comparing logistic regression and neural networks. *Photogramm Eng Remote Sens 67*, 73–81.

Freund, Y. & Schapire, R.E. (1997). A decision-theoretic generalization of on-line learning and an application to boosting. *J. Comput. Syst. Sci. 55*(1), 119–139.

Iliadis, L., Papastavrou, A. & Lefakis, P. (2002). A computer system that classifies the prefectures of greece in forest fire risk zones using fuzzy sets. *Forest Policy Economics 4*, 43–54.

Kearns, M.J., Littman, M.L. & Singh, S.P. (2001). Graphical models for game theory. In *UAI '01: Proceedings of the 17th Conference in Uncertainty in Artificial Intelligence*, pp. 253–260. Morgan Kaufmann.

Konijnendijk, C., Nilsson, K. Randrup, T. & Schipperijn, J. (2005). *Urban Forests and Trees A Reference Book*. Springer.

Kuensch, H., Geman, S. & Kehagias, A. (1995). Hidden markov random fields. *Ann. of App. Prob. 5*, 577–602.

Lapucci, A., Lombardo, S. Petri, M. & Santucci, A. (2005). A kdd based multicriteria decision making model for fire risk evaluation. In *AGILE 2005, 8th Conference on Geographic Information Science*, pp. 143–152. Toppen F, Painho M. (eds).

Li, L., Song, W. Ma, J. & S.K. (2009). Artificial neural network approach for modeling the impact of population density and weather parameters on forest fire risk. *International Journal of Wildland Fire 18*, 640–647.

Martinez, J., Chuvieco, E. & Martin, P. (2004). Estimation of risk factors of human ignition of fires in spain by means of logistic regression. In *Proc. of the Second International Symposium on Fire Economics, Planning, and Policy*.

Pez, A. & Scott, D. (2005). Spatial statistics for urban analysis: A review of techniques with examples. *GeoJournal 61*, 53–67.

Reineking, B., Weibel, P. Conedera, M. & Bugmann, H. (2010). Environmental determinants of lightning-v. human-induced forest fire ignitions differ in a temperate mountain region of switzerland. *International Journal of Wildland Fire 19*, 541–557.

Vega-Garcia, C. (1993, Feb). Geographic and temporal factors that seem to explain human-caused fire occurrences in whitecourt forest, alberta. In *GIS93*, Volume 1, pp. 115–119.

Urban and Regional Data Management – Zlatanova, Ledoux, Fendel & Rumor (eds)
© 2012 Taylor & Francis Group, London, ISBN 978-0-415-67491-1

A decision-making support model for systemic vulnerability assessment in urbanized areas

A. Sole, R. Albano, S. Pascale & L. Giosa
Department of Environmental Engineering and Physics, University of Basilicata, Potenza, Italy

F. Sdao
Department of Structures, Geotechnics, Engineering Geology, University of Basilicata, Potenza, Italy

Å. Sivertun
Swedish National Defence College, Stockholm, Sweden

ABSTRACT: This paper aims to analyze the systemic vulnerability of a complex urban area when a combined flood-landslide event occurs. The proposed approach of vulnerability evaluation differs from traditional methods as it provides an analysis of "Systemic Vulnerability" measuring the relationship between extreme event magnitude and direct and indirect consequences of this event on the territorial system. GIS technology has been introduced for the proposed systemic vulnerability assessment because it allows implementation of the spatially and thematically distributed models used. The software also consents to the automated application of the method by creating a model that processes, analyzes and displays data in the form of maps. The model proposed was shown to be able to support the territorial planning and the appropriate management of an emergency due to natural catastrophes such as floods and landslides. The priority intervention elements can be defined as those which were more affected by systemic vulnerability making the system less flexible.

1 INTRODUCTION

Landslides and flood phenomena are some of the highest risk factors for human life, and environmental and economic activities. They prove to be extremely dangerous when they coincide each other or with other natural events. Landslides can, for example, block the normal flow of a river to produce a dam-breaking effect downstream. The rapid urbanization and the construction of buildings and infrastructures increase the territorial vulnerability because during a natural catastrophe, such as a landslide or flood, the vulnerability depends not only on the features of the territorial elements but also on the interdependence between them. For example, the collapsing of a bridge, apart from the intrinsic loss, could bring about the interruption of an entire traffic zone, or interrupt the functions of a hospital with consequences on the health system in general as well as on mobility, communication, efficiency of civil protection forces, etc.

The Systemic Vulnerability model proposes a generalized method applicable to each specific urban area subject to any kind of natural disaster or combination of two or more events. This is possible because it considers the different interconnections among the elements of the system which can be many and various depending on the degree and form of the urbanization of the territory. One study which has looked at the functional relationships between interconnecting elements of a territorial system subject to natural hazards is Tamura et al. (2000), Ezell et al. (2000), Haimes and Jiang (2001) and Minciardi et al. (2006). The approaches mentioned above do not appear suitable for the analysis of the consequences of landslide-flood events, because they deal with modeling aspects which are too specific.

For the purposes of this study, a more general methodology is proposed, so that the end user can make decisions regarding a defined territorial system without being an expert in risk modelling. The proposed model, through GIS technology, gives planners an automatic and easy to understand tool to take into account the effect that the fragility of a node could produce in the overall functioning of the whole territorial system. Through the consultation of the maps, elaborated in the model, the planners can recognize the priorities of interventions, i.e., the elements more affected by Systemic Vulnerability, and can give the community a more efficient support to mitigation strategies, to delineate the emergency plans and to respond to calamitous events with civil defense actions.

2 A GIS MODEL FOR SYSTEMIC VULNERABILITY ASSESSMENT IN URBANIZED AREAS EXPOSED TO COMBINED LANDSLIDE-FLOOD RISK

The model has been created in GIS environment using the tool Model Builder and consists of the following steps:

- the first phase deals with the topological characterization of the territorial system, represented through a network, and consists in the identification of its components on one hand and their connections and interactions on the other. Then it focuses on the evaluation of the scenarios at risk (in this case landslide-flood). The overlay of the territorial network and the risk scenarios highlights the elements directly affected by the event;
- in the second phase we use the data created in the previous step in order to analyze the direct after-effects of an event on the urban system. In fact, in this step, the functional vulnerability of each element is evaluated in function of the intensity of the event. For example, for events with the same intensity, a greater vulnerability functional is assigned to elements that have important functions in the system;
- the last step focuses on the estimate of the systemic vulnerability of the elements of the system taking into account the results of the previous step, the interconnection and the influence between the elements.

Each step represents a module which will be reported in detail in the following.

2.1 *Module 1: Characterization of the system*

A complex territorial system is composed of many subsets with specific functions, (Minciardi et al., 2006, Pascale et al., 2009 and 2010; Sdao et al., 2011, Pascale S et al., 2010): built-up areas, health centers, civil and environmental protection structures, different infrastructures, energy distribution centers etc. These can be either connected by mutual interrelations in order to guarantee a proper and effective functioning of the complex system.

To identify the interconnections of the territorial system, the latter is represented by the network G (E, A), where E is the set of nodes and A is the set of the arcs (link between nodes). The nodes, belonging to the E set, can represent specific territorial points (such as hospitals or fire stations) or areal infrastructures (such as industrial areas or sites) and the arcs, belonging to the A set, can represent linear elements (such as road infrastructures, railways, subways etc.).

The damage or malfunctioning of only one of these elements can have a remarkable influence on the proper functioning of the entire territorial system.

After the characterization of the system, such as a dense interconnection network, the flood-landslide risk scenario is overlapped on this network to create a map that visualizes the potential elements at risk (Fig. 1).

2.2 *Module 2: Analysis of external stress on the elements*

The scenario, outlined in the previous step, is studied to evaluate the intensity of the event, and it will be useful to estimate the functional vulnerability of elements at risk, identified in

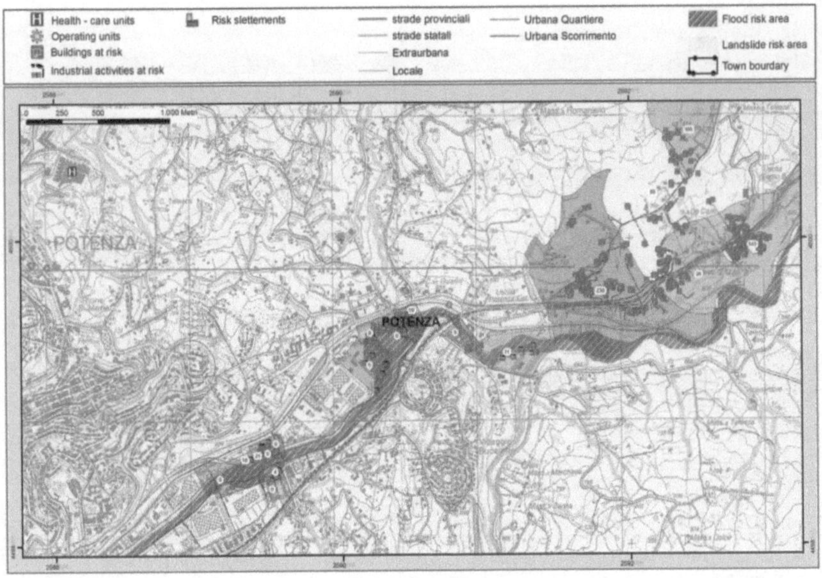

Figure 1. Potenza city network overlapped with the scenario to show elements directly involved by a flood–landslide event.

the last step. In fact, the vulnerability assessment is grounded on the event magnitude and on the function that the element plays in the proposed network.

2.2.1 *Flood and landslide intensity maps*

Magnitude of floods is measured by means of a matrix (Pascale et al., 2009, Sole et al., 2011) based on several parameters: return time, water depth and flow speed. The parameter values are observed on single points and then interpolated in order to achieve a common vision of the entire course of the river, so you can see a map for each parameter. Those maps show some numerical or textual values which are not comparable to one another, so that no objective total intensity value can be derived from these. Consequently, the maps are classified by subdividing the parameter into value intervals to which a 1-to-10 score is assigned by the command "Reclassify" of the tool "Spatial Analyst", depending on the specific interval and the classification of the different parameters. Then the score relating to each parameter is added and subdivided further into proper magnitude classes by the above mentioned tool. A flood magnitude value ranging from 1 to 10 and of an intensity ranging from very weak (I) to extremely strong (X) is assigned to each interval.

The same approach from the landslide magnitude is measured by means of a matrix (Pascale et al., 2010, Sdao et al., 2011) constituted by various parameters which characterize the landslide and express its intensity including: volume, velocity, run out, depth, area involved, deformation and typology.

Therefore, by using the vector landslide and flood data we can draw up the landslide-flood intensity maps of the municipality of Potenza (Fig. 2). These two maps are joined and compared with the Mosaic to New Raster tool so as to create a map which also provides information about a possible combined risk.

2.2.2 *The socio-economic and environmental vulnerability maps*

The drawing-up of the event intensity map will be useful for estimating to what extent the calamitous event affects the elements at risk. In order to evaluate the element vulnerability, it is important to consider the magnitude of the event and the element functionality, estimated on the basis of the typology as well as the population density in the territory where they can be found.

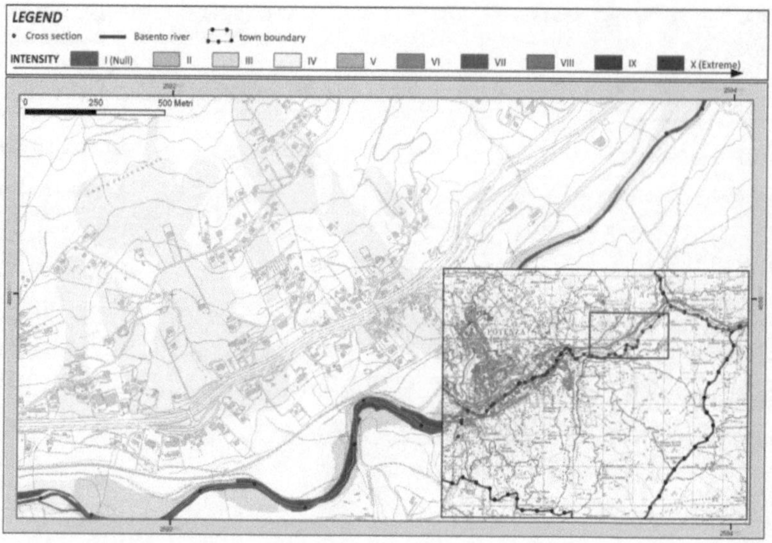

Figure 2. Intensity map of combined flood–landslide event.

The typology of the element is important for the role it plays in the case of an emergency and for the socio-economic and environmental damage it can cause. For instance, if a hospital is damaged the whole system is affected by an increase in the rescue workload on other forms of assistance. The risk elements with different physical, social and economic features are set in Categories A, B and C. The importance of these features moving from Category A to C:

- Category A includes all the most important elements in the case of an emergency such as hospitals, fire stations and civil protection stations. These are all those elements that give assistance when catastrophic events occur, as well as link roads such as highways.
- Category B includes all the major socio-economic and environmental elements such as factories, which can also deal with dangerous materials, and big shopping centers as well as all other public buildings including Universities, Libraries and Churches. All of these can contain a large number of people and can also be important from a historical, artistic and cultural point of view. This Category includes secondary roads (main and provincial roads), as well.
- Category C includes private buildings, small business activities and local roads.

A number has been assigned to each Category: number 1 to Category C, number 2 to Category B and number 3 to Category A.

The population density is important because it provides a priori information on the number of people who may need rescue and on how to act in the case of danger when an appropriate emergency plan can be prepared which foresees the right kind of intervention necessary where the population density is higher.

The population density is classified into three categories (high, medium and low), following the "Equal Interval" criterion: the value interval is divided into sub-fields of an equal dimension so as to specify the interval number. This procedure facilitates an adjustment of the situation to the model giving the flexibility and application multiplicity of the model itself to any territorial system. A score is successively assigned to each Category thanks to a proper reclassification: the "Low Density" Category is characterized by a score equal to 10, the "Medium Density" by a score equal to 20 and the "High Density", having a higher density, equal to 30.

The density map reclassified values have been added cell by cell to those reclassified in the territorial element typology map, thus obtaining a new raster giving information on density and functionality. By summing up these values, 11, 12, 13 and 21, 22, 23 and 31, 32, 33 are obtained. The first figure read from left to right gives information on density (from low 1 to high 3), while the second gives information on the element functionality. Each of these values has been extrapolated from the new raster created by drawing up a new map through the IN operator of the Map Algebra tool. In fact, this operator is able to extrapolate the 9 different values (11, 12, 13 and 21, 22, 23 and 31, 32, 33) from the raster which is the sum of the functionality and density values, thus creating 9 new rasters, that is 9 maps.

Then to draw up the functional vulnerability map of the landslide-flood risk elements, the value of the event intensity map, of the population density map and also of the territorial element typology map are combined in different equations.

2.3 *Module 3: Analysis of systemic vulnerability*

The last sub-model is the one which permits an evaluation of the real systemic vulnerability where the interconnections between the different elements and the territorial system flexibility in the case of a calamitous event are studied.

2.3.1 *Maps of influence*

This is the specific phase which studies the interconnections between the territorial system elements. The first operation consists in evaluating the at-risk element influence on all the elements constituting the territorial system.

The way each at-risk element affects the network organization and flexibility can be observed and estimated. For instance, if there is only one road connecting the fire station to an at-risk building, the network is not so efficient in the case of an emergency being very rigid and not so dynamic, therefore the positive influence of the element on the system will be very low. As a consequence, the planners can try to solve such problems emphasized by the model and they can imagine new connecting roads and ramifications so as to make the building affected by the calamitous event safe, or else plan other interventions. The connection typology must also be taken into consideration: if the road given in the example above is classified as a highway, the system acquires more flexibility because the connection is faster in the case of an emergency, and so that particular building will have more influence on the network. The better the territorial network is planned and prepared for the emergency, the higher the influence is since it ranges between 0, that is the element without any interconnection with the others, and 1, that is the element with strong interconnections giving the system the maximum flexibility level (Pascale et al., 2010).

2.3.2 *Systemic vulnerability maps*

In order to make the element vulnerability comparable and interdependent with their influence on the network, firstly the influence value scale is inverted by the equation (Pascale S et al., 2010):

$$\alpha_{ij} = -9,9 \times \mathrm{I}_{\mathrm{a,b}} + 10 \tag{1}$$

where $\mathrm{I}_{\mathrm{a,b}}$ is the influence index defined for each couple of categories of territorial elements; and secondly this is put into a relation with functional vulnerability through the equation (Pascale S et al., 2010):

$$w_{ij}\left(x_j\right) = \left(1 - 0.1\alpha_{ij}\right) \frac{\left(1 - e^{-\alpha_{ij} x_j^2}\right)}{\left(1 - e^{-\alpha_{ij}}\right)} + 0.1\alpha_{ij} \tag{2}$$

Each equation is inserted into the Map algebra command window in order to obtain the respective results.

Equation 1 enables the passage from the element influence meant as the system flexibility increasing to the element influence meant as the critical level system. The former considers

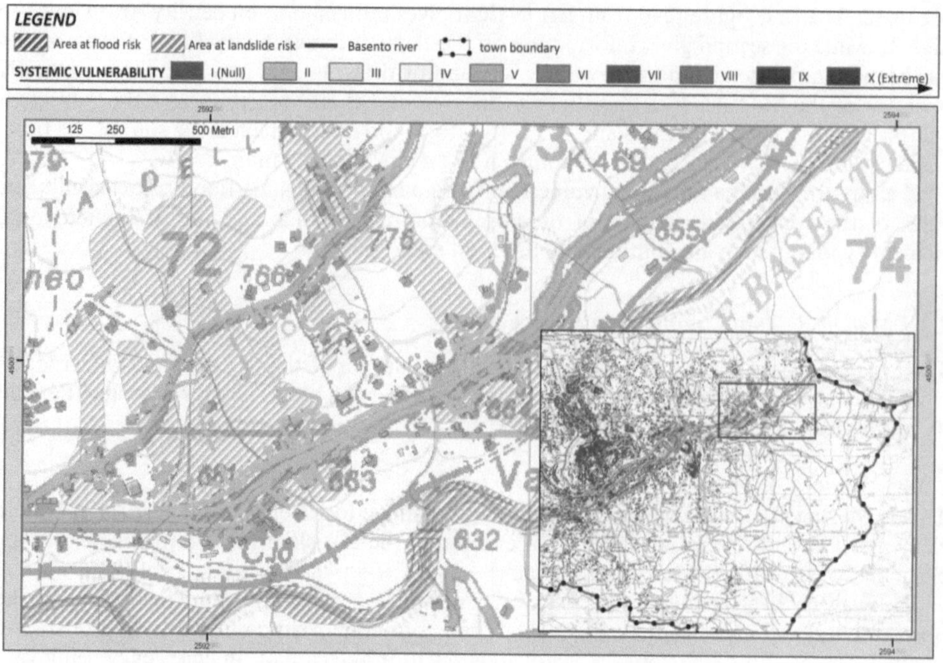

Figure 3. Map of systemic vulnerability.

how decisive the element was for the system in the case of an emergency, the latter considers how the element can put the system dynamicity at risk.

The range of values which was between 0 and 1 is now between 1 and 10 where the lowest system flexibility represented by value 0 changes into the element with a maximum critical level in the network, that is to say 10, and the maximum flexibility which was 1 now expresses the complete element homogeneity and amalgamation in the network associated with values 0.

With Equation 2, the vulnerability expressing the degree of critical level for each single element (damage on one single element), is compared with the parameter of the previous equation which expresses the critical level element for the system, (damage on the entire system), thus obtaining a new parameter called influence level.

The new parameter (influence level) shows that an element is critical for the system not only due to the damage which can cause individual suffering, but also because of the damage it can cause as an element of a territorial system.

The influence level ranges from 0 to 1, where 0 represents the element having no criticality, and it is entirely integrated in the system, while 1 represents the maximum criticality of the element in the system, and constitutes a node in which the network is completely rigid. Therefore, as not to underestimate any evaluations made on the critical level element, the vulnerability map is compared once again with that of the influence level taking the maximum value cell by cell with the Mosaic to New Raster tool. This tool compares the values of the two maps and considers the highest one to be in a condition of maximum safety in order to create the Systemic Vulnerability map (Fig. 3) with these values.

3 CASE STUDY

The model for the evaluation of systemic vulnerability in combined flood-landslide risk areas has been validated in the urban area of Potenza, Southern Italy. The city of Potenza covers approximately 174 km^2. The area is a part of the Apennine Chain and has an average annual rainfall of about 800 mm. Elevation ranges from 593 to 1339 m a.s.l.

It was selected as suitable for the evaluation of the model because it is subject to instability phenomena due to its geological, lithological and geo-morphological features. A total of 920 landslides have been found in the area, equal to a surface of about 46 km^2, 26% of the entire area of the town of Potenza. The latter was suitable because it is the Regional Capital of the Basilicata (96,000 inhabitants). It is the city with the greatest concentration of communication means, productive activity and residential areas and it is intersected by the “Basento” River, one of the most important rivers in the Region. This is a torrential Mediterranean river which flows into the Ionian Sea. The river bed consists of gravel and stone. In this area, the river has an inclination of 0.6% and an average depth of 1.5 meters. Along the river there is a high concentration of residential and business settlements and most of the industrial activities of the city. Therefore, if this river floods it could cause extensive damage to this sector.

3.1 *Data*

In the case study, the scenario is not the entire area at flood-landslide risk but only a specific zone (Fig. 4). The latter has been chosen as suitable for testing the model for its wide urbanization and its territorial characteristics, the coexistence and the interdependence between the area at flood risk and the landslide area. Infact, the landslide scenario considered in this study involves the area of “Varco d’Izzo”, that is an area which has been recently subject to extensive urbanization and is characterized by the diffuse presence of human activity and several important transport routes (“S.S. 407 Basentana”, “Railway Potenza-Metaponto”). The flood scenario involves the flooded area of “Basento” River with Tr = 500 years (Return Time). This landslide area is generally fan-shaped and reaches the “Basento” channel which consequently is meandered with evident lateral deviations.

All the scenario data have been produced by the organization responsible for protection against hydrological risks of the territory considered, the Basilicata Inter-regional Basin Authority (AdB of Basilicata).

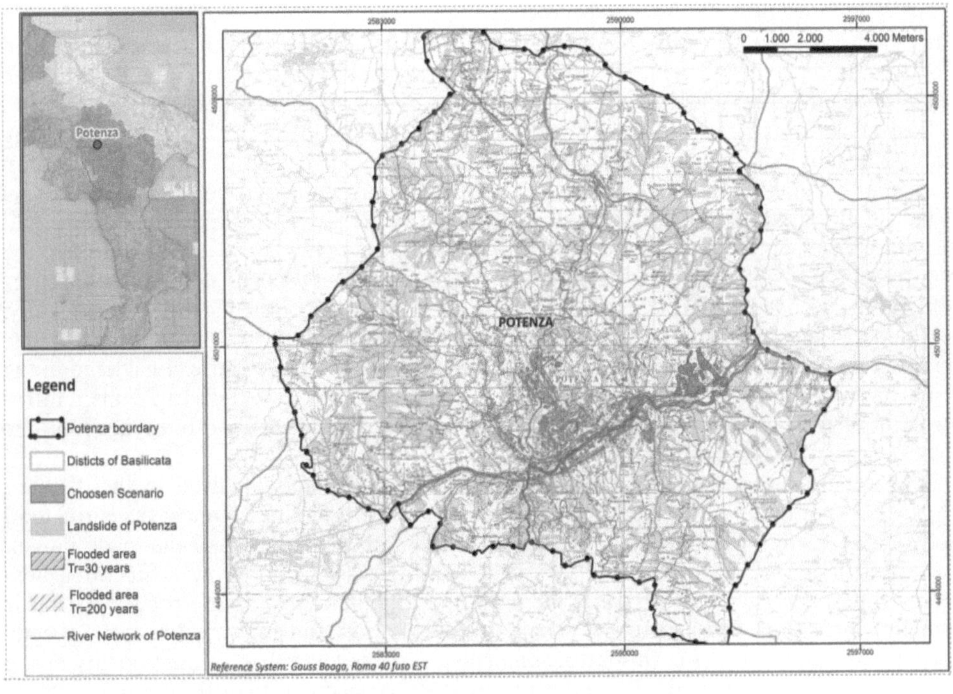

Figure 4. Scenario utilized in the case study.

We can use the following set of parameters from AdB database: volume, velocity, run out, depth, involved area, deformation and typology to estimate the intensity of the landslide.

The AdB has selected reference flood events with return periods, "Tr", equal to 30, 200 and 500 years for the identification of flooded areas and, by means of mathematical models of a flood simulation, was able to define the areas of the territory directly involved in the flood event, the maximum water depth, the maximum velocity of the water in the river and in other flooded areas. This study only deals with the event with a return period of 500 years, characterized by a maximum flow value of 1015 m^3/s, because its maximum flooded area is very similar to those with return periods of 30 and 200 years in the fluvial area intersecting the town of Potenza.

3.2 *Resolution*

The first step identified the Potenza territorial network, characterized by 18883 elements, 1283 areal elements (residential districts), 5000 lines (infrastructures: streets, roads, railways and so on) and 21000 points (crossroads, public buildings such as schools, local health centers, hospitals, military and fire stations, industries and so on).

Then the scenario was overlaid on the territorial network to highlight the 140 elements at risk (Fig. 1). From the data of AdB the intensity of the calamitous event was estimated (Fig. 2): the water depth, the maximum velocity of the water and the return time parameters were utilized to estimate the flood intensity and volume, velocity, run out, depth, involved area, deformation and typology to estimate the landslide magnitude.

Therefore, the intensity of the combined flood-landslide event with the functional categories of the elements of the territorial network have been necessary to evaluate the functional vulnerability map.

Finally, this map was combined with the map of influence, that shows the interdependence of a couple of categories of elements within the system, and the systemic vulnerability map was obtained, (Fig. 3), with the equation of the mathematical model, (Pascale S et al., 2010).

3.3 *Results*

This study has put forward an approach for estimating the consequences of combined flood-landslide events drawn up by means of the utilization of a territorial model based on a network of influences and an estimation procedure of systemic vulnerability.

The reliability of the proposed approach has been tested on a case study regarding flood-landslide risk in the district of Potenza and has led to identify the infrastructural elements that mainly affect the functionality of the territorial system, i.e., the National Road 407 Basentana and the Railway Naples-Taranto. Although the same conclusion could also have been achieved by the analysis carried out by an expert in the territory (due to the relative simplicity of the territorial system considered), it is to be considered as an important result, because it highlights the reliability of the obtained results and, consequently, of the decisions to come. The significance of such analyses in the ambit of the particular territory considered must also be emphasized. Analogous events to those considered in the study have occurred in the Potenza area and have caused serious damage to the main infrastructural systems (water pipelines and roadways) as well as direct and indirect economic losses on most of the surrounding area.

The scenario utilized has highlighted 140 territorial elements involved in the disaster: 35 areas, 31 points (18 public buildings and 8 industries) and 74 lines. Almost the entire road network is directly or indirectly involved in the disaster: more than 30% of the roads, in fact, are affected by systemic vulnerability. All the considered elements show a functional loss within the system of at least 10% and, therefore, not negligible, and, in the specific, the linear elements more than 20%: these are urban streets, subways and railways (close or directly involved in the area at risk). This reduction depends on the low value of functionality of the streets within the system and it is not due to a loss due to the intensity of the event, but rather to a loss due to the roads connected to them. Infact, the flood-landslide event does not have a wide areal extension and the flood event is quite rare (Tr = 500 years).

4 CONCLUSION

The proposed model analyzed the environmental problems of Potenza (South Italy), simulating a combined flood-landslide event, and showed that the Systemic Functionality may increase because of the stress that affects the single element but also due to its losses of functionality induced by the interconnection with the other elements.

It is a generalized methodology, applicable to every territorial system and each kind of event or combination of events.

The model permits us to formulate hypotheses rapidly, to test and verify them to improve comprehension of the studied phenomenon and decisional processes. After the long reconstruction of the territorial system, that can be implemented with other information such as social status of the population, speed limit on the street, etc., the planner is able to modify the scenario or the territorial network to choose the best planning, and the best environmental and economic decisions.

The GIS technology allows us to evaluate and verify the influence that local damage produces on the overall functioning, and to visualize the results, through maps, as a realistic representation and to identify and manage the process through an easily comprehensible flow-chart programming. The model is a valid decision-making support for an efficacious and also participated planning, that allows all the people, specialized or not, to be receptors and actors of their own territory.

REFERENCES

Ezell, B.C., Farr, J.V. & Wiese I. 2000. Infrastructure risk analysis model, *Journal of infrastructure system*: 114–117.

Galli, M. & Guzzetti, F. 2007. Landslide Vulnerability Criteria: A Case Study from Umbria, Central Italy, *Environ Manage*: 40, 649–664.

Geoprocessing in ArcGIS, Environmental Systemic Research Institute, Inc.

Haimes, Y.Y. & Jiang, P. 2001. Leontief-based model of risk in complex interconnected infrastructures, *Journal of infrastructure system*: 1–12.

Khatibi, R. Systemic nature of, and diversification in system exposed to, flood risk, *First International Conference on Flood Recovery, Innovation and Response*: 91–101, 2008.

Minciardi, R., Sacile, R. & Trasforini, E. 2006. Assessing the efficiency and the criticality of the elements belonging to a complex territorial system subject to natural hazards, *Natural Hazards and Earth System Sciences*: 6, 21–32.

Network Analyst, Environmental Systemic Research Institute, Inc.

Pascale, S., Giosa, L., Sdao, F. & Sole, A. 13–15 May 2009. Assessment of systemic vulnerability in flood prone areas, *Fourth International Conference on Sustainable Development and Planning*, Cyprus.

Pascale, S., Sdao, F. & Sole, A. 2010. A model for assessing the systemic vulnerability in landslide prone areas, *Nat. Hazards Earth System Science*: 10, 1575–1590, DOI: 10,5194/nhess-10–1575.

Sdao, F., Sole, A., Pascale, S. & Giosa, L. 2011. Systemic vulnerability in urban areas exposed to combined risk of landslide and flood, *Italian Journal of Engineering Geology and Environment*, in press.

Sole, A., Sdao, F., Sivertun, A., Albano, R., Pascale, S. & Giosa, L. 2011. A GIS implementation of a model of systemic vulnerability assessment in urbanized areas exposed to combined risk of landslide and flood, *Geographic Information Analysis for Sustainable Development and Economic Planning: New Technologies, IGI Global*, (in press).

Tamura, H., Yamamoto, K., Tomiyama, S. & Hatono, I. 2000. Modeling and analysis of decision making problem for mitigating natural disaster risks, *Eur. J. Oper. Res.*, 122, 461–468, 2000.

Using ArcGIS Spatial Analyst, Environmental Systemic Research Institute, Inc.

Using Model Builder, Environmental Systemic Research Institute, Inc.

Part IV:
Remote sensing for urban applications

Urban and Regional Data Management – Zlatanova, Ledoux, Fendel & Rumor (eds)
© 2012 Taylor & Francis Group, London, ISBN 978-0-415-67491-1

LiDAR VR generation with point-based rendering

M. Nakagawa
Shibaura Institute of Technology, Tokyo, Japan

ABSTRACT: We have developed a point-based rendering application with high-speed spatial interpolation using a point cloud to generate virtual reality (VR) data called LiDAR VR. This application not only transforms from a point cloud to VR data but can also generate missing-point data in the rendered result from a selected viewpoint. Moreover, we have focused on the following features of the LiDAR VR generation. First, panorama imagery can be generated automatically from arbitrary viewpoints from one observation data set. Second, it is possible for a spatial interpolation with rendering results to reduce the processing task. These features have been verified using case studies.

1 INTRODUCTION

Accurate 3-D scanning techniques such as terrestrial laser scanning are used for various applications such as geographical information systems (GIS) data acquisition in urban areas or archaeological digital archiving. A laser scanner can acquire a 3-D point cloud by measuring the distance to a surface and the scanning angles. Moreover, a calibrated digital camera with a laser scanner can be used to acquire color information with the point-cloud data.

A point cloud created using these techniques has no limitations on its visualization in 3-D space. When we compare a point-cloud representation with a Triangulated Irregular Network (TIN) or computer-aided design (CAD) model representation, the point-cloud representation has the advantage of high presence. However, a viewpoint translation in point-cloud rendering reduces the visualization quality because of noticeable occlusion exposure and a noticeably uneven point distribution. Although the point cloud preserves accurate 3-D coordinate values, the phenomenon that transparent far points exist among near points reduces the visualization quality for users. In addition, recent drastic increases in scanning speed and scanning data volume require a large video memory and high-frequency CPU.

In contrast, image-based rendering or panorama VR such as Google Street View by Google Inc. or QuickTime VR (Shenchang 1995) by Apple Inc. can provide a good presence for users, even though the data processing is very simple. File sizes for the panorama VR are considerably smaller than for those of point clouds, because the panorama VR consists of a projected panorama image or a projected mosaic image and a simple screen. However, panorama VR data are not actual 3-D data. In addition, this methodology requires users to capture images one by one from different viewpoints.

In recent research, simulated panorama imagery was generated from different viewpoints using digital cameras [2, 3]. These approaches require very accurate camera translation and rotation parameters. Generally, these parameters are calculated through a camera orientation algorithm using a position data taken from RTK-GPS and rotation data taken from a gyro-sensor as initial values. When the camera orientation is omitted, simulated panorama image generation may fail to overlap images correctly. In addition, increasing time differences in image acquisitions may also cause failures in overlapping images because of moving objects in the acquired images.

2 OBJECTIVES

An advantage of 3-D point-cloud data is that it allows accurate display from an arbitrary viewpoint. By contrast, panorama imagery has the advantage of providing higher attractiveness with less data. In addition, panoramic image georeference (Dykes 2000) and distance-value-added panoramic image processing (Verbree, Zlatanova and Dijkman 2005) show that both advantages can be combined for 3-D-GIS visualization. Thus, we focus on the possibility that these advantages can be combined by a point-cloud projection into panorama space. Splat-based ray tracing (Linsen, Muller, Rosenthal and Cohen 2007) is a methodology to generate a photo-realistic curved surface on a panoramic view using normal vector from point-cloud data. A problem is the large amount of time required for surface generation in 3-D work space. Furthermore, the curved-surface description is inefficient at representing urban and natural objects as GIS data. In particular, we consider that a simpler filtering algorithm is important to achieve high volumes of point-cloud processing with high speeds. Thus, our objective is to develop a point-based rendering application with a simpler filtering algorithm to generate VR data, which we call LiDAR VR data. We have also focused on the following two features of LiDAR VR imagery generation.

The first feature is LiDAR VR imagery generation with a translated viewpoint. In general, an image from a viewpoint should be created from an observation. Moreover, when panorama images are taken from several viewpoints, position data from RTK-GPS and rotation data from a gyrosensor are required to calculate relative positions and rotations between cameras. LiDAR VR images from many viewpoints can be simulated from observation data through the approach proposed in this research. Moreover, the proposed methodology can simulate camera translation and generate new simulated imagery without the position and rotation sensors.

The second feature is real-time spatial interpolation using a point cloud. In general point-cloud interpolation processing, missing points can be interpolated by a traditional 3-D filter from points in a space. This type of 3-D interpolation requires huge amounts of processing capability. Therefore, we focus on the possibility that spatial interpolation with rendered results can reduce the processing task. We have also developed a pixel-selectable averaging filter. This function can replace spatial filtering in 3-D space by spatial filtering in 2-D space with real-time processing. Thus, any quality reduction caused by viewpoint translation can be solved in real time.

3 METHODOLOGY

The processing flow of the proposed methodology in this research is described below. First, sensors acquire a point cloud with additional color data such as RGB data or intensity data. The sensor position is defined as an origin point in a 3-D work space. If color data cannot be acquired, distance values are attached to a color index. Thus, for example, we can use a laser scanner, stereocamera, or time-of-flight camera. Second, a LiDAR VR image from the simulated viewpoint is generated using the point cloud. Finally, the generated LiDAR VR image is filtered to generate missing points in the rendered result using distance values between the viewpoint and objects.

3.1 *LiDAR VR image generation using the point cloud*

The colored point cloud is projected from 3-D space to panorama space. This transformation simplifies viewpoint translation, filtering, and point-cloud browsing. The LiDAR VR data consist of a panorama model and range data. The panorama space can be a cylindrical model, a hemispherical model, or a cubic model. Here, the spherical model is described. The measured point data are projected onto a spherical surface, and can be represented as range data as shown in Figure 1. The range data can preserve measured point data such as X, Y, Z, R, G, B, and intensity data in the panorama space in a multilayer style. Azimuth and elevation angles from the viewpoint to the measured points can be calculated using 3-D vectors generated from the view position and the measured points. When azimuth angles and elevation angles are converted to column counts and row counts in the range data with adequate spatial angle resolution, a spherical panorama image can be generated from the point cloud.

Based on this spherical panorama, the range data are generated using the point cloud with a translated viewpoint, as shown in Figure 2. When points from P1 to P10 are projected into a panorama space generated from a viewpoint Xo, these points are arranged continuously from P1 to P10 in the range data. An azimuth or elevation angle from a viewpoint Xo to a measured point P1 is denoted as Ro. When the same scene is captured from a different viewpoint Xt, the angle from the viewpoint Xt to the measured point P1 is denoted as Rt. The position of the projected point in the range data moves according to the change in angle from Ro to Rt.

3.2 *Filtering using distance values between a viewpoint and measured points*

Three types of filtering are executed after the viewpoint translation, as shown in Figure 2. The first filtering is an overwriting of the occluded point. When the viewpoint is translated to Xt, the projected point P1 becomes an occluded point behind P2. Therefore, P1 is overwritten by P2.

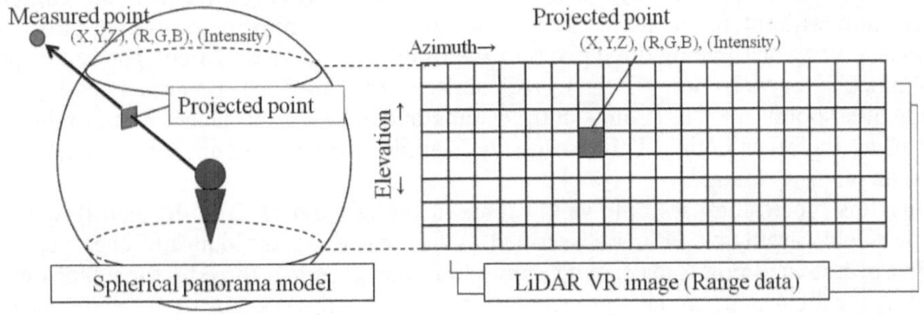

Figure 1. LiDAR VR consisting of a spherical panorama (left side in the figure) and range data (right side in the figure).

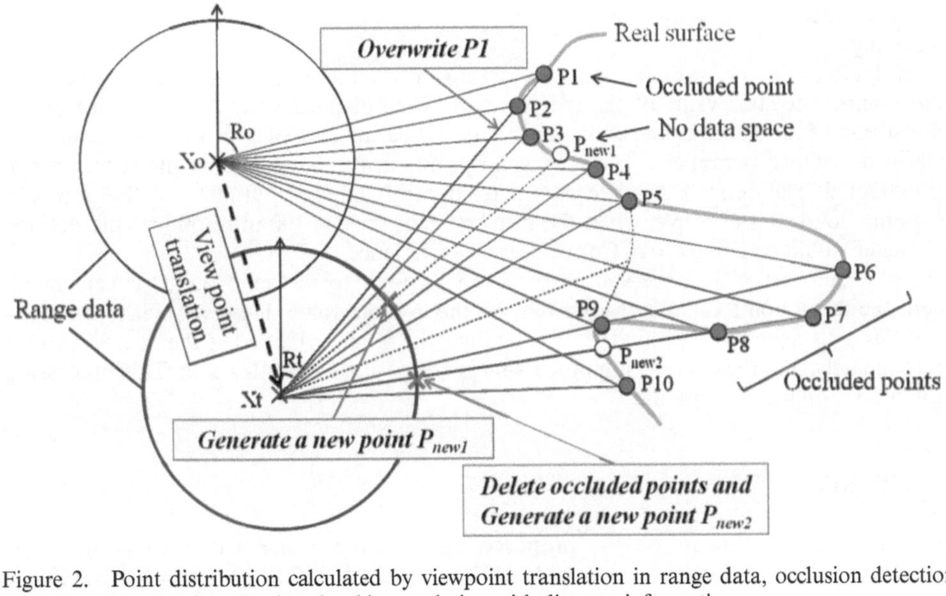

Figure 2. Point distribution calculated by viewpoint translation in range data, occlusion detection using the point cloud, and point-cloud interpolation with distance information.

The second filtering is a generation of new points in no data space. This occurs when the viewpoint is translated to Xt and no data space exists between the projected points P3 and the projected points P4. Thus, Figure 2 shows that P_{new1} is generated.

The third filtering is detection of occluded points and generation of new points instead of detected occluded points. When the viewpoint is translated to Xt, the point P8 exists between P9 and P10 after the first filtering. However, the actual point P8 should be occluded because the point P8 exists behind the real surface. Therefore, the occluded point P8 should be given a new distance value as P_{new2}, calculated with interpolation processing using the distance values of points P9 and P10. In addition, new points are generated with a pixel-selectable averaging filter developed in this research as follows.

3.3 *Pixel-selectable averaging filter*

In general, when an image is transformed, each pixel in the image has its color data resampled by using pixel values around it. Points projected into the panorama space are also processed using a similar technique to improve the quality of the range data. However, general resampling techniques such as nearest interpolation reduce the quality of the range data because valid, erroneous, and missing data are blended in the resampling. Therefore, a focused pixel-selectable averaging filter is applied to this problem. The filtering processing uses only valid pixels around a pixel for the resampling. This processing is equivalent to missing-point regeneration without reducing geometrical accuracy by a uniform smoothing effect. For example, a block of three by three pixels in the range data set is prepared. The center point in the block is the focus point in the range data set. When the focus point is a missing point and the other points are valid points in the point cloud, a new pixel value is given to the focus point using the other points. This processing is applied to each channel in the RGB image and to an intensity image.

Thus, this processing uses only valid points in the range data. The detailed flow of the pixel-selectable averaging filter is described as follows. First, the data are checked to see whether or not valid points exist. When more than one pixel in a three by three block exists, including the focus point and the points around it, processing proceeds to the next step. If no point exists in the block, the next processing step is omitted.

Second, the number of valid pixels in the block is counted. When there are more than two valid pixels, the processing proceeds to the next step. If there is only one valid pixel and the extracted point is the focus point, it is deleted as spike noise. If there is only one valid pixel and the extracted point is not the focus point, the focus point remains without the filtering processing.

Third, after these point-extraction steps, a range of search in distances is given to extract valid points. The start value of the search range is the distance from the viewpoint to the nearest point found among the extracted points, while the end value is the start value plus a defined distance parameter. Then, all valid points in the block within the search range are extracted. The defined distance parameter depends on the continuity of the points in the point cloud. For example, when the measurement scale is 100 m to 500 m, the defined parameter would be 1 m to 10 m from empirical experience.

Finally, an average distance value from the viewpoint to the valid points is calculated. Then, the focus point value is overwritten by the average value. However, when the focus point has a distance value within the search range, the point is defined as the nearest surface approximately, and the overwriting processing is not performed in this case. This processing sequence is applied to all points.

4 EXPERIMENT

We conducted experiments on the proposed methodology using a colored point cloud taken from a terrestrial laser scanner LMS-Z420i produced by RIEGL Laser Measurement

Systems GmbH. This scanner can acquire panoramic distance data and corresponding color data over 360° in the horizontal direction. The test field was a campus that has good visibility, including flat ground planes, individual trees, puddles, and buildings behind trees. The input data set after 3 cm^3 spatial filtering consisted of 1.5 million points. Figure 3 shows the result projected into panorama space. This panorama space was set as 1800×350 pixels and 0.2° spatial angle resolution. Part of the panorama image after a viewpoint translation from the sensor point to a point 2.5 m distant, and part of the panorama image after iterative filterings (three times) using the panorama imagery and a viewpoint translation are shown in Figure 4. The processing time for the conversion coded in C from the input data to the panorama image data set was 0.70 s with an Intel Core 2 Duo processor (2.80 GHz, single thread) for one-shot generation. Moreover, the processing time for the filtering was 0.11 s with the same environment.

We also conducted experiments using a point cloud taken from a terrestrial laser scanner VZ-400 produced by RIEGL. The test field was a park. The input data set consisted of 1.7 million points after 3 cm^3 spatial filtering. Part of the panorama image after filtering using the panorama imagery and a viewpoint translation from the sensor point to a point 2.5 m distant is shown in Figure 5. The average processing time for the panorama image

Figure 3. Input data (panorama image of LiDAR VR).

Figure 4. Part of the panorama image generation result (1): left image is the result after a viewpoint translation, and right image is the result after filtering.

Figure 5. Part of the panorama image generation result (2): left image is the result after a viewpoint translation, and right image is the result after filtering.

Figure 6. Part of the panorama image generation result (3): left image is the result after a viewpoint translation, and right image is the result after filtering.

conversion and iterative filterings (three times) by parallel programming in C was 0.13 s (Intel Core i7 (2.80 GHz, eight threads)) for one-shot generation in multiple viewpoints without file I/O. In addition, part of the panorama image after filtering using the panorama imagery and a viewpoint translation from the sensor point to a point 7 m distant is shown in Figure 6.

5 DISCUSSION

We can verify the features of our LiDAR VR generation as follows. The first feature is panorama-image generation with an arbitrary viewpoint. We have verified that the proposed methodology can successfully generate LiDAR VR images from many viewpoints automatically with one observation data set from one viewpoint. Moreover, we have verified that camera movement and rotation from a viewpoint can be simulated successfully to generate a new image without a position sensor such as RTK-GPS or a direction sensor such as a digital compass. In addition, we have confirmed that continuous camera trajectory simulation can achieve LiDAR VR imagery generation from accurate viewpoints.

The second feature is real-time spatial interpolation using the point cloud. We have confirmed that spatial interpolation with rendering can be processed in real time (approximately 8 fps with parallel processing) including panorama-image generation. First, a panorama projection is performed successfully as shown in Figure 3. Next, a viewpoint translation simulates a new LiDAR VR image with some missing point data as an interleaved pixel array. Then, the interpolation processing for missing points in the panorama space generates new point data from the new viewpoint. Thus, almost all of the missing data are recovered after filtering. As a result, a higher-quality image has been generated, as shown in Figure 4. Finally, although the panorama projection requires a long processing time in proportion to the number of points, we have confirmed that the proposed overall processing can be achieved in real time. Moreover, even if missing points remain, they can be removed via two or more iterative filterings in most cases. However, larger translations of the viewpoint expose more occlusions in the point cloud. Therefore, even if iterative processing is applied to a generated image from a more distant viewpoint, the interleaved pixels may not be correct, as shown in Figures 5 and 6. However, this is not a critical problem because a common interpolation is undesirable in a case with data.

6 CONCLUSION

We have developed a point-based rendering application called LiDAR VR with high-speed spatial interpolation using a point cloud to generate VR data. This application has achieved not only a transformation from point cloud to VR data but also automatic regeneration of missing points in the rendered result from a chosen viewpoint. Moreover, we have focused on two features of the LiDAR VR generation. The first is that panorama imagery can be generated automatically from arbitrary viewpoints with one observation data set. The second is the possibility that spatial interpolation with rendering results can reduce the processing tasks. Finally, we have presented results that confirm these features in case studies.

ACKNOWLEDGEMENT

This work is supported by Strategic Information and Communications R&D Promotion Programme (SCOPE) of the ministry of internal affairs and communications, Japan.

REFERENCES

Chen, S.E. 1995. Quicktime VR—An Image-Based Approach to Virtual Environment Navigation, *Proceedings of ACM SIGGRAPH'95*: 29–38.

Dykes, J. 2000. An approach to virtual environments for visualization using linked geo-referenced panoramic imagery. *Computers, Environment and Urban Systems* 24, 127–152.

Kopf, J., Chen, B., Szeliski, R. & Cohen, M.F. 2010. Street Slide: Browsing Street-Level Imagery. *ACM Transactions on Graphics (Proceedings of SIGGRAPH 2010)*, 29(4), Article No. 96.

Katsushi, I., Masao, S., Hiroshi, K. & Imari, S. 2004. Constructing Virtual cities by Using Panoramic Images. *International Journal of Computer Vision*, Vol. 58, No. 3: 237–247.

Linsen, L. Muller, K. & Rosenthal, P. 2007. Splat-based Ray Tracing of Point Clouds. *Journal of WSCG*, Vol. 15, Issue: 1–3: 51–58.

Verbree, E., Zlatanova, S. & Dijkman, S. 2005, Distance-value-added panoramic images as the base data model for 3D-GIS. *Panoramic Photogrammetry Workshop*.

Urban and Regional Data Management – Zlatanova, Ledoux, Fendel & Rumor (eds)
© 2012 Taylor & Francis Group, London, ISBN 978-0-415-67491-1

Quasi-real-time 3D reconstruction from low-altitude aerial images

J. Roters & K. Hinrichs
Visualization and Computer Graphics Research Group, Institute of Computer Science, University of Münster, Germany

F. Steinicke
Immersive Media Group, Institute of Human-Computer-Media, University of Würzburg, Germany

ABSTRACT: Generating large-scale virtual 3D models from real-world images is of major interest for many application domains, like urban planning and surveying, and has attracted the computer vision and photogrammetry communities since decades. However, most of the approaches that have been developed are rather restricted since they involve time-consuming manual modeling. Therefore, often the reconstructed scenes are already outdated when they become available for deployment. One recent approach to address this issue is to use *Miniature Unmanned Aerial Vehicles* (*MUAVs*) equipped with digital cameras. MUAVs can be used to capture sequences of low-altitude aerial images, which may cover areas under multiple viewing angles with high ground resolution. Using advanced network technology these images are transferred continuously in *quasi*-real-time, i.e., within a few seconds, to a mission control ground station where they can be processed for immediate display. For this new approach novel photogrammetric processing techniques are needed in order to handle such dynamic and continuously updated low-altitude images.

In this paper we present our multiple view 3D reconstruction workflow and introduce a modification to the traditional image registration, which support incremental refinement of the 3D structure and aim at meeting the performance requirements.

1 INTRODUCTION

Today aerial images play an essential role in surveying, the creation of urban maps, city planning, solar potential analysis and many other applications. The automatic generation of virtual 3D models from aerial images and geo-referenced basis data is of major interest for many different application domains (Snavely, Seitz, and Szeliski 2006). In traditional aerial photography images are acquired by airplanes or satellites. While capturing a region an airplane usually follows an exact trajectory and time protocol. A common pattern is to acquire images with an overlap of approximately 60%. In most cases calibrated cameras and exact on-board measurement devices are used in order to simplify the registration and generation process. These constraints enable the extraction of three-dimensional (3D) points from the aerial images based, for instance, on epipolar or trifocal geometry (Kraus 2007). Since image acquisition by airplanes is very expensive, the usage of more flexible and efficient approaches to obtain aerial images has been in the focus of photogrammetric communities.

Miniaturized unmanned flying robots, also referred to as MUAVs (Miniature Unmanned Aerial Vehicles), open up new vistas in this context. Because of the emergence of MUAVs, the nowadays widespread use of digital cameras and the progress of photogrammetric technologies, low-altitude image acquisition attracts major interest in photogrammetric areas. In most approaches a digital camera is attached to a single MUAV, which simplifies the acquisition process. However, the processing of aerial photographs, regardless of whether acquired by airplanes or MUAVs, requires a lot of manual work, and usually it takes a long time (up to several weeks) until the recorded data is completely processed and a corresponding 3D representation can be visualized.

The AVIGLE project[1] has the goal to develop an avionic digital service platform consisting of a swarm of partly autonomous, interconnected flying MUAVs which can be equipped with different sensing technologies. These MUAVs can be used to capture low-altitude images with high ground resolution from which a virtual 3D model has to be reconstructed in quasi-real-time, i.e., within a couple of seconds. Currently, the mostly used low-altitude platforms for image data acquisition are helicopters, remote controlled model aircraft or tethered balloons (Huang and Lin 2003). One of the many examples of image acquisition using MUAVs is a project at ETH Zurich (Püschel, Sauerbier, and Eisenbeiss 2008) that uses a workflow which involves a significant amount of manual data processing. Furthermore, these projects focus on a single UAV and do not consider a swarm of UAVs.

Although first proposals to address the challenges of photogrammetric processing of such low-altitude images have been made (Zhang 2008), acquisition of low-altitude image sequences with MUAVs and their photogrammetric processing are still in an early stage. Until now, in the context of MUAVs none of the existing solutions provides sufficient concepts to allow the reconstruction of a 3D model of the captured area in quasi-real-time. One reason for this is that in contrast to traditional approaches prescribed relationships between adjacent images acquired with MUAVs cannot be guaranteed. In particular, there may be significant variations in overlap and angular alignment between adjacent images.

Our research focus within the AVIGLE project lies in the area of 3D reconstruction and exploration. Compared to traditional approaches like using helicopters or airplanes higher quality results at lower costs should be achieved with the aerial platform envisioned in the AVIGLE project. Cameras mounted on MUAVs (see Fig. 1) acquire aerial photographs of a scene or area. These image data are transferred to a mission control ground (MCG) station where they are further processed in order to build a 3D model. This incrementally constructed model is used as a basis for a virtual environment showing the 3D world in the area covered by the MUAVs. Since the system is intended to be used in desaster management scenarios, it is crucial that the results are displayed as soon as possible, and that this incrementally growing virtual environment can be explored interactively. Therefore, an initial 3D model of the area covered by the MUAVs needs to be created in quasi-real-time.

In Section 2 we present our 3D reconstruction workflow, which is optimized for low-altitude images acquired by MUAVs. In particular, our algorithm performs in quasi-real-time and supports incremental refinement of the 3D models. This is achieved by modifications to the traditional image registration algorithm. In Section 3 we describe an evaluation of the developed concepts and conclude in Section 5.

Figure 1. MUAV equipped with a camera that captured the aerial photographs for this paper.

[1]AVIGLE project: http://www.avigle.de

2 PHOTOGRAMMETRIC PROCESSING

There are different ways to extract 3D information from images. Single view approaches are often semi-automatic (Zhang, Dugas-Phocion, Samson, and Seitz 2002), or when they are fully automatic numerous prerequisites have to be met to obtain 3D information (Hoiem, Efros, and Hebert 2005), for example strong vanishing lines or adequate segmentations. Alternatively, reconstruction can be achieved using multiple view approaches which explore point-to-point references of several images captured from different viewpoints in order to derive 3D information (Hartley and Zisserman 2004). Using a multiple view approach for reconstructing 3D information from low-altitude aerial photographs is challenging for several reasons. With regard to the small camera distance to the objects degeneracies may occur when only one plane was captured (Torr, Fitzgibbon, and Zisserman 1999), for instance a side of a building or the ground. Small viewpoint changes may affect the quality of the reconstruction due to occlusions of objects of interest, for example by trees. These occlusions may reduce correspondences between images and thus limit the quality of the reconstruction.

For the fully automatic 3D reconstruction method presented in this section we have developed a two-view geometry approach that enables us to provide incremental, robust and automatic 3D reconstructions from aerial images taken by a calibrated camera and especially addresses our main objective, namely interactivity. In order to visualize 3D information as soon as possible, the first reconstructions consisting of 3D points should be available within a few seconds.

Our approach aims at a fast and incremental 3D reconstruction from low-altitude aerial images, hence our primary focus is not on designing the best algorithm with respect to precision. Thus, the discussed computer vision methods aim at the creation of an approximate 3D model in quasi-real-time.

When using two-view geometry the following prerequisites have to be satisfied. First, the captured areas must have a certain amount of overlap. Second, the different views have to be captured from different camera positions, i.e., it is not sufficient to just rotate the camera at a fixed position (Torr, Fitzgibbon, and Zisserman 1999). Based on the GPS and *inertial measurement unit* (*IMU*) sensor data the approximate overlap of images can be determined in order to restrict the reconstruction to images that have a certain minimum amount of overlap.

The current photogrammetric processing workflow (see Fig. 2) within the AVIGLE project is performed in three stages: (1) image processing (see Sec. 2.1), (2) computation of the 3D point cloud (Euclidean geometry) using a two-view approach (see Sec .2.2), and (3) triangulation and texturing.

In the first stage it is important to find stable feature points in each image. Between two overlapping images these points are matched. These matches potentially identify 3D points which are computed in the next stage.

The second stage performs the 3D reconstruction. To establish the basis for the epipolar geometry the essential matrix is estimated. This matrix represents the transformation between the camera poses of two images capturing the same scene. The corresponding camera rotation and translation have to be extracted from this matrix. Thereafter, the Euclidean geometry consisting of the 3D points is extracted directly from the current image pair using normalized image coordinates and the previously determined camera pose. At the end of the two-view reconstruction the computed 3D pointsare refined in several ways. For instance, the coherent 3D point clouds are combined into larger point clouds. After these refinements a 3D model consisting of 3D points is ready for rendering or export.

The third stage which is not further discussed in this paper consists of the following steps:

1. Mesh Generation: triangulate the point cloud in a smart way using heuristics involving edge and line detection and textures of the captured images (Strothoff, Feldmann, Steinicke, Vierjahn, and Mostafawy 2011).

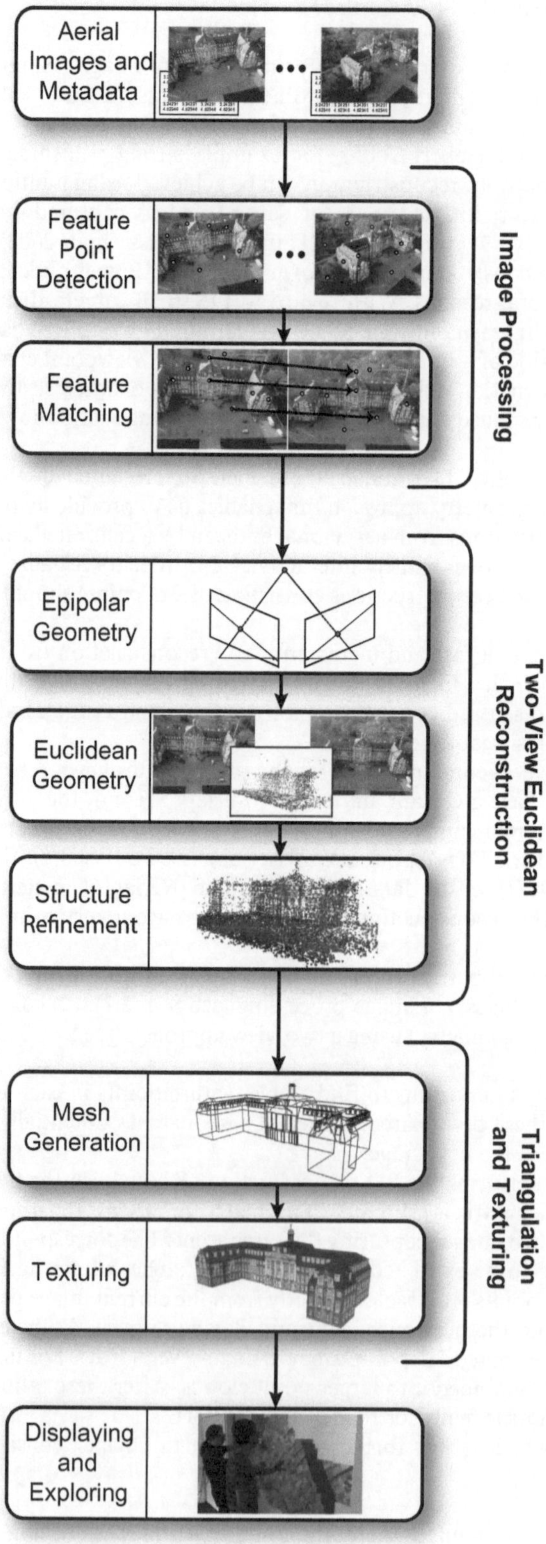

Figure 2. Photogrammetric processing pipeline for low-altitude aerial images.

2. Texture Mapping: map parts of the aerial images as textures onto parts of the 3D mesh (Feldmann Steinicke, and Hinrichs 2011).

After these steps the current 3D model consisting of textured meshes is ready for rendering or export.

Since in our project setting we get images with a high resolution (4256×2832 pixels) which have to be processed in quasi-real-time, our approach is structured in two phases: (1) initial estimation (to get an approximate initial result in quasi-real-time) and (2) incremental improvement (to improve the results and to densify the 3D information). While the initial estimation is computed with every new incoming aerial photograph, the incremental improvements are computed when the system is idle.

2.1 *Image registration*

At the beginning of the image registration workflow several images with meta data are received by the MCG station (see Fig. 2). We assume that these images overlap and are captured from different viewpoints. The image registration has a large impact on the 3D reconstruction performance and quality. For this reason we had to develop a robust registration method that supports incremental updates, since new images are continously captured by the MUAVs and received by the MCG.

2.1.1 *Feature point detection*

The feature point detection is the first basic step in image registration. A feature point is given by coordinates in an image and an additional feature descriptor. Such a point should be recognizable with respect to its descriptor in other images which have been taken from different perspectives and in which the point is also visible. These points and their corresponding descriptors are required for succeeding processing steps and are therefore computed for every new image of the sequence.

Since the feature points have to be determined on low-altitude aerial images the viewpoint change and the distance to the objects have a big impact on the projected 3D structure. Several feature point detectors are available, but due to the expected viewpoint changes, our feature detector has to be invariant against changes of scale, rotation and viewpoint. The most interesting detectors for our approach are the *Speeded Up Robust Features* (*SURF*) (Bay, Ess, Tuytelaars, and Gool 2008) and the *Scale Invariant Feature Transform* (*SIFT*) (Lowe 2004). While SURF provides a faster feature point detection, we have observed that the results from SIFT are more reliable, especially when it comes to viewpoint changes. Therefore, we decided to use SIFT for extracting features for image registration. Furthermore, Lowe's default parameter setting is used, since it provides a stable registration.

Four steps are required to locate the scale invariant feature points and to compute their descriptors (Lowe 2004):

1. Detection of extrema in *Gaussian Scale Space* (*GSS*)
2. Localization and refinement of feature points
3. Orientation assignment to the feature points
4. Generation of feature descriptors

As a major drawback SIFT requires significant computing power, since the creation of the GSS (step 1) is very expensive, especially for large images. Furthermore, for each feature point a descriptor has to be generated which requires many orientation calculations (steps 3 and 4). Note that in our project setting we expect images with high resolution (about 12 MP). Our SIFT implementation usually provides more than 40 000 feature points per image.

To address this drawback we introduce a modification to step 1, the extrema detection in the GSS. This modification has significant impact on the number of feature points, so that further steps (based on the feature points) in image registration as well as in 3D reconstruction are computed in less time. The Gaussian Scale Space in which minima and maxima have to be detected is obtained by cascade filtering (Lowe 2004). It consists of several octaves, each describing a scale space (see Fig. 3). Octave 0 (O_0) represents the scale space of the original image size. Each following octave is down-sampled by a factor of 2. Thus, the image size of octave k is given by 2^{-k} times the original image size. To cover almost all possible scales we compute n octaves limited by the minimum size of 8×8 pixels. With our current project setting with image sizes of approximately 4000×3000 we compute $n = 9$ octaves.

On the one hand, it is expensive to compute octaves for large image sizes. On the other hand, in general these octaves provide more feature points than the ones for smaller image sizes. As mentioned above, our approach is structured in two phases in such a way that the initial result can be computed in quasi-real-time. To achieve this goal, our approach considers only the octaves between some *initial octave* O_m and the octave O_{n-1} with the smallest image size during the initial estimation phase (see Fig. 3). The octaves $O_0, \ldots, O_{m-1}$ with larger image sizes are considered individually during the incremental steps. As long as new images are received the initial phase is active, and the computation of the incremental steps is suspended. Based on our experiments we choose $m = 3$. Thus, incremental steps may upgrade the feature detection results up to 3 times until the final result is reached.

Note that the initial result of our feature point detection is used for the remaining process. As a conclusion this reduces the number of features which also speeds up the remaining 3D reconstruction pipeline. The suspended octaves could be computed afterwards, i.e., during idle time.

2.1.2 *Feature matching*

The feature matching between two images is reasonable if there is sufficient overlap between the captured images. To compute an approximate overlap ratio the vertices of the view frustums could be projected onto the ground. For this computation we would need (1) the

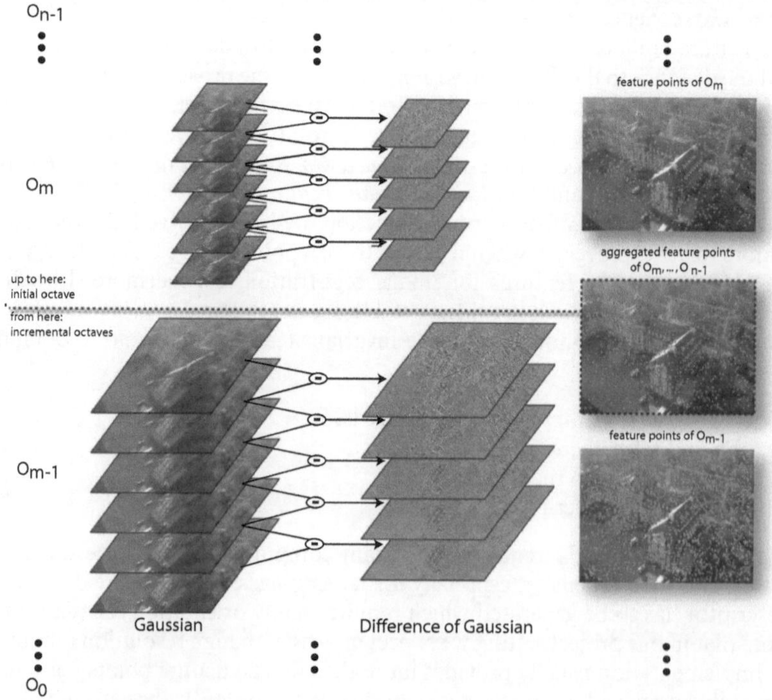

Figure 3. Representation of the Gaussian Scale Space. Each octave is represented by 6 images obtained by applying Gaussian cascade filtering. Adjacent images are subtracted to produce the difference images in which the local extrema have to be detected. The detected feature points of the octaves are shown on the right.

current position and the view direction of the MUAV given through GPS and IMU sensors and (2) the height of the ground which is for example available through the Shuttle Radar Topography Mission[2].

To compute the matching between the feature points of two images we have to determine nearest neighbor pairs among their feature descriptors. In our project setting each feature descriptor is a vector with 128 components. Since we expect the SIFT algorithm to detect many feature points (>40,000) for octaves with large image size, the brute-force nearest neighbour algorithm would be too expensive to compute the matches in quasi-real-time.

To speed up the matching process we use an approximate nearest neighbor algorithm which uses multiple randomized kd-trees to determine the nearest neighbor pairs (Muja and Lowe 2009). Since this algorithm provides only approximate results outliers may be obtained, i.e., false matches. In succeeding steps these outliers may be removed based on additional constraints.

Several feature points in one image may be matched with the same feature point in the other image. Thus, we cannot decide which is the correct match. The quality of a match between two feature points is determined by the Euclidean distance between their feature descriptors. Hence we only accept a match if the distance between the descriptors of the best match is at least 50% larger than the distance between the descriptors of the second best match. Otherwise the matches (and consequently the corresponding feature points) are rejected.

At the end of the image registration process the overlap of the images could be checked again with respect to the found matches. We reject the complete feature matching between two images if

$$m \leq c \cdot \min(pts_1, pts_2) \tag{1}$$

where pts_1 (resp. pts_2) is the number of feature points in the first (resp. second) image and m is the number of matches between them. In other words, if there are not enough matches between the two images, they may not overlap sufficiently. Based on our experiments we set $c = 0.15$ in order to have at least 15% matches (of all possible matches).

2.2 *Two-view euclidean reconstruction*

As shown in Figure 2, the first step of the Euclidean reconstruction is the determination of the two-view (epipolar) geometry which is the basis of our 3D reconstruction approach. In the calibrated case it is represented bythe essential matrix from which the camera matrices can be extracted. Then the Euclidean geometry (represented as a 3D point cloud) is computed by using these matrices and the point matchings. Finally, we refine the point clouds to improve the results.

The epipolar geometry describes the geometric relations between two camera views of a 3D scene captured as 2D images. For a known epipolar geometry the epipolar constraint

$$\vec{x}^T E \vec{x}' = 0 \tag{2}$$

is satisfied if $\vec{x} = (x_1, x_2, 1)^T$ is a normalized image coordinate of a point in one image and $\vec{x}'$ is the corresponding coordinate in the other image. Furthermore, E denotes the essential matrix that describes the epipolar geometry withrespect to the epipolar constraint.

In our workflow the essential matrix is estimated by the use of the 8 Point Algorithm (Longuet-Higgins 1981). Due to further improvements, which reduce the impact of noise (Hartley 1997), it has become a very robust algorithm.

As mentioned before, the epipolar geometry is the basis of our 3D reconstruction algorithm. Thus, it is very important for further processing steps that the essential matrix is estimated as precisely as possible. Due to the use of an approximate nearest neighbor algorithm for the feature point matching the data contains outliers. Hence, RANSAC (Fischler

[2]NASA SRTM: http://www2.jpl.nasa.gov/srtm

and Bolles 1981) is used to support the robust estimation of the essential matrix and to remove outliers. RANSAC chooses 8 matching feature point pairs randomly and computes the epipolar geometry, which is afterwards validated using the remaining matches.

To compute the Euclidean geometry we need to extract the camera matrices from the essential matrix. A short description of this procedure is given in (Hartley and Zisserman 2004). With the known camera matrices and the feature matches we compute the Euclidean geometry using linear triangulation.

Due to small measurement errors of the feature point detector the rays through the camera centers and the feature points (given in normalized image coordinates) do not meet in 3D space. Thus, the epipolar constraint (see Eqn. 2) is not satisfied exactly and results in an increased error of the 3D point. Therefore we suggest a correction of the image points using the optimal triangulation approach (Hartley and Sturm 1997) in an optional step.

Some techniques for structure refinement are used to improve the (initial and incremental) results. On the one hand, outliers in the 3D point cloud are removed, and on the other hand, several point clouds are combined into coherent 3D point clouds.

In our approach an outlier is defined through the Euclidean distance to the median value m of the 3D point cloud in the following way. Let σ be the standard deviation of the 3D coordinates from the point cloud of two corresponding images. Then a point is rejected if its distance to m is larger than $3 \cdot \sigma$.

Due to our incremental architecture, we do not optimize the structure with bundle adjustment while receiving new images, i.e., in the initial phase. On the one hand, bundle adjustment would optimize the structure and cameras with respect to the minimization of the reprojection error (Triggs, McLauchlan, Hartley, and Fitzgibbon 2000), but on the other hand it would reduce the interactivity since in each new incremental step the reprojection error has to be minimized again.

However, in order to get a coherent 3D point cloud we try to estimate a rigid body transformation (with an additional scaling term) for each pair of 3D point clouds using three feature matches which occur in three overlapping images.

At the end of the photogrammetric processing workflow there is a 3D model consisting of coherent 3D point clouds corresponding to a certain set of aerial photographs. In the case of good matching and overlapping images only one coherent point cloud is reconstructed representing all captured photographs. Note that two MUAVs capturing completely different areas or objects provide disjoint 3D models which cannot be combined.

3 EVALUATION

As mentioned above, one of the major goals of the AVIGLE project is to provide an approximate 3D reconstruction as fast as possible, which then has to be incrementally refined. Such an initial approximation of the 3D model can be used, for example, to avoid collisions between MUAVs and physical objects such as buildings.

In order to evaluate if our adaptations allow to reconstruct a 3D model of sufficient quality, we conducted an evaluation based on a real-world image sequence. The image sequence we have chosen to evaluate our approach consists of 20 aerial photographs. Figure 4 shows some example images of the sequence. The images were captured with a calibrated 12 MP (4032×3024 pixel) camera. With this image size both Lowe's SIFT as well as our adapted version compute a total of 9 octaves for the final result.

Figure 4. Example images from the sequence of 20 images.

4 RESULTS

Figure 5 shows a comparison of the evaluation (left) using the unmodified SIFT algorithm against (right) our modified version showing the final results up to octave O_0. As it can be seen, both versions of the feature detectors identified almost the same number of feature points for the final result; SIFT detected on average 43 403 feature points, while the modified incremental version detected 43 986 feature points, i.e., approx. 1.34% more. The unmodified version of SIFT successfully matched in each image on average 22.69% of the detected feature points against feature points in the adjacent image, while our adapted version matched on average 25.93%.

The 3D reconstructions of our adapted version, i.e., the initial result and each incremental step, are shown in Figure 6. When it comes to computation time in the initial phase each image is processed on average 2.39 seconds, so that the first 3D reconstruction between two overlapping images is available within 5 seconds. Such a first 3D reconstruction is shown in Figure 7.

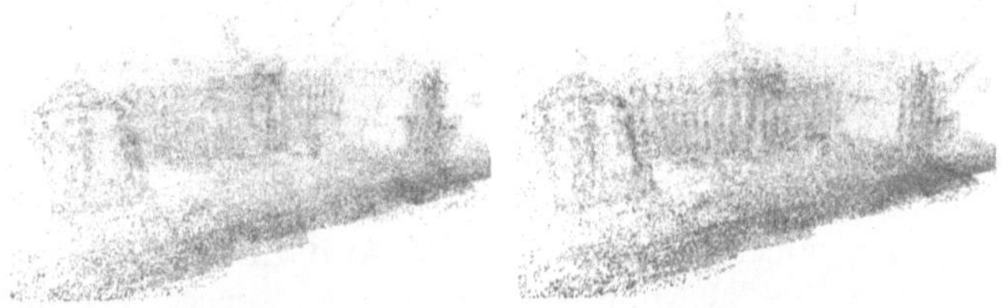

version:	normal	modified
point count:	72332	73359
computation time:	1 587s	1 808s

Figure 5. Final result of the unmodified version (left) and final result of the adapted incremental version (right) up to octave O_0.

computation up to octave:	O_3	O_2	O_1	O_0
octave size:	504 × 378	1008 × 756	2016 × 1512	4032 × 3024
number of points:	1645	6100	20786	73359
computation time:	48s	145s	456s	1 808s

Figure 6. Results of the modified version of the photogrammetric processing workflow of 20 images. On the left the final result of the initial octave O_3 is shown. From left to right final results of the octaves are presented with the aggregated point clouds.

Figure 7. Example of a result of the 3D reconstruction between 2 images computed within 5 seconds in contrast to a 3D model.

5 CONCLUSION

In this paper we present an incremental method for 3D reconstruction from low-altitude images, which generates first results within a few seconds. We describe a 2-phase architecture to compute an initial structure during the first phase that can be refined incrementally during the second phase in order to improve the 3D point cloud. For this purpose we developed an incremental version of SIFT that has significant impact on the computation time by reducing the number of feature points. We conducted an evaluation of our modifications, which illustrated that our adaptations perform in quasi-real-time while providing sufficient quality.

In the future we will further improve the 3D reconstruction process. For instance, we plan to optimize the implementation, which is currently realized in Matlab. An implementation using a multi-core CPU or a GPU would decrease the computing time significantly, so that the initial octave may be increased to get initial results with more detail. Moreover, further aspects (e.g., bundle adjustment, triangulation, texturing) of the entire 3D reconstruction workflow have to be adapted to an incremental. In addition, we plan to perform a qualitative evaluation, which analyses the visual quality of each incremental model considering perceptual aspects.

ACKNOWLEDGEMENTS

Our work has been conducted within the project AVIGLE, which is part of the Hightech. NRW initiative funded by the Ministry of Innovation, Science, Research and Technology of the German State of North Rhine-Westphalia. AVIGLE is conducted in cooperation with several industrial and academic partners. We thank all project partners for their work and contributions to the project. Furthermore, we thank Cenalo GmbH for their image acquisition.

REFERENCES

Bay, H., Ess, A., Tuytelaars, T. & Gool, L.V. (2008). Surf: Speeded up robust features. *Computer Vision and Image Understanding (CVIU) 110*(3), 346–359.

Feldmann, D., Steinicke, F. & Hinrichs, K.H. (2011). Flexible clipmaps for managing growing textures. In *Proceedings of International Conference on Computer Graphics Theory and Applications (GRAPP)*.

Fischler, M. & Bolles, R. (1981). Random sample consensus: A paradigm for model fitting with applications to image analysis and automated cartography. *Communications of the ACM 24*(6), 381–395.

Hartley, R. (1997, Jan.). In defense of the eight-point algorithm. *IEEE Transactions on Pattern Analysis and Machine Intelligence 19*(6), 580–593.

Hartley, R. and Sturm, P. (1997, Jan.). Triangulation. *Computer Vision and Image Understanding 68*(2), 146–157.

Hartley, R.I. & Zisserman, A. (2004). *Multiple View Geometry in Computer Vision* (Second ed.). Cambridge University Press, ISBN: 0521540518.

Hoiem, D., Efros, A.A. & Hebert, M. (2005). Automatic photo pop-up. In *ACM SIGGRAPH 2005 Papers*, SIGGRAPH '05, New York, NY, USA, pp. 577–584. ACM.

Huang, H. and Lin, S. (2003). Digital photogrammetry using tethered balloon. *Journal of Photogrammetry and Remote Sensing of Taiwan 8*(2), 55–68.

Kraus, K. (2007). *Geometry from Images and Laser Scans* (2nd ed.). Number 1 in Photogrammetry. Walter de Gruyter.

Longuet-Higgins, H. (1981, Sep). A computer algorithm for reconstructing a scene from two projections. *Nature 293*(5828), 133–135. 10.1038/293133a0.

Lowe, D. (2004). Distinctive image features from scale-invariant keypoints. *International journal of computer vision 60*(2), 91–110.

Muja, M. & Lowe, D.G. (2009). Fast approximate nearest neighbors with automatic algorithm configura-tion. In *International Conference on Computer Vision Theory and Application (VISSAPP'09)*, pp. 331–340. INSTICC Press.

Püschel, H., Sauerbier, M. & Eisenbeiss, H. (2008). A 3d model of castle landenberg (ch) from combined photogrammetric processing of terrestrial and uav-based images. *The International Archives of the Photogrammetry, Remote Sensing and Spatial Information Sciences 37*, 93–98.

Snavely, N., Seitz, S.M. & Szeliski, R. (2006). Photo tourism: Exploring photo collections in 3d. In *SIGGRAPH Conference Proceedings*, New York, NY, USA, pp. 835–846. ACM Press.

Strothoff, S., Feldmann, D., Steinicke, F., Vierjahn, T. & Mostafawy, S. (2011). Interactive generation of virtual environments using muavs. In *Proceedings of the International Symposium on Virtual Reality Innovations (ISVRI)*.

Torr, P.H., Fitzgibbon, A.W. & Zisserman, A. (1999). The problem of degeneracy in structure and motion recovery from uncalibrated image sequences. *International Journal of Computer Vision 32*, 27–44.

Triggs, B., McLauchlan, P., Hartley, R. & Fitzgibbon, A. (2000). Bundle adjustment—a modern synthesis. In B. Triggs, A. Zisserman, and R. Szeliski (Eds.), *Vision Algorithms: Theory and Practice*, Volume 1883 of *Lecture Notes in Computer Science*, pp. 298–372. Springer-Verlag.

Zhang, L., Dugas-Phocion, G., Samson, J.-S. & Seitz, S.M. (2002). Single view modeling of free-form scenes. *Journal of Visualization and Computer Animation 13*(4), 225–235.

Zhang, Y. (2008). Photogrammetric processing of low altitude image sequences by unmanned airship. *The International Archives of the Photogrammetry, Remote Sensing and Spatial Information Sciences 37*, 751–758.

Urban and Regional Data Management – Zlatanova, Ledoux, Fendel & Rumor (eds)
© 2012 Taylor & Francis Group, London, ISBN 978-0-415-67491-1

The vision of mapping the global urban footprint using the TerraSAR-X and TanDEM-X mission

H. Taubenböck, A. Roth, T. Esch, A. Felbier, A. Müller & S. Dech
German Remote Sensing Data Center (DFD), German Aerospace Center (DLR), Wessling-Oberpfaffenhofen, Germany

ABSTRACT: How much of the Earth's landmass is urban, and where are these urbanized areas located? Because of the key role of cities in sustaining human civilization and their impact on the environment, there is a substantial need for and interest in global mapping of human settlements. Satellite sensors provide one of the few globally existing spatial data sources of observations. The German space mission TanDEM-X will acquire two coverages of the entire landmass for 2011 and 2012. This study presents a developed pixel-based classification approach aiming to extract urban and non-urban structures from single radar imagery. The intended product, the so called 'urban footprint' is presented for selected cities, thematically defined and accuracy assessment is carried out. Beyond that sample applications are discussed and one example, the mapping and monitoring of urban growth is applied.

1 INTRODUCTION

We live in a world where the population is expected to increase continuously from currently 6.7 billion to 9.3 billion in 2050. Trends imply that almost all the expected world population growth will be absorbed by urban areas (UN, 2007). This immense urbanization is a world-wide phenomenon, but what spatial consequences do and will come along with it?

How much of the Earth's landmass is urban, and where are these urbanized areas located? Before 2000, if one asked this question, the only global map available was a digital mosaic of maps and charts from the 1960s through the 1990s (Digital Chart of the World (DCW)) or (Vector Map Zero) (Potere & Schneider, 2009). Mapping, monitoring, quantification, analysis and visualization of the global process of massive urbanization are crucial to raise awareness, to identify and localize problems or trends and to develop strategies for sustainable development. Thus, these high current and expected dynamics of urbanization require the increased utilization of fast and operational global monitoring systems. A series of satellite based sensors provide Earth observation data well suited for this purpose.

In the last decade Earth observation sensors developed to a stage where new maps have been made possible by the availability of global coarse resolution data (250 m–2 km). Even though higher resolution systems are available the achievement of a global coverage is not an easy task. Limitations like cloud coverage and illumination, on board storage capacity, sensor utilization and sharing of the same source with other EO projects cause a several years lasting acquisition period and a relatively coarse resolution. Also the data costs and the processing effort are significant.

In 2011 and 2012 the German space mission TanDEM-X will acquire two coverages of the entire landmass of the world that will be processed to a digital elevation model with 12 m posting (Huber, Younis & Krieger, 2010). The vision of mapping the global urban footprint makes use of this immense data source. This paper presents the global urban footprint project carried out at the German Aerospace Center at the German Remote Sensing Data Center (DFD), including the data, the techniques, the product and its accuracy supplemented by sample applications or research ideas.

2 THE VISION OF MAPPING THE GLOBAL URBAN FOOTPRINT

2.1 *Data*

The classification of the global urban footprint map bases upon data from the German radar satellite TerraSAR-X launched on June 15th 2007 as well as on data from the German TanDEM-X satellite launched on 21st of June 2010. Both satellites record high detailed radar images of the Earth independent from cloud coverage and daytime based on SAR technology (synthetic aperture radar).

Being the first bistatic SAR-satellite mission **TanDEM-X** (**T**erraSAR-X **a**dd-o**n** for **D**igital **E**levation **M**easurement) opens a new era in space borne radar remote sensing. Both almost identical satellites are flying in a closely controlled formation with typical distances between 250 m and 500 m generating a stereo like view to the Earth's surface. Primary mission objective is the generation of a consistent global Digital Elevation Model (DEM) with an unprecedented accuracy. Applying SAR interferometry the 3D-coordinates of each image element are determined. Beyond that, TanDEM-X provides a highly reconfigurable platform for the demonstration of new SAR techniques and applications.

The TanDEM-X mission (Bartusch et al., 2010) is implemented as a public private partnership (PPP) between the German Aerospace Centre DLR and EADS Astrium GmbH. DLR is the owner of the TerraSAR-X and TanDEM-X satellites and data and coordinates their scientific utilisation. Infoterra GmbH a subsidiary of EADS Astrium will exclusively distribute the TanDEM-X products to commercial customers.

Three different image modes with a resolution up to 1 m are available (Fritz et al., 2007). The standard mode for the DEM production is the 'Stripmap' single polarized mode with approximately 3 m spatial resolution. These data will also be input to the urban footprint extraction. The swath width is 30 km, one strip is up to several 100 s of km long. The input product for the mapping of the built-up area is the so called CoSSC, the single look complex format. Single look imagery is well suited for the speckle analysis on which the footprint classification is based on. Additionally the CoSSC is an intermediate product created during the standard TanDEM-X-processing (Fritz et al., 2010) and provides an ideal interface to the global urban footprint processor. Figure 1 shows stripmap data for the example of the city of Islamabad in Pakistan as well as the calculated speckle divergence. The technique is introduced below.

2.2 *Technique*

A pixel-based classification approach aims at extracting urban and non-urban structures from a single TerraSAR–X image. The basic concept is to extract reliable urban features, which are represented by bright point scatterers (corner reflectors).

The idea of the classification approach is a texture analysis of the local development of speckle to derive the so called 'speckle divergence'. Speckle noise is a granular noise that inherently exists in and degrades the quality of the active radar and synthetic aperture radar (SAR) images. Speckle noise in conventional radar results from random fluctuations in the return signal from an object that is no bigger than a single image-processing element. It increases the mean grey level of a local area (Tso & Mather, 2009).

A possibility to quantify the development of speckle and therewith the heterogeneity in SAR images is the coefficient of variation. The local coefficient of variation is defined as the ratio between the local standard deviation and the local mean. In homogenous image regions typical for land-cover classes such as grassland or water the local coefficient of variation equals the fading texture. In these regions the speckle is randomly distributed. Compared to homogenous image regions the presence of distinct structures within the resolution cell leads to an increased proportion of directional backscatter. As a result the speckle characteristics are no longer randomly distributed. Especially in highly structured urban areas this leads to an increased value of the local coefficient of variation. Thus, the 'speckle divergence' measures the difference between randomly distributed speckle and image heterogeneity resulting from directional backscatter (Esch et al., 2011).

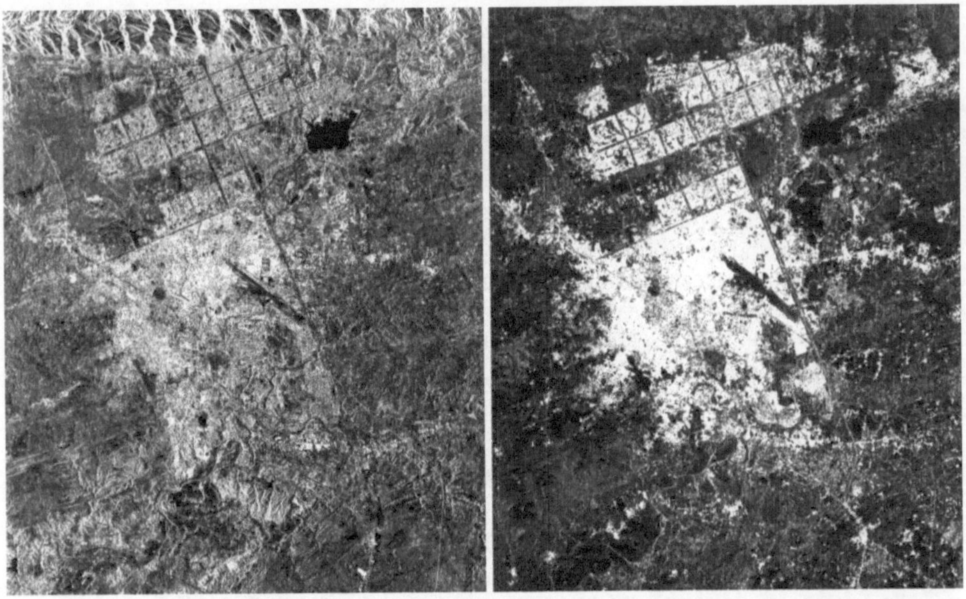

Figure 1. TerraSAR-X stripmap data (left) and 'speckle divergence' (right) for the example of the city of Islamabad in Pakistan.

For this reason thresholds are applied on the intensity values of the radar image and on the speckle divergence texture image. These 'urban seeds' are defined as bright point scatterers (corner reflectors) in a heterogeneous environment and depict the reflection information of vertical structures. However, the 'urban seeds' are rated in five different probability classes. As example, if the intensity value and the speckle divergence texture value are both very high, the probability of being an urban seed is assumed to be high. One example approach to derive the final 'urban footprint' classification is a generalization technique. This technique closes the gaps between the detected 'urban seeds' with respect to intensity and speckle divergence texture values. For it, a threshold to the mean filtered 'urban seeds' (kernel size of 15×15) and to the texture image is applied (cp. Fig. 1) (Esch et al., 2010).

The 'Global Urban Footprint' processor is part of the TanDEM-X processing chain. Purpose of the first processing stage is to be fully inline with the incoming data. In this first stage every radar image of the TanDEM-X mission is used to derive an amplitude image, the speckle divergence texture layer and the 'urban seeds'. A first idea to derive the final urban footprint from the urban seeds on a global scale is presented above by the generalization technique.

However, the classification procedure is still in the development process, where manifold parameters will be included to improve the classification result and minimize false classification. TanDEM-X will acquire two coverages of the entire landmass for 2011 and 2012. Thus, two global urban seeds masks will be derived. The combination of both classification results will be merged to derive one final global urban footprint mask. The main problem regarding the classification approach has been identified in very steep mountainous regions due to foreshortening and layover effects. The slope map derived from the global TanDEM-X digital height model will be included to reduce misclassifications induced by these SAR effects. The idea is to detect critical areas and exclude misclassification using a probability-based approach and a plausibility check.

Beyond that rocky, icy or heterogeneous tropical rain forests have also already been identified as critical areas. However, tests already show promising results that the spatial distribution and the probability of these critical areas clearly differ from detected urban seeds. In addition, it is also tested whether external pre-classifications from the GlobeCover product will support the classification accuracy.

One goal of the TanDEM-X mission is to record the landmass of the entire Earth in just one year. For this reason a number of approximately 400 CoSSCs data sets per day have to be processed, which corresponds to an amount of data of approximately 1 TB per day. This means that entire Europe can be classified in 25 days. A smaller area like the Iberian Peninsula requires just 1.5 days of processing.

All processed layers are slightly distorted due to the radar image geometry. Furthermore storage limitations have to be taken into account. For this reasons multi looking techniques are used to transform all layers to the target spacing of 12 meters (Eineder et al., 2004). Based on intensity and texture information and the urban seeds the final classification result is performed by using generalization techniques. In a third stage the classification result is geocoded and mosaicked to the final product. Figure 2 shows the urban footprint classification result for the example of Islamabad, Pakistan.

2.3 *The product*

The term 'footprint' has often been used in literature as metaphor to capture and extend our conception of the human impact on the ecosphere (Rees, 1992; Wackernagel, 1994). To define

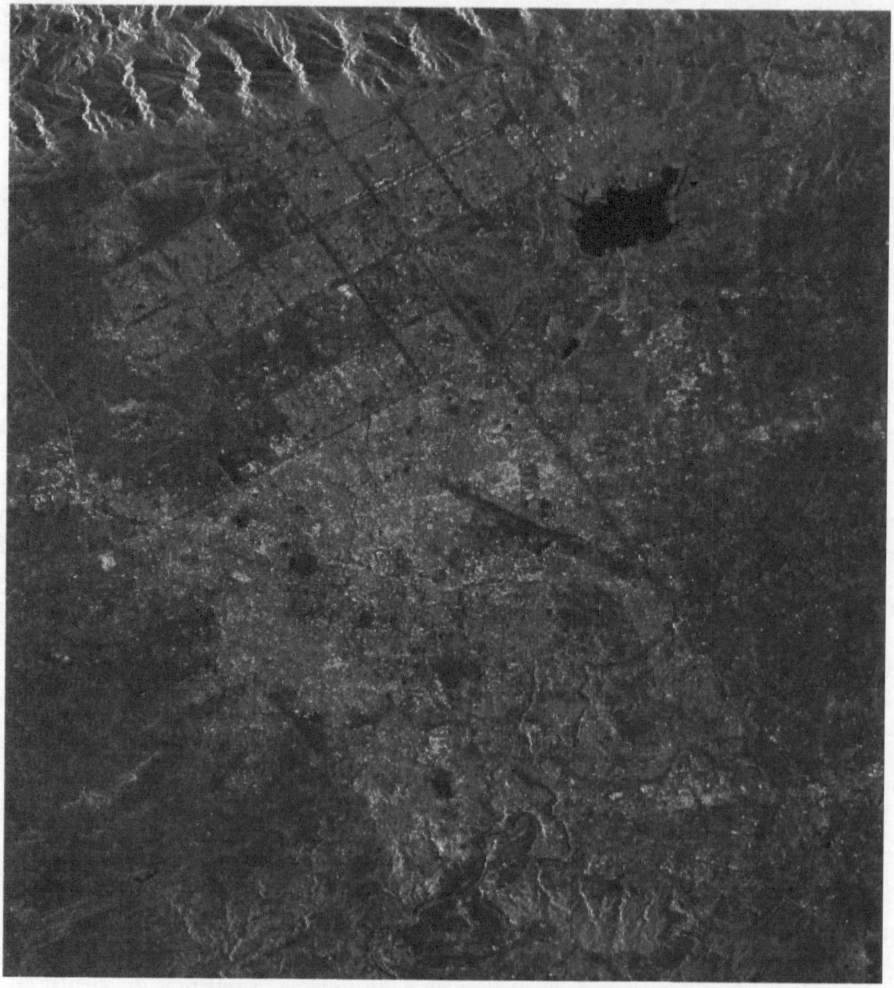

Figure 2. Urban footprint classification for the city of Islamabad in Pakistan derived from TSX-stripmap data.

the term 'urban footprint' a wide range of social, economic, demographic, administrative or political indicators could be used; however, there is no consensus on how to construct a consistent definition based on any single set of attributes (Montgomery et al., 2003); e.g., administrative boundaries are not reliable means of defining urban space, because they change frequently over time, are not comparable across cities and they over- or underestimate the urban land (Potere et al., 2009).

The term 'urban footprint' is widely used in literature and basically refers to the spatial extent of urbanized areas on a regional scale; however, it is a fuzzy definition. It raises e.g., the question, whether a front garden of a house is part of the human impact on the ecosphere, or not? Cahn (1978) understood the term 'urban footprint' as the land directly occupied by a particular structure. Thus, his definition is represented by a 'settlement mask', defined by buildings, streets and impervious surfaces. With an intended geometric resolution of 12 meters a concept behind classifying a pixel as 'urban' is as follows: Pixels of the classification refer to the landcover type urban, if a pixel is dominated by the built environment, which includes human-construct elements, roads, buildings, runways and industrial facilities. When open land (e.g., vegetation, bare soil) dominates a pixel, these areas are not considered urban, even though they may function as urban space (Potere et al., 2009).

Based on this theoretical concept, the definition for the intended 'urban footprint' product derived from TerraSAR-X and TanDEM-X data needs to come directly from the algorithm itself. As defined in section 2.2 the developed algorithm primarily detects vertical structures, which consist of buildings, but also cars, street signs, etc. In addition the subsequent generalization of the detected 'urban seeds' by the algorithm allows identifying highly structured areas around the seed points. As a result, in line with the settlement mask defined above, the physically based definition of the 'urban footprint' is a logical consequence for our 'urban footprint' map. However, it has to be stressed that the 'urban footprint' therefore reflects the distribution of man-made structures with a vertical dimension—e.g., buildings, poles, traffic signs, etc. In turn, flat areas such as streets, runways etc. are not included.

Figure 2 displays the result of the urban footprint mapping product for the city of Islamabad in Pakistan. It clearly reveals the algorithms capabilities to detect the urbanized, highly structured areas and at the same time detecting open land within the urban environment. Thus, the urban footprint product provides insight into dimension, directions, patterns and location of the city on a regional scale.

2.4 *Accuracy*

The accuracy of the intended global urban footprint product has been tested for different urban areas around the globe using manifold methods and reference data sets. In any assessment of map accuracy it is important to have a clear definition of the map attributes under review (cp. the physical definition from section 2.3).

Esch et al. (2010) checked the classified urban footprint comparing it to the original TerraSAR-X stripmap data by a visual verification process. Using a random distribution of check points the accuracy has been calculated for 10 structural different cities around the globe (e.g., Buenos Aires, Munich or Padang). On average the overall accuracy resulted in 90% and higher. Thus, the urban footprint proves high accuracy values on a regional scale. However, in a different study the derived urban footprints have been compared to geometrically and thematically highly resolved reference data: a 3-D city model with a geometric detail of 1 meter and a thematic resolution of 8 land-cover classes, namely 'buildings', 'streets', 'sealed areas', 'grassland', 'forest', 'bare soil' and 'water' and an accuracy of 97% is used (Taubenböck et al., 2009).

With it, a multi-scale comparison between the regional urban footprint product and the high resolution 3-D city model was applied. The binary layer of the "urban footprint" was related to two different definitions of the urban footprint: First to the built-up areas consisting of buildings, streets and sealed areas (settlement mask). Second, to an alternative definition—the building mask only comprising vertical artificial elements.

The square measures comparison reveals that the urban footprint derived from TerraSAR-X stripmap data relates closely to the spatial extension of the settlement mask. The overall accuracy resulted in 79.2% to the settlement mask and 74.28% to the building mask. This accuracy assessment allows defining in more detail the often unclear usage of the word 'urban footprint' (Taubenböck et al., 2011). The accuracy shows that due to the physical classification concept of the radar data an exact and direct reference to thematic classes as shown with the building mask or the settlement mask (cp. Section 2.4—accuracy) is partly misleading. However, the settlement mask derived from the 3-D city model proved to be a good approximation of what the urban footprint represents. With respect to the measured accuracy (cp. Section 2.3), the derived urban footprint classifies structured areas.

Furthermore, a dependence of the accuracy has been measured with respect to differing urban structures: e.g., with rising built-up densities the over-classification of the urban footprint increases as well. Thus, it is expected that the accuracy of the global product will not be consistent as the urban structures vary immensely around the globe.

2.5 *Sample applications*

The data and techniques presented above as well as other remote sensing techniques have already shown their value in mapping and monitoring urban areas at various scales, and as data sources for the analysis of urban land cover change (Donnay et al., 2001; Batty & Howes 2001; Herold et al., 2002; Esch et al., 2010; Taubenböck & Dech, 2010; Rashed & Jürgens, 2010). With the intended global urban footprint product a reference data set for the year 2011 and 2012 would be available for manifold applications and scientific questions.

The question from the introduction "How much of the Earth's landmass is urban, and where are these urbanized areas located" can be answered for the current situation. This global question can be resolved down to different administrative units—to a national or even regional scale. Monitoring tasks, reporting and statistical analysis will be possible on the issue of "urbanization". As example, questions regarding newly developed and developing urban corridors can be answered spatially.

Recent research has used remotely sensed images to quantitatively describe the spatial structure of urban environments and characterize patterns of urban morphology on urban footprint level. With methods like landscape metrics, gradient analysis or zonal statistics the urban patterns have been quantified (Jat, Garg & Khare, 2008; Luck & Wu, 2003; Sudhira, Ramachandra & Jagadish, 2003; Taubenböck et al., 2009). The consistent global data sets will also enable comparative studies between cities. Furthermore, multi-sensoral approaches have already been applied to analyse the spatial dimension, configuration and fragmentation over time (Schneider, Seto & Webster, 2005; Taubenböck et al., 2010). In addition, research is ongoing towards estimation of building density utilizing the intensity data in combination with the classified urban footprint (Schmidt et al., 2010). Thus, a higher thematic detail intends to allow internal structural pattern analysis.

Figure 3 shows the process of spatial urbanization for the megacity Seoul in South Korea using multi-sensoral remotely sensed data from the Landsat Sensors MSS in 1975, TM in 1988, ETM+ in 2001 and TerraSAR-X in 2010. The analysis visualizes the immense dimension of urban sprawl and its complex developing urban patterns. While in 1975 Seoul was still compact city with small suburbs and one major harbour city it developed over time as polycentric, axial and complex megacity.

The spatial and dimensional knowledge of urbanized areas on a global scale will support different disciplines. As examples, risk and vulnerability analysis is depending on the location and amount of exposed urbanized areas as one of many factors. Or energy-relevant questions need spatial knowledge on distribution of supply and demand.

Beyond that interdisciplinary research can be triggered using these data sets for value-adding due to data, method and knowledge combination. Correlating remote sensing data with demographic data is a common research field. Coarse resolution imagery has been used in improving global population density assessments (Dobson et al., 2000). Amaral et al.

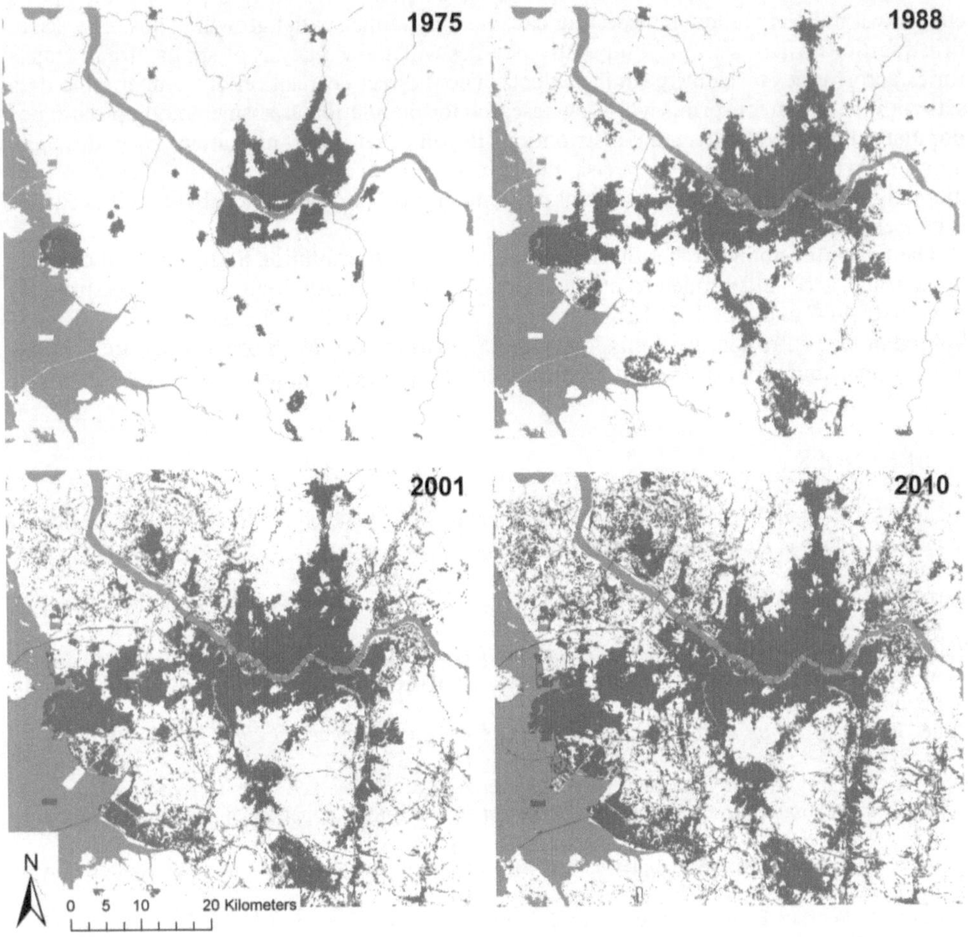

Figure 3. Spatio-temporal analysis of urban sprawl at megacity Seoul, South Korea using multi-sensoral remotely sensed data sets from Landsat and TerraSAR-X.

(2006) or Sutton et al. (2001) developed extrapolation methods of known population data of sample cities onto a nationwide or global scale using stable light night-time imagery. The availability of a global urban footprint map aims at research adopting and developing methods for population assessments.

3 CONCLUSIONS AND OUTLOOK

The new German satellite missions of TerraSAR-X and TanDEM-X allow a never before seen geometric capability for mapping urbanized areas on a global scale. Thus, the German Remote Sensing Data Center of DLR intends to provide a global urban reference data set. The term 'urban footprint' refers to a physical definition described by man-made, highly-structured patterns—urbanized areas. This global urban footprint product will become publicly available with a geometric resolution of three arc-seconds.

The urban footprint map will allow analysis on the percentage of urbanized areas, their location and spatial configuration on a global, continental, national or regional level. With high accuracies, to date tested on sample cities across the world; the urban footprint product will provide a reliable, independent, area-wide and up-to-date geo-information base.

The measurement and quantification of urban extents, urban corridors or spatiotemporal effects will indirectly allow interpreting causes of dynamic spatial growth. Thus, new scientific questions will arise: e.g., comparative studies will allow analysis of spatial urban trajectories across cities to identify similar effects. The indirect correlation of demographic data sets with the urban footprints will allow research for population assessment and up-dating of population numbers and spatial distribution. Beyond that other interdisciplinary ideas and approaches (e.g., urban planning, risk and vulnerability, energy) will add to the usage of this mapping product. But also transdisciplinary integration into political debates and decisions is conceivable.

The technique applied relies on the availability of high resolution SAR data and can easily be transferred and applied to other sensors. This has already been tested successfully for RADARSAT-2 and ALOS. Different missions providing similar radar data such as CosmoSkyMed or the future Sentinel 1 mission from ESA within the GMES program assure continuation to map and monitor the systematically urban extent on a global scale.

REFERENCES

Amaral, S., Monteiro, A.M.V., Camara, G. & Quintanilha, J.A. (2006). *DMSP/OLS night-time light imagery for urban population estimates in Brazilian Amazon.* International Journal of Remote Sensing, Vol. 27, No. 5, 855–870.

Bartusch, M. Miller, D. & Zink, M. (2010). *TanDEM-X Mission Overview and Status.* Proceedings of EUSAR 2010.

Batty, M. & Howes, D. (2001). *Predicting temporal patterns in urban development from remote imagery.* In J.P. Donnay, M.J. Barnsley, P.A. Longley (Hrsg.), Remote Sensing and urban analysis (pp. 185–204). London: Taylor and Francis.

Cahn, R. (1978) *Footprints on the planet: A search for an environmental ethic.* ISBN-: 0876633246, p. 277.

Dobson, J.E., Bright, E.A., Coleman, P.R., Durfee, R.C. & Worley, B.A. (2000). *LandScan: A global population database for estimating populations at risk.* Photogrammetric Engineering and Remote Sensing 66, 849–857.

Donnay, J.P., Barnsley, M.J. & Longley, P.A. (2001). *Remote sensing and urban analysis.* London: Taylor and Francis.

Eineder, M., Boerner, E., Breit, H., Holzner, J., Fritz, T., Palubinskas, G. & Balss, U. (2004). *TerraSAR-X Payload Ground Segment—TMSP Design.* pp. 64–71.

Esch, T., Schenk, A., Ullmann, T., Thiel, M., Roth, A. & Dech, S. (2011). *Characterization of Land Cover Types in TerraSAR-X Images by Combined Analysis of Speckle Statistics and Intensity Information,* IEEE Trans. Geosci. Remote Sens., IEEE Xplore 10.1109/TGRS.2010.2091644, p. 15.

Esch, T., Thiel, M., Schenk, A., Roth, A., Mehl, H. & Dech, S. (2010). *Delineation of urban footprints from TerraSAR-X data by analyzing speckle characteristics and intensity information,* IEEE Trans. Geosci. Remote Sens., vol. 48, no. 2, pp. 905–916, Feb. 2010.

Esch, T., Taubenböck, H., Heldens, W., Thiel, M., Wurm, M., Klein, D., Dech, S., Roth, A. & Schmidt, M. (2009). *Monitoring and assessment of urban environments using space-borne earth observation data—selected applications.* In: Krek, Rumor, Zlatanova & Fendel (eds). "Urban and Regional Data Management", Taylor & Francis Group, London, ISBN 978-0-415-55642-2. pp. 385–398.

Fritz, T., Mittermayer, J., Schättler, B., Balzer, W., Buckreuß, S. & Werninghaus, R. (2007). *TerraSAR-X Ground Segment—Level 1b Product Format Specification.* p. 257, http://sss.terrasar-x.dlr.de/

Fritz, T., Breit, H., Balss, U., Yague-Martinez, N. Rossi, C. Niedermeier, A., Lachaise, M. & Rodriguez, F. (2010): *Processing of Interferometric TanDEM-X Data.* Proceedintgs of EUSAR 2010.

Herold, M., Scepan, J., & Clarke, K.C. (2002). *The use of remote sensing and landscape metrics to describe structures and changes in urban land uses.* Environment and Planning A, 34, 1443–1458.

Huber, S., Younis, M. & Krieger, G. (2010). *The TanDEM-X mission: overview and interferometric performance.* International Journal of Microwave and Wireless Technologies, Seiten 1–11. Cambridge University Press. DOI: 10.1017/S1759078710000437.

Jat, M.K., Garg, P.K. & Kahre, D. (2008). *Monitoring and modelling of urban sprawl using remote sensing and GIS techniques.* In: International Journal of Applied Earth Observation and Geoinformation, vol. 10, issue 1, pp. 26–43.

Luck, M. & Wu, J. (2002). *A gradient analysis of urban landscape pattern: a case study from the Phoenix metropolitan region, Arizona, USA.* In: Landscape Ecology, vol. 17, pp. 327–339. Kluwer Academic Publishers. Richardson 1980.

Montgomery, M., Stren, R., Cohen, B. & Reed, H. (2003). *Cities Transformed: Demographic Change and Its Implications in the Developing World* (Washington D.C.: National Academies Press).

Potere, D. & Schneider, A. (2009). Comparison of Global Urban Maps.—Gamba, P. & Herold, M. (Eds.): *Global Mapping of Human Settlements: Experiences, Data Sets, and Prospects.* Taylor & Francis Group: 269–308.

Potere, D., Schneider, A., Angel, S. & Civco, D.L. (2009). *Mapping urban areas on a global scale: which of the eight maps now available is more accurate?.* International Journal of Remote Sensing, Vol. 30, No. 24, 2009. pp. 6531–6558.

Rashed, T. & Jürgens, C. (2010). *Remote Sensing of Urban and Suburban Areas.* Springer, 1st Edition, XII, p. 352 ISBN 978-1-4020-4371-0.

Rees, W.E. (1992). *Ecological footprints and appropriated carrying capacity: What urban economics leaves out.* In: Environment and Urbanization. Vol. 4, No. 2, pp. 121–130.

Schmidt, M., Esch, T., Klein, D., Thiel, M. & Dech, S. (2010). *Estimation of building density using TerraSAR-X-Data.* In: Proceedings of IGARSS 2010, Honolulu, USA, 25–30 July 2010.

Schneider A., Seto K.C. & Webster, D.R. (2005). *Urban growth in Chengdu, Western China: application of remote sensing to assess planning and policy outcomes.* Environment and Planning B: Planning and Design 32(3) 323–345.

Sudhira, H.S., Ramachandra, T.V. & Jagadish, K.S. (2003). *Urban sprawl: metrics, dynamics and modelling using GIS.* In: International Journal of Applied Earth Observation and Geoinformation. Volume 5, Issue 1, February 2004, Pages 29–39.

Sutton, P., Roberts, D., Elvidge, C. & Baugh, K. (2001). *Census from Heaven: an estimate of the global human population using night-time satellite imagery.* International Journal of Remote Sensing, Vol. 22, No. 16, 3061–3076.

Taubenböck, H., Wegmann, M., Roth, A., Mehl H. & Dech, S. (2009). *Urbanization in India—Spatiotemporal analysis using remote sensing data.* Computers, Environment and Urban Systems 33. pp. 179–188 DOI information: 10.1016/j.compenvurbsys.2008.09.003.

Taubenböck, H., Wegmann, M., Wurm, M., Ullmann, T. & Dech, S. (2010). *The global trend of urbanization—Spatiotemporal analysis of megacities using multitemporal remote sensing, landscape metrics and gradient analysis.* SPIE Europe, Toulouse, France. p. 20.

Taubenböck, H., Esch, T., Felbier, A., Roth, A. & Dech, S. (2011). *Pattern-based accuracy assessment of an urban footprint classification using TerraSAR-X data.* In: IEEE Geoscience and Remote Sensing Letters. Vol. 8; No. 2; pp. 278–282.

Taubenböck, H. & Dech, S. (Hrsg.) (2010). *Fernerkundung im urbanen Raum—Erdbeobachtung auf dem Weg zur Planungspraxis.* Wissenschaftliche Buchgesellschaft Darmstadt. August 2010. p. 192.

Tso, B. & Mather, P. (2009). *Classification Methods for Remotely Sensed Data* (2nd ed.). CRC Press. pp. 37–38. ISBN 1420090720.

United Nations (2007). *World Urbanization Prospects—The 2007 revision.* New York.

Wackernagel, M. (1994). *Ecological Footprint and Appropriated Carrying Capacity: A Tool for Planning Toward Sustainability.* (PhD thesis), Vancouver, Canada: School of Community and Regional Planning. The University of British Columbia. OCLC 41839429.

Urban and Regional Data Management – Zlatanova, Ledoux, Fendel & Rumor (eds)
© 2012 Taylor & Francis Group, London, ISBN 978-0-415-67491-1

Author index